Disaster Management and Risk Reduction

Role of Environmental Knowledge

Disaster Management and Risk Reduction

Role of Environmental Knowledge

Editors

Anil K. Gupta
Sreeja S. Nair
Florian Bemmerlein-Lux
Sandhya Chatterji

Narosa Publishing House

New Delhi   Chennai   Mumbai   Kolkata

Disaster Management and Risk Reduction
Role of Environmental Knowledge
316 pgs. | 97 figs. | 39 tbls.

Editors
Anil K. Gupta
Sreeja S. Nair
National Institute of Disaster Management
Ministry of Home Affairs, Government of India
IIPA Campus, New Delhi

Florian Bemmerlein-Lux
Ifanos Concept and Planning Germany
Vor der Cramergasse 11
90478 Nürnberg, Germany

Sandhya Chatterji
Ifanos Concept and Planning India
C9/9405 Vasant Kunj, New Delhi

NAROSA PUBLISHING HOUSE PVT. LTD.

22 Delhi Medical Association Road, Daryaganj, New Delhi 110 002
35-36 Greams Road, Thousand Lights, Chennai 600 006
306 Shiv Centre, Sector 17, Vashi, Navi Mumbai 400 703
2F-2G Shivam Chambers, 53 Syed Amir Ali Avenue, Kolkata 700 019

www.narosa.com

*The book is an outcome of the Indo-German cooperation project ekDRM implemented by GIZ,
Germany (IGEP New Delhi) and NIDM New Delhi.*

Printed from the camera-ready copy provided by the Editors.

ISBN 978-81-8487-251-4

Published by N.K. Mehra for Narosa Publishing House Pvt. Ltd.,
22 Delhi Medical Association Road, Daryaganj, New Delhi 110 002

Printed in India

Foreword

Mega- and mini-disasters are becoming more frequent. Coastal tsunamis and storms as well as inland floods and drought are also occurring at an accelerated pace. Although the precise reasons for the growing occurrence of extreme weather events are not known, the impact of climate change can no longer be ignored. Floods, drought, rise in sea level and similar climate related events need greater attention and pro-active action. It is therefore timely that Dr Anil Kumar Gupta and a group of international experts have compiled this book, based on the outcome of the Indo-German Environment Partnership Programme (IGEP). We owe a deep sense of gratitude to GIZ of Germany for sponsoring a detailed study on Environmental Knowledge for Disaster Risk Management. The book provides guidelines for the management of disasters, both natural and man made. Thus chemical mishaps as well as the adverse effect of climate change are covered in the book. The authors of the 21 chapters contained in the book are all eminent persons and hence the book contains the stamp of their knowledge and professional authority.

We need to develop proactive measures to mitigate the hardships arising from disasters. For example, in 1972, I delivered a lecture under the auspices of the Indian Society of Agricultural Statistics on how to face drought without food imports. In that lecture I pointed out that just as grain reserves are important for food security, seed reserves are important for crop security. 2013 marks the 70th anniversary of the Bengal Famine which resulted in the death of an estimated 1.5 to 3 million children, women and men during 1942-43. A constellation of factors led to this mega-tragedy, such as the Japanese occupation of Burma, the damage to the am an (kharif) rice crop both due to tidal waves and a disease epidemic caused by the fungus Helminthosporium oryzae, panic purchase and hoarding by the rich, failure of governance particularly in relation to the equitable distribution of the available food grains, disruption of communication due to World War II, and the indifference of the then UK Government to the plight of the starving people of undivided Bengal.

Famines were frequent in colonial India and some estimates indicate that 30 to 40 million died out of starvation in Tamil Nadu, Bihar and Bengal during the later half of the 19th century. This led to the formulation of elaborate Famine Codes by the then colonial government, indicating the relief measures that should be put in place when crops fail. This book will help us to develop for each agro-ecological region a Disaster Prevention and Management Code.

I hope that this book will be widely read by professionals, policy makers, public and research scholars. On the basis of the suggestions contained in the book, every state in our country can prepare a road map for disaster prevention, mitigation and management. We owe a deep sense of gratitude to Prof Gupta and his colleagues for their labour of love for the cause of a disaster free India.

Prof. M.S. Swaminathan, FRS

Preface

Disasters have grown especially over the past decades, both in frequency and in the intensity of their impact on human and natural resources. Since the turn of the millennium, more than one million people have been killed and 2.3 billion others have been directly affected by natural disasters around the world (CRED, 2012). A statistical study of weather-related disasters published by Netherlands Environmental Assessment Agency in 2012 shows that hydro-meteorological disasters affected an average of 140 million people, caused 41,000 deaths and economic losses of USD 57 billion annually during 1980–2010 period. The growing challenges of hydro-meteorological disasters coupled with the complexities of land-use within industrial settings are being noticed world-over. At the same time in developing nations while high damage probabilities exist on one hand, so do opportunities for proactive risk reduction on the other hand. Vulnerabilities of the people, land-habitations, resources and the infrastructure are governed and modified by the availability of natural resources, bio-productivity, technological know-how, people's practices and behaviour.

Environmental changes, in particular the climate-change, natural resource degradation and land-use changes are known to cause new hazards besides aggravating the impacts of existing hydro-meteorological disaster risks. Management of environmental resources has significant role to play in reducing risk and vulnerability especially of the climatic disasters and in coastal and mountain environments. Knowledge of environment is crucial for all the stages of disaster management cycle including pre-disaster prevention and mitigation, and during post-disaster response, relief, reconstruction and recovery. Decline in environmental services has been identified as a key concern in meeting the needs for life, livelihoods, health and overall wellbeing of people.

The solution to dealing with these complexities and risks lies is the appropriate utilization of environmental knowledge. In essence, environmental knowledge – technology based, traditional or perceptional - plays a key role in understanding hazards, vulnerabilities and risks. These experiences and knowledge are crucial in risk management planning, preparedness, mitigation and recovery, or in other words, for defining the overall frameworks for disaster risk reduction. 'Utilizing environmental knowledge for disaster risk management' has gained attention of policymakers, professionals and communities at large, given the increasing awareness of climate-change related risk, and more intensively in the backdrop of industrial-chemical mishaps, e.g., Bhopal gas tragedy, Jaipur fire, British petroleum oil spill, Mumbai oil spill and the more recent Fukushima nuclear disaster..

The book is the collaborative effort of the Environmental Knowledge for Disaster Risk Management (ekDRM) project team with contributions from experts and scholars at national and international levels. An international conference on Environmental Knowledge for Disaster Risk Management was held in May 2011 at Vigyan Bhavan New Delhi, to systematically capture some of the available knowledge on the subject and identify gap areas. The Conference was part of the project on Environmental Knowledge for Disaster Risk Management implemented under Indo-German collaboration by the National Institute of Disaster Management (NIDM), Ministry of Home Affairs, and the Deutsche Gesellschaft fuer Internationale Zusammenarbeit (GIZ) GmbH. The ekDRM project focused on capacity building, training, research, documentation and knowledge management.

It was felt that, although a rich domain, environmental knowledge is polarized into technical knowledge and local knowledge, and the inputs are not integrated while planning and implementing various environment and disaster management and risk reduction programmes. Contributions from the experts, professionals and scholars along with the select papers of the conference, are thus, presented in the book to supplement knowledge gaps in the field of environmental knowledge and disaster risk management. The term "Environmental Knowledge" as envisaged here, covers a broad spectrum of information and knowledge collated from a wide range of sources ranging from space based inputs to field surveys carried out at community level. Forty authors from technical and scientific organizations, academic institutions, as well as researchers and policy makers provided manuscripts which were reviewed by experts in India and Europe. The diversity in knowledge and experience of authors and reviewers contributed to the richness in contents.

The book is divided into five parts. Section I provides a detailed Introduction and Overview on environmental knowledge. Section II on Environmental Statistics and Decision Support System covers more technological aspects with papers focused on weather forecasting, monitoring, hazard mapping, vulnerability and risk assessment and their applications in developing efficient Decision Support Systems. Section III is devoted to Spatial Planning for Disaster Risk Management. Papers in this section discuss the role of spatial planning in reducing disaster risk with case studies on technological and natural disasters from India and Germany. Section IV covers the Legal and Policy Framework for Disaster Risk Management. Section V of the book addresses the issues of Natural Resource Management for Disaster Risk Reduction. Papers on traditional practices and ecosystem based approaches in various geo-environmental settings like coastal areas, islands, dry land areas, urban ecosystems and agro-ecosystems, were incorporated in this section.

The book covers a wide range of aspects related to environment and disaster management and highlights possible solutions, with case studies from developed and developing world, and outlines concepts, tools and methodologies that can be ingested into the policy making, planning and implementing process. Prevalent practices adopted by using scientific and technological tools coupled with traditional practices and examples from various part of the globe are also described.

The editors are thankful to the National Institute of Disaster Management (Ministry of Home Affairs, Govt. of India), Deutsche Gesellschaft fuer Internationale Zusammenarbeit (GIZ) GmbH, the Ministry of Environment and Forests and other Ministries and Departments of the Government of India for their generous support in developing this volume. We are grateful to M/s. Narosa Publishing House, New Delhi and in particular to Mr. N. K. Mehra, Managing Director, for accepting the proposal and agreeing to publish the treatise. A number of scholars and officials in various professional and personal capacities helped making of the edited book a reality. The editorial support extended by Ms. Sunanda Day, ekDRM Project Associate and Ms. Swati Singh, Project Associate, IGEP and formatting and designing inputs from Ms. Vidya Satija worth special mention.

We hope the book will be of direct use to practitioners in the field of disaster management and natural resource management and also for researchers and post-graduate students of natural resource management, environment, agriculture, disaster management and social work.

Anil K. Gupta
Sreeja S. Nair
Florian Bemmerlein-Lux
Sandhya Chatterji

Acknowledgements

Editors acknowledge with gratitude the support and cooperation of following persons and organizations in the successful organization of the international conference 'ekdrm at Vigyan Bhavan, New Delhi and/or in the preparation of the book:

- Mr. M. Shashidhar Reddy, Vice-chairman, National Disaster Management Authority, India
- Prof. M.S. Swaminathan, FRS, Member of Parliament (Rajya Sabha), Chairman – MSSRF, Chennai
- Mr. Christian-Mathias Schlaga, Charge d'Affairs, Embassy of the F. R. Germany, Delhi
- Dr. Dieter Mutz, Director, Indo-German Environment Partnership Progamme (IGEP-GIZ), Delhi
- Dr. Akhilesh Gupta, Secretary, University Grants Commission (then Advisor, DST, GoI)
- Major Gen. (Dr.) K. Bansal, Member, National Disaster Management Authority, India
- Prof. D.P. Singh, Vice-chancellor, Banaras Hindu University (presently VC- Indore University)
- Dr. Anil K. Singh, Vice-chancellor, Agriculture University Gwalior (then Dy. DG –NRM, ICAR)
- Sr. Prof. Vinod K. Sharma, IIPA New Delhi & Vice-chairman – State DM Authority Sikkim
- Prof. N.R. Madhava Menon, Founder Director, National Judicial Academy, Govt. of India
- Prof. Mohd. Yunus, Chairman, CES (formerly Head, Dept. of Env. Sc., BBA Central Univ. Lucknow)
- Mr. P.G. Dhar Chakrabarti, IAS, Secretary, NDMA (& Addl. Secretary Centre-State Relations)
- Mr. S.K. Das, Director General, Central Statistical Organization, GoI, New Delhi
- Major Gen. (Dr.) R. Siva Kumar, Head, Natural Resource Data Management System, DST, GoI
- Mr. K.C. Gupta, Former DG, National Safety Council, Ministry of Labour, GoI
- Dr. S.P. Gautam, Chairman, Central Pollution Control Board, New Delhi
- AVM Rtd. (Dr.) Ajit Tyagi, DG India Meteorology Department (presently WMO Professor)
- Dr. P.S. Roy, Geospatial Chair Professor, Hyderabad Central University (formerly Director, IIRS)
- Prof. Dilanthi Amaratunga, Head, Centre for Disaster Resilience, University of Salford, UK
- Dr. Indrani Chandrasekhran, Sr. Advisor (E&F), Planning Commission, Govt. of India
- Mr. Sanjay S. Gahlout, Dy. Director General, National Informatics Centre, GoI
- Mr. R.K. Srivastava, IAS, JS (DM), Ministry of Home Affairs, GoI (presently JS Kashmir)
- Mr. Hem Pandey, IAS, JS, Ministry of Environment & Forests, GoI (In-charge, EPDRM project)
- Dr. Satendra, IFS, Executive Director, NIDM and Director SAARC DM Centre, New Delhi
- Prof. Anjana Vyas, Dean, Faculty of Geomatics, CEPT University, Ahmedabad
- Dr. K.J. Ramesh, Advisor, Ministry of Earth Science, GoI
- Mr. N.M. Prusty, Chairman, Strategy Centre, New Delhi

- Dr. Luther Rangreji, Senior Legal Officer, Ministry of External Affairs, GoI
- Mr. N. Raghu Babu, Indo-German Environment Partnership Programme, (IGEP-GIZ), Delhi
- Dr. Shital Lodhia, Associate Professor, Sardar Patel Institute, Ahmedabad, India
- Mr. Jaiganesh Murugeshan, DRR Specialist, UN-HABITAT Myanmar
- Ms. Vidya Satija, GIS Executive, Global Hydrogeological Solutions, New Delhi
- Ms. Sunanda Dey, Research Associate with ekDRM (GIZ-NIDM) project, New Delhi
- Ms. Richa Arya (Research Scholar of AKG), presently Lecturer (Science), Govt. of UP
- Ms. Reetika Goyal, Research Associate with ekDRM (GIZ-NIDM) project, New Delhi
- Ms. Prerna Kaushik, Health Programme Officer, Butterfly, New Delhi
- Faculty & colleagues at NIDM Prof. Santosh Kumar, Prof. C. Ghosh, Dr. Surya Parkash, Dr. K. J. Anandha Kumar, Mr. Arun Sehdeo, Mr. Santosh Tiwari, Mr. Hemant Kumar, Mr. J. N. Jha
- Editor's family members including Mr. Laxmi Narayana Addanki, Dr. Alka Gupta, Vanya and Vibhu
- Indo-German Environmental Partnership Programme, GIZ

- *National Institute of Disaster Management, New Delhi*
- *Flood Information Centre, Bavarian Environment Agency, Germany*
- *United Nations University Institute of Environment & Human Security, Bonn, Germany*
- *Gorakhpur Environmental Action Group (Resource Agency of MoEF), Gorakhpur, India*
- *German Commission of Process Safety, Bonn, Germany*
- *Institute of Social & Environment Transition (ISET-US), Colorado*
- *Indian Agriculture Research Institute, ICAR*
- *United Nation Environment Programme (Disaster Management Branch), Geneva*
- *Consortium for Environment & Sustainability (CES India)*
- *Potsdam University Centre for Climate Research, Potsdam, Germany*
- *International Centre for Integrated Mountain Development (ICIMOD)*
- *UNDP-Disaster Risk Management Programme, India*
- *Ministries of GoI–Environment & Forests, Earth Sciences, Agriculture, Water Resources*
- *UNESCO Cluster office for South Asia, New Delhi, UNESCO-International Hydrology Programme*
- *Central Statistical Office, Govt. of India, New Delhi*
- *Ifanos C&P Germany & Ifanos India*
- *Narosa Publishing House, New Delhi*

Contents

INTRODUCTION AND OVERVIEW

ENVIRONMENT STATISTICS AND DECISION SUPPORT SYSTEMS

List of Figures

List of Tables

List of Contributors

Alfons Vogelbacher, (Dr.)
Head, Flood Information Centre, Bavarian Environment Agency, Munich, Germany

Anandita Sengupta
Faculty of Geo-Information Science and Earth Observation (ITC), University of Twente, The Netherlands

Anil K. Gupta, (Dr.)
Associate Professor, National Institute of Disaster Management, IIPA Campus, New Delhi, India

Anjali Singh
Sr. Research Fellow, Division of Agricultural Physics, Indian Agricultural Research Institute, New Delhi, India

Anne Van der Veen, (Prof)
Faculty of Geo-Information Science and Earth Observation (ITC), University of Twente, The Netherlands

B.D. Bharat
Scientist, Human Settlement and Analysis Division, Indian Institute of Remote Sensing, 4, Kalidas Road, Dehradun, Uttarakhand, India

B.S. Sokhi
Scientist, Human Settlement and Analysis Division, Indian Institute of Remote Sensing, 4, Kalidas Road, Dehradun, Uttarakhand, India

Balaji, S.
Department of Coastal Disaster Management, Pondicherry University, India

Cees J van Westen (Dr.)
Associate Professor, Faculty of Geo-information Science and Earth observation (ITC), University of Twente, The Netherlands

Charles Kelly
Freelance International Disaster Management Consultant.

Christian Jochum (Prof.)
Chairman, German Commission on Process Safety Director, European Process Safety Centre.

Claudia Bach
Researcher, United Nations University, Institute for Environment and Human Security (UNU-EHS)

Debanjan Bandopadhyay
PhD Scholar, Faculty of Geo-Information Science and Earth Observation (ITC), University of Twente, The Netherlands

Hasem Roney
Department of Geography & Environment, University of Dhaka

Jai Shree
Assistant Professor, Government PG College, Chamba, Himachal Pradesh

Jörn Birkmann (Prof.)
Director, United Nations University Institute of Environment and Human Security (UNU-EHS), Bonn, Germany

Kathyln Kissy H. Sumaylo
Project Manager of IOM Cambodia's Building Resilience to Natural Hazards in North-East Cambodia project, Cambodia

Kesavan, P.C. (Prof.)
Emeritus Professor, IGNOU, New Delhi & Distinguished Fellow of M.S. Swaminathan Research Foundation, Chennai, Tamil Nadu, India

M.S. Bhat (Dr.)
Department of Geography and Regional Development, University of Kashmir, J&K, India

N.M. Prusty
Chief Mentor cum Director, Centre for Development and Disaster Management Support Services, A Strategy Center, New Delhi, India

Nilanjan Paul
Manager, SENES Consultants India Pvt. Ltd., Kolkata, India

P.K. Joshi, (Prof.)
Professor & Head, Department of Natural Resources TERI University, New Delhi, India

Prasad Dharmasena
Research Scholar, Department of Geography and Regional Development, University of Kashmir, Srinagar, J&K, India

Priti Attri
Research Scholar, Institute of Environmental Studies, Kurukshetra University, Haryana, India

Shreya Roy
Junior Research Fellow, North Eastern Space Application Centre (NESAC), Dept. of Space, Umiam, Meghalaya, India

Shrikant Maury
Department of Coastal Disaster Management, Pondicherry University, India

Smita Chaudhary (Prof.)
Institute of Environmental Studies, Kurukshetra University, Haryana, India

Sreeja S. Nair
Assistant Professor, National Institute of Disaster Management, IIPA Campus, New Delhi, India.

Subrat Sharma
G.B. Pant Institute of Himalayan Environment and Development, Almora, Uttarakhand, India

Sunanda Dey
Research Associate, EkDRM Project, National Institute of Disaster Management, IIPA Campus, New Delhi, India

Swaminathan, M. S. (Prof.)
Member of Parliament (Rajya Sabha),Chairman, M.S. Swaminathan Research Foundation,Third Cross Street, Taramani Institutional Area, Chennai, Tamil Nadu, India

Swati Singh
Programme Associate, Indo-German Environment Partnership (IGEP) Programme, Deutsche Gesellschaft für Internationale Zusammenarbeit (GIZ) GmbH, New Delhi, India

T.V. Ramachandra (Dr.)
Energy & Wetlands Research Group, Centre for Ecological Sciences & Centre for Sustainable Technologies, Centre for infrastructure, Sustainable Transportation and Urban Planning (CiSTUP), Indian Institute of Science, Bangalore, India

Thara, K.G. (Prof.)
Head, Disaster Management Centre, Institute of Land & Disaster Management, Trivandrum & Member, Kerala State Disaster Management Authority, Kerala, India.

V. B. Negi (Dr.)
Assistant Registrar, IGNOU Regional Centre, Chauhan Niwas Khalini, Shimla, Himachal Pradesh, India

Vijita Agarwal (Dr.)
Associate Professor, University School of Management Studies, GGS Indraprastha University, Delhi, India

Vinay Kumar Sehgal (Dr.)
Sr. Scientist, Division of Agricultural Physics, Indian Agricultural Research Institute, Pusa Campus, New Delhi, India

Vinod K. Sharma (Prof.)
Professor, Indian Institute of Public Administration, New Delhi and Vice Chairman, Sikkim State Disaster Management Authority.

1

Environmental Knowledge for Disaster Risk Management:
Reference to Hydro-meteorological Disasters

Anil K. Gupta, Sreeja S. Nair and Vinod K. Sharma

1.1 INTRODUCTION

In recent years, 90 percent of natural disasters worldwide have been related to water and climate; floods account for nearly 70 percent of the people affected in Asia. During the period 2000 to 2006, 2,163 water-related disasters were reported globally in the EM-DAT database, killing more than 290,000 people, afflicting more than 1.5 billion people and inflicting more than US$422 billion in damages. The United Nations University Institute for Environment and Human Security (UNU-EHS) warns that unless preventative efforts are stepped up, the number of people vulnerable to flood disasters worldwide is expected to mushroom to two billion by 2050 as a result of climate change, deforestation, rising sea levels and population growth in flood-prone lands.

In general, all water-related disasters events increased between 1980 and the end of the twentieth century. Floods and windstorm events increased drastically from 1997 to 2006, but other types of disaster did not increase significantly in this period. Floods doubled during the period 1997 to 2006 and windstorms increased more than 1.5 times. Drought was severe at the beginning of the 1980s and gained momentum again during the late 1990s and afterwards. The numbers of landslides and water-borne epidemics were at their highest during the period 1998–2000 and then decreased. Waves and surges increased between 1980 and 2006. Figure 1 based on the data from the World Meteorological Organization shows that floods accounted for highest number of disasters (37%) followed by windstorms (28%) and droughts and famines (8%) during 1993-2002.

United National International Decade for Disaster Reduction (IDNDR) devoted its 1997 year campaign on the theme 'water......too much....too little, ... leading cause of natural disasters' is still relevant even after 15 years of worldwide efforts of understanding and reducing risk of natural disasters. Environmental changes, in particular – climate change, natural resource degradation and land-use changes, are known to cause new hazards besides aggravating impact of existing hydro-meteorological disaster risks. Loss of environmental resources and ecosystem

Disaster Management and Risk Reduction: Role of Environmental Knowledge; Editors Anil K. Gupta, Sreeja S. Nair, Florian Bemmerlein-Lux and Sandhya Chatterji; Copyright © 2013, Narosa Publishing House, New Delhi

resilience noted worldwide and especially in the developing regions like India, has been identified as a key concern in managing the livelihoods, agricultural productivity, health, and thus, the overall vulnerability toward impact of water and climate related disasters. Management of land-resources and land-use has significant role in reducing risk and vulnerability especially of the climatic disasters and in coastal and mountain environments. Coastal zone management is recognized with serious concern for reducing the risk of major disasters like catastrophic cyclone, tsunami, coastal erosion, flooding, salt water intrusion, etc. In mountain areas the approach of ecosystem management in particular the watershed protection, rangeland management, joint forest management etc., are directly associated with resilience against climatic disasters.

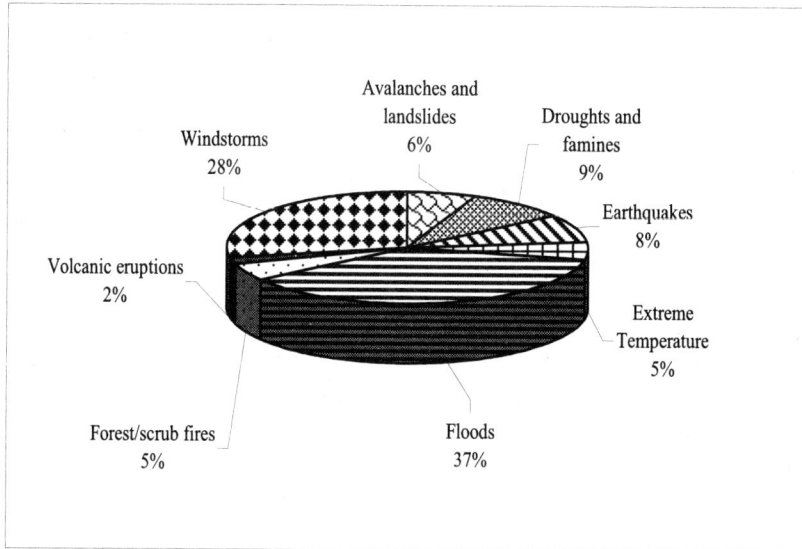

Figure 1.1: Global natural hazards by categories during 1993-2002. (WMO)

Irony as witnessed from the experience is that disaster managers often fail to recognize the environmental dimensions of disaster risk, vulnerability and also of the post-disaster actions, whereas environmental managers seldom focus on evaluating disaster risk aspects within their studies. There is also a need to analyze and/or to develop approaches, tools, techniques and methodologies, legislative framework and decision support systems of environmental management for their role in disaster risk management. "Development is neither a simple, nor a straightforward linear process. It is a multi-dimensional exercise that seeks to transform society by addressing the entire complex of interwoven strands, living impulses, which are part of an organic whole" (Haqqani, 2003). With increasing number of environmental hazard-induced disasters leading to ever increasing human tragedy and economic costs, the international community is calling for the substantial reduction of disaster risk by 2015. The Hyogo Framework for Action sets out the path for the International Strategy for Disaster Reduction (ISDR), and since its adoption in 2005, disaster risk reduction (DRR) has come a long way with respect to concept development and practical application.

United Nations agencies including UNEP, UNDP, IUCN, UN-ISDR and UNU Institute of Environment and Human Security, jointly with many international organizations like ADPC, WWF, GFMC, ProAct Network, SEI, and the Council of Europe have formed a Partnership for Environment and Disaster Risk Reduction (PEDRR) in year 2008 with headquarter at Geneva. UN-OCHA has set up a joint Environment Unit with UNEP to emphasize environmental aspects of disasters and their management. UN-ISDR and IUCN have come up with a number of publications on linkage of environment and disasters with particular focus on hydro-meteorological risks. However, at national levels the initiatives for emphasizing the linkages between the two are yet to be institutionalized in India. The recent understanding of the environment-development interface that sets background for disaster risks and opportunities for climate change adaptation is depicted in the Figure 1.2.

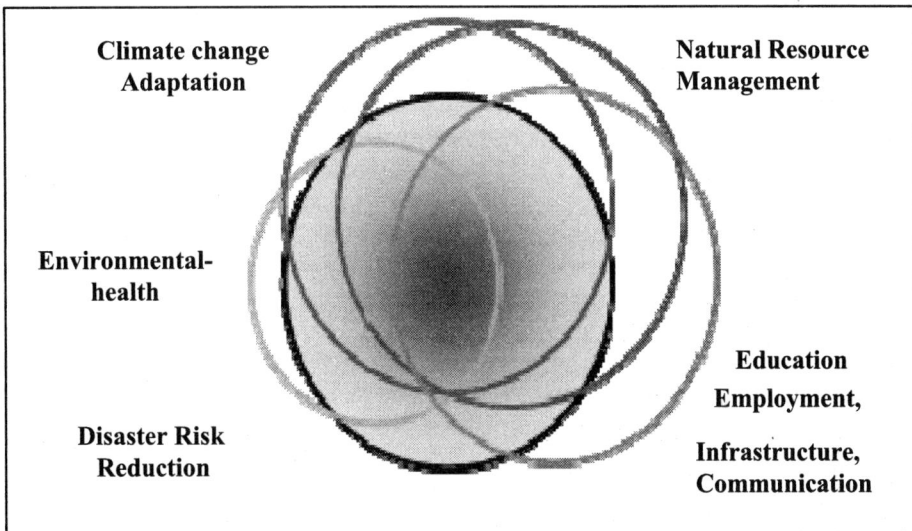

Figure 1.2: Environment-development interface for disasters and adaptation

1.2 ENVIRONMENT AND DISASTERS

Environmental degradation is the deterioration of the environment through depletion of resources such as air, water and soil; the destruction of ecosystems, habitat and the loss of natural homeostasis. It is defined as any change or disturbance to the environment perceived to be deleterious or undesirable, be it quantitative or qualitative.

Environmental degradation is one of the Ten Threats officially cautioned by the High Level Threat Panel of the United Nations including World Resources Institute, UNEP, UNDP and the World Bank. Disasters are the events of environmental extremes which are inevitable entities of this living world. Over the years, with the increasing human intrusion in the natural systems and the changing global environment, the frequency and impact of the disasters is augmenting. Following are the three major faces of environmental change primarily responsible for causing or aggravating hydro-meteorological disasters (Figure 1.3).

1. Climate change
2. Natural resources
3. Land-use changes

Environment and disasters are inextricably linked, particularly in context of hydro-meteorological hazards, climatic risks and epidemiological challenges. It is now widely recognized in terms of following interface:

(i)   Environmental degradation leads to water and climate related disasters: Environmental degradation can occur owing to natural processes or as impact of human activities. Alteration of natural systems and processes, destruction of habitats, loss of environmental quality and exploitation of resources are the broad indicators of environmental degradation. Climate-change and natural resource degradation are known to generate or aggravate disasters especially of the hydro-meteorological origin. Increasing trend in these disasters like floods, drought, cyclone, pest-attack, fires, (and erosion, slides or epidemics) as secondary impacts), worldwide and especially in continents of Asia and Africa is a serious concern for governments and communities.

(ii) Environmental degradation causes vulnerability: Besides causing new hazards and aggravating such precursors of disaster events, degradation of environment increases socio-economic vulnerability by reducing water, food and nutrition, sanitation and health, livelihoods, housing, bio-productivity, entrepreneurship, and thus overall economics jeopardizing the coping capacity of people. Low capacities result in high exposures to hazards, unsafe locations and conditions of high disaster risk. Poor environmental quality and degraded natural resources also result in social conflicts and political instability.

(iii) Disasters impact environment: Disaster events are also known for causing serious impacts on environment affecting natural processes, natural resources and ecosystems, and thereby creating conditions for secondary or future disasters including complex emergencies. Natural disasters can also trigger chemical or technological disasters. Environmental sustainability is compromised during disaster management operations and recovery process due to emergency engineering, improper disposal of disaster and relief wastes, acute exploitation of natural resources and inappropriate land-use/landscape modifications.

1.3 ENVIRONMENTAL MANAGEMENT AND DISASTER MANAGEMENT CYCLE

Knowledge of environment is crucial in all stages of disaster management cycle including pre-disaster prevention and mitigation, and during post-disaster response, relief, reconstruction and recovery (figure1.3). Experience of the past disasters indicated that environmental services like shelter, water, food safety, sanitation and waste management form crucial components in emergency relief especially in case of water and climate related disasters. On the other hand, concern of disaster risk and mitigation is equally important in all stages of environment management from prevention of hazards and environmental degradation, control, impact minimization, remediation, rehabilitation overall sustainability in environmental systems.

A well-managed environment can act as a buffer against disasters. This can happen in two ways. A healthy or well-functioning ecosystem can regulate or mitigate the hazard itself, thus preventing a disaster from taking place or reducing disaster's impacts. In addition, healthy ecosystems reduce people's vulnerability to disasters by increasing the resilience of communities through meeting basic needs (water, food, health, fuel, etc.) and supporting sustainable local livelihoods and economies.

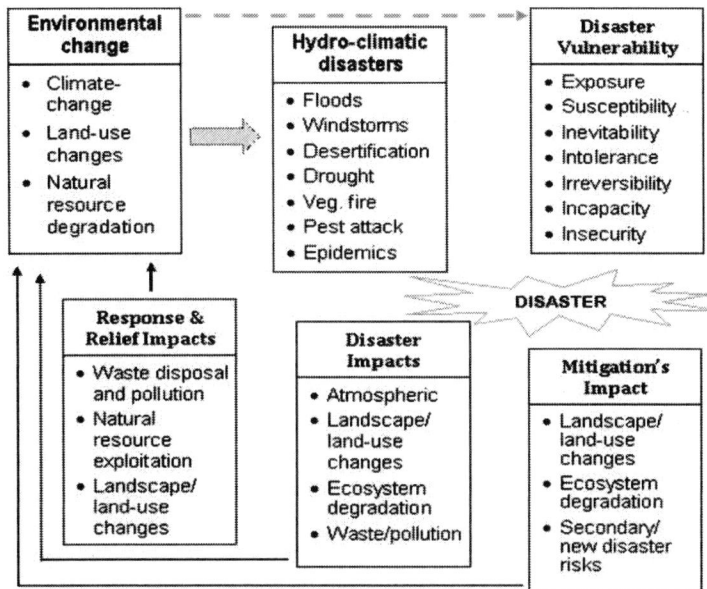

Figure 1.3: Environmental aspects in various stages of disaster management

Opportunities of integrating environmental management and disaster risk management together, hence, are a prime concern emerged globally. Environmental laws and strategic instruments, viz. EIA, Risk & Vulnerability Assessment, Ecological modelling and predictions, Auditing, Environmental Laws facilitate at key stages of disaster risk management.

1.4 HYDRO-METEOROLOGICAL DISASTERS: IMPLICATIONS OF CLIMATE-CHANGE

Hydro-meteorological hazards are the natural processes or phenomena of atmospheric, hydrological or oceanographic nature, which may cause loss of life or injury, property damage, social and economic disruption, or environmental degradation. Hydro-meteorological hazards include: floods, debris and mud floods; tropical cyclones, storm surges, thunder/hailstorms, rain and windstorms, blizzards and other severe storms; drought, desertification, vegetation fires, temperature extremes, sand or dust storms; permafrost and snow or ice avalanches. Hydro-meteorological hazards can be single, sequential or combined in their origin and effects.

India is among the world's most disaster prone areas. India supports 1/6th of the world's population on just 2 percent of it landmass. Nearly 59 percent of India's land area is prone to earthquakes of moderate to very high intensity; over 40 million hectares (12 % of land) is prone to floods; close to 5700 km. of its 7516 km. coast line (about 8%) is cyclone prone and exposed to tsunamis and storm surges. 2 percent of land is landslide prone and 68 percent of India's arable land is affected by droughts (GOI, 2004).

These trends are likely to exacerbate in future with climate change. The projected increase in precipitation and rainfall, the glacial meltdown and rising sea levels will affect India particularly severely, creating conditions for more hazardous events and will lead to increase in incidences of floods, cyclones, and storm surges. Though it is not possible to predict the future frequency or timings of extreme events but there is evidence that the risk of drought, flooding, and cyclone damage is increasing and will continue to do so.

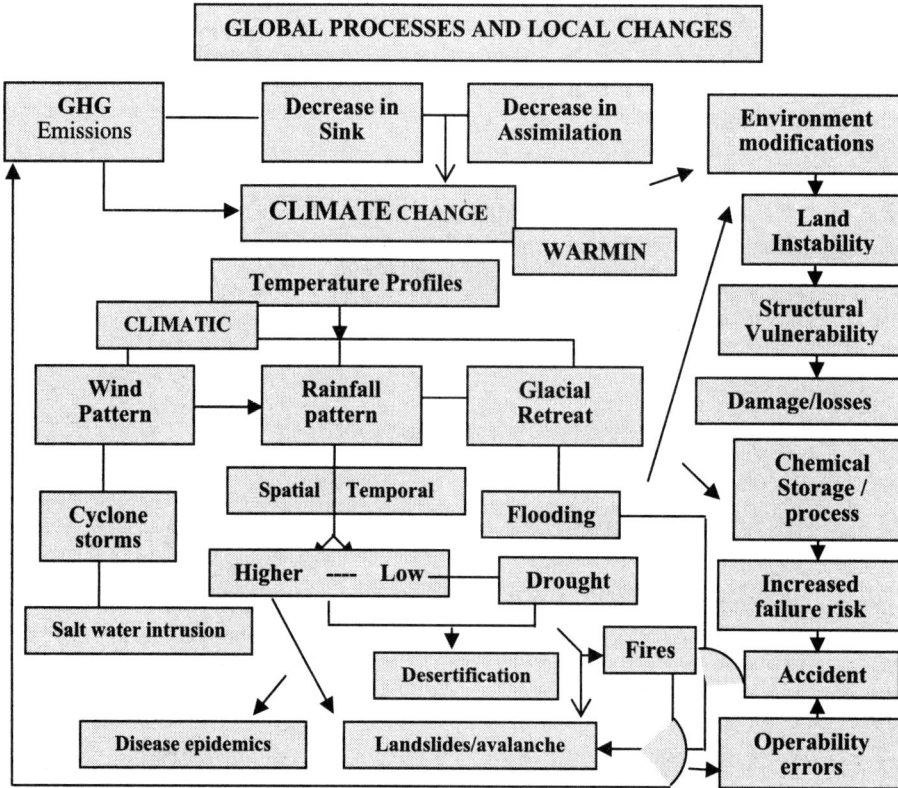

Figure 1.4: Climatic and environmental implications on disaster risks (Gupta, 2010)

Climate change is also likely to threaten India's food security, increase water stress, and increase occurrences of diseases especially malaria. Implications of global processes and local changes driven by climate change that results or aggravates disasters are shown in figure 1.4. Lack of availability and access to technological and financial resources coupled with a high dependence on climate sensitive sectors-agriculture, fisheries, forestry have made India highly vulnerable to climate change. A large and growing population densely populated and a low-lying coastline, and an economy closely tied to its natural resource base further intensifies this vulnerability.

1.5 ECOSYSTEM APPROACH TO ADAPTATION AND DRR

Ecosystem-based Adaptation (EbA) options are often more accessible and affordable to the poor than adaptation interventions based on infrastructure and engineering. It is consistent with community-based approaches to adaptation; can effectively build on local knowledge and needs; and can provide particular consideration to the most vulnerable groups of people, including women, and to the most vulnerable ecosystems. Improved ecosystem management practices need to be developed and promoted to strengthen adaptation to climate change impacts (Figure1.5). As examples, at the local specific level, these include appropriate agricultural and water management practices, breeding techniques for introduction of drought-tolerant, salt-tolerant, and standing-water tolerant crops and tree species; improved livestock management and fodder management practices. At the systems level, this might include alternative cropping or land-use model, strengthening the extension system, or supporting more effective decentralized planning processes for watershed management that can enhance more locally specific adaptive processes. A more systematic approach of integrated natural resource management for cyclone risk management is shown in Figure 1.6.

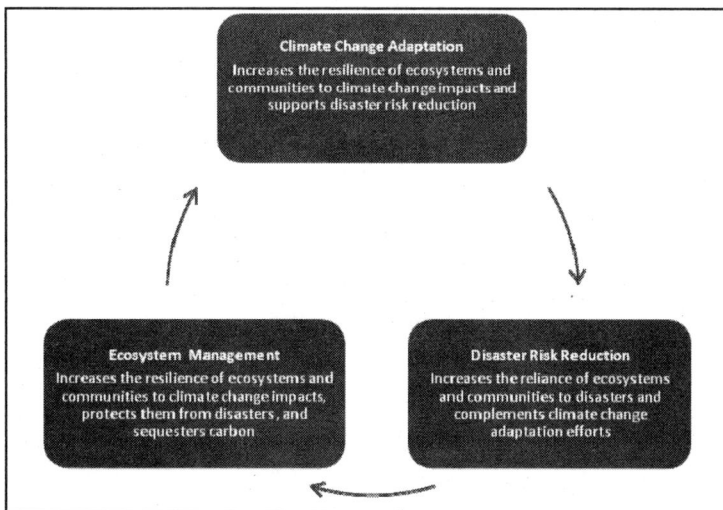

Figure 1.5: Linking climate change adaptation, ecosystem management and disaster risk reduction
(Source: UNEP 2009)

Environment Impact Assessment (EIA) as a decision making tool helps in identifying viable and sustainable option for structural or non-structural measure of disaster mitigation, whereas post-disaster EIA forms a part of damage, loss and needs assessment that enables planning for reconstruction and recovery. Although most EIAs implicitly have focus on natural disaster risk in the project site context and include Disaster Management as part of EMP, the information contained therein the state of environment section and predictions under the section on environmental impacts can be used for carrying out detailed risk assessment and disaster management planning as well. However, there is a need to carry out a systematic analysis of these policies and laws for applying these legal provisions for various activities under disaster management. Role of environmental knowledge in disaster risk management is depicted in Figure 1.7.

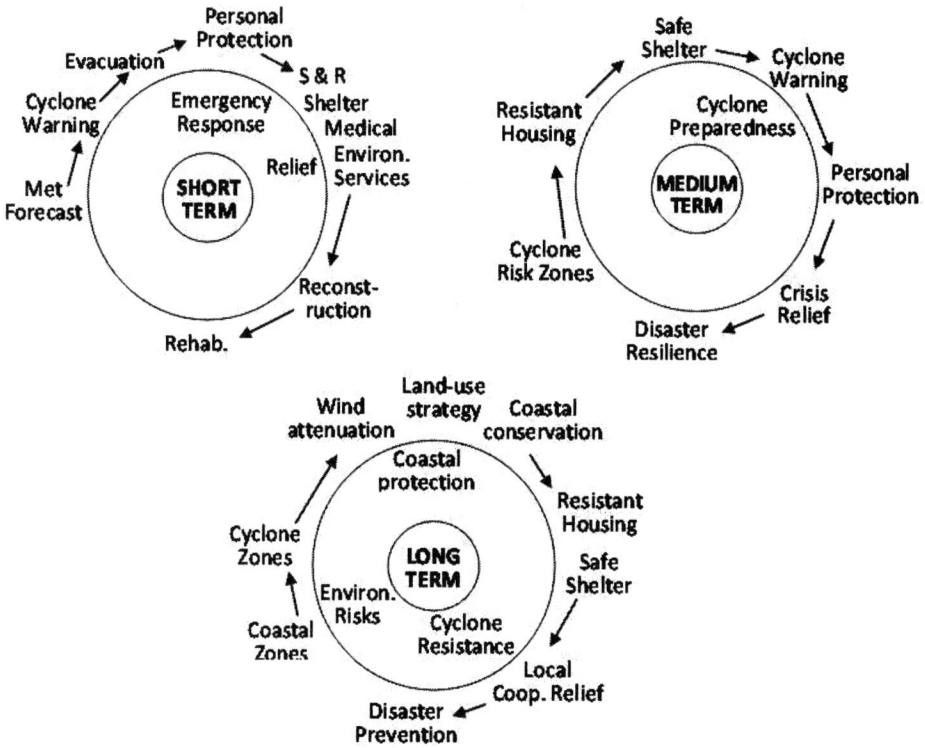

Figure 1.6: Coastal management framework for short-term and long-term mitigation for cyclone disaster
(Gupta and Nair, 2012)

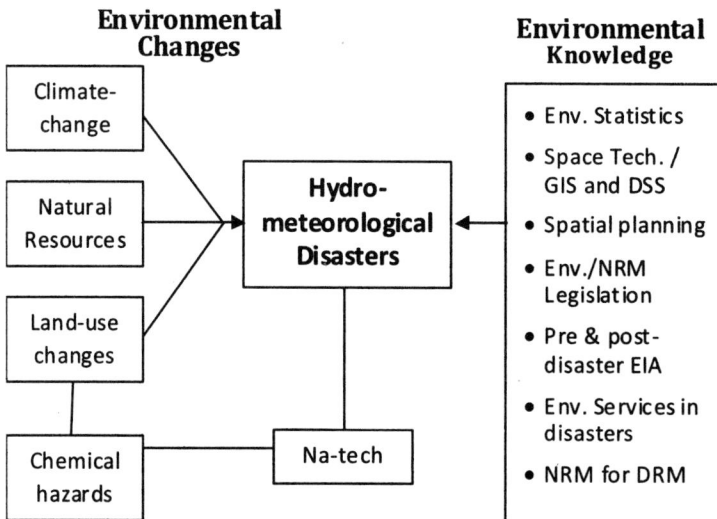

Figure 1.7: Central concept of environmental knowledge for disaster risk management (ekDRM)
(Gupta and Nair, 2011)

Disaster management has primarily been a concern for emergency response and a post-disaster focused approach until the realization of paradigm shift from 'response and relief' to 'prevention and preparedness'. The climate-change awareness globally has brought-in a greater understanding on role of global, regional and local environmental aspects in disaster management (risk assessments, mitigation, early warning and effective response). In many countries, the framework of disaster management has been functional in total separation from the systems that deal with environmental protection and natural resources management. However, at international level the guiding documents like Agenda-21, Hyogo Framework of Action, Millennium Ecosystem Assessment, IPCC 4th Assessment Report, Ramsar Convention, Convention on Desertification, and many other strategic documents have recently emphasized environmental aspects of disaster management and vice versa with intense focus.

At national levels as well, for example in India, the Disaster Management Act and policy have defined 'substantial damage to the environment' as a disaster, and has focused on environmental compatibility and sustainable development as strategic issue in disaster management cycle. On the other hand, policies and legal framework on environment – water, forests, agriculture, land-use, atmosphere and climate change, waste management besides constitutional provisions on environment, have provided options for reducing hazards and vulnerability in context of disasters. For example, the National strategy on climate change actions and missions thereof, policy statement on conservation of natural resources, coastal zone regulation, area specific notifications like one on Doon valley, are of key importance in disaster risk management.

In India the National Disaster Management Authority is the apex national organization for developing guidelines and plans on various aspects of disaster management, whereas Ministry of Environment & Forests is nodal agency for environmental protection including dealing with issues of climate change, forest and habitat conservation, environmental quality, EIA, coastal zone, river conservation, Himalayan ecosystem, etc. Various aspects of land-use and natural resources are dealt by different Ministries/Departments like Rural Development (Dept. of Land Resources), Water Resources, Agriculture, Earth Sciences, Science & Technology, Biodiversity Board, etc. National Institute of Disaster Management has also set-up a specialized 'Environment, Climate change and Disasters Cell' for interdisciplinary activities under Hydro-meteorological disasters division of the institute. Similarly Indian Institute of Public Administration has institutionalized a Centre for Environment, Climate change and Drought Management for focused activities of research and capacity building.

Since the environmental degradations resulting into disaster risk are either slow onset process or have wider geographic extents, local administrators and planners' understanding usually is not sound enough to analyze cause-consequence relationship and may result in inappropriate strategies for disaster mitigation and response. However, many good case examples of utilizing environmental knowledge and management into disaster risk management and also disaster risk reduction's significance in sustainable environment are available particularly in context of hydro-meteorological disasters in mountain and coastal areas and also in context of urban flooding, epidemics and agro-ecosystems.

References

DEFRA/Government of India (2005b). 'Climate Change Impacts on Sea Level in India. Key Sheet 4'. National Institute of Oceanography, Goa, India. In *Investigating the Impacts of Climate Change in*

India. Report by Department of Environment, Food and Rural Affairs (DEFRA), UK and Ministry of Environment and Forests (MoEF), Government of India (GoI), 2005. Retrieved from http://www.defra.gov.uk/environment/.

German Advisory Council on Global Change (WBGU) (2008). *Climate Change as a Security Risk*. UK: Earthscan.

Government of India (2004). *Disaster Management in India - A Status Report*. National Disaster Management Division, Ministry of Home Affairs, Government of India, MHA/GOI/28/06/2004, Jun 2004.

Gupta, A. K., Yunus, M., and Misra, J. (1998). Disaster Reduction and Sustainable Development. *Environews 4*, 06.

Gupta, A. K. (1999). Science, Sustainability and Social Purpose: Barriers to Effective Articulation, Dialogue and Utilization of Formal and Informal Science in Public Policy. *International Journal Sustainable Development, 2* (3), 368-371.

Gupta, A. K., Suresh, I. V., Misra, J., and Yunus, M. (2002). Environmental risk mapping approach: Risk minimization tool for development of industrial growth centres in developing countries. *Journal of Cleaner Production, 10*(3), 271-281. doi: 10.1016/S0959-6526(01)00023-3.

Gupta, A. K., and Yunus, M. (2004). India and the WSSD (Rio + 10), Johannesburg: Issues of national concern and international Strategies. *Current Science, 87*(1), 37-43.

Gupta, A. K., Nair, S. S., and Shard, S. (Eds.) (2009). Chemical Disaster Management, Proceeding Volume of the National Workshop, 30 Sept.-01 Oct. 2008. New Delhi: Ministry of Environment & Forests (GoI) and National Institute of Disaster Management.

Gupta, A. K., Nair, S. S., and Sehgal, V. K. (2009). Hydro-meteorological disasters and climate change: conceptual issues and data needs for integrating adaptation into environment and development framework, *Earth Science India, 2* (I), 117 – 132.

Gupta, A. K., Nair, S. S., Chopde, S., and Singh, P. K. (Eds.) (2010). *Risk to Resilience: Strategic Tools for Disaster Risk Management*. Proceedings of the International Workshop, 2-3 Feb 2009. New Delhi: NIDM, ISET-Colorado.

Gupta, A. K., Nair, S. S., and Sharma, V. K (2011). Environmental Knowledge for Management of Hydro-meteorological Disasters. In Anil K.Gupta & Sreeja S. Nair (eds) *Environmental Knowledge for Disaster Risk Management* (pp.106-117). New Delhi: NIDM and GIZ.

Gupta, A. K., and Nair, S. S. (2012). *Environmental Extremes Disaster Risk Management – Addressing Climate-change*. New Delhi: NIDM.

Haqqani, A. B. (ed.) (2003). *The Role of Information and Communication Technologies in Global Development: Analyses and Policy Recommendations*. New York: United Nations Information and Communication Technologies Task Force.

Kumar, S. (2006). Hazard, risk, vulnerability and climate change in India. In *Proceedings of the International Conference on Adaptation to Climate Variability and Change, Jan 5-7, 2006,*

Institute for Social and Environment Transition (ISET), Boulder, Colorado, USA & Winrock International (WII), . New Delhi, India (pp. 22-27).

Lal, M. (2001). Future Climate Change: Implications for Indian Summer Monsoon and its Variability. *Current Science 81*(9), 1196-1297.

Nair, S. S., and Gupta, A. K. (2010). Industrial siting in multi hazard environment – Role of GIS and MIS. *International Geoinformatics Research and Development Journal, 1*(2), 1-14.

Patwardhan, A., Narayanan, K., Parthasarathy, D., and Sharma, U. (2003). Impacts of Climate Change on Coastal Zones. In: P. R. Shukla, Subodh K. Sharma, N. H. Ravindranath, A. Garg and S. Bhattacharya (Eds). *Climate Change and India; Vulnerability Assessment and Adaptation*. (pp. 326-359).Universities Press (India), Hyderabad, India.

UNDP (2007b). Managing Risks of a Changing Climate to Support Development. In *Report of the Asia Regional Workshop,* 23-26 Apr 2007, Kathmandu, Nepal, UNDP Regional Center Bangkok, Thailand. Retrieved from http://regionalcenterbangkok.undp.or.th.

UNEP (2008). *Environmental Needs Assessment in Post-Disaster Situations - A Practical Guide for Implementation.* Retrieved from http://oneresponse.info/crosscutting/environment/-publicdocuments /UNEP_PDNA_draft.pdf.

2

Sustainable Rural Development for Disaster Risk Reduction

P.C. Kesavan and M.S. Swaminathan

2.1 INTRODUCTION

The word 'disaster' is variously defined in the literature and in this paper it is defined as a natural or man-made event that negatively affects life, property, livelihood or employment often resulting in "permanent or almost irreversible changes" to human societies, ecosystems and environment. Our solar system is estimated to have been formed about 4.56 billion years ago with a violent gravitational collapse of a small part of a giant molecular cloud, which in turn was the result of an earlier explosion (13 to 20 billion years ago) of a subatomic unit. The 'Big Bang Theory' explains it. Most of the collapsing mass collected in the centre, forming the sun, while the rest flattened into a proto-planetary disc out of which the planets, moons, asteroids and other celestial bodies formed. Collisions between galactic bodies are still going on in a cycle of annihilations and creations. Earth formed from such violent events is inherently violent. Over the billions of years, after several major cataclysmic events, Earth has become substantially but not entirely pacific. The geophysical disasters (e.g. earthquakes, and volcanoes) and the water and weather-related (hydro-meteorological) disasters which occur at varying frequencies and intensities establish not only the violent past, but also the violent present of our planet. Even more rarely, but without doubt, astro-physical disasters (e.g. a meteorite colliding with our space ship Earth) have also occurred. One view is that dinosaurs became extinct after a meteorite impact on Earth about 60 million years ago.

The astro-physical, geo-physical and hydro-meteorological extreme events are classified as 'natural disasters', whereas major accidents such as Chernobyl nuclear reactor accident (26th April 1986), and the Bhopal gas (methyl isothiocyanate) disaster (night of 2nd and early hours of 3rd December 1984) are the man-made disasters. The March 2011 disaster in Japan is unique in the sense that a major earthquake (M~8.9) induced a powerful tsunami and the combined impact of these two caused serious structural damage to the nuclear power reactors in Fukushima Daiichi and consequently explosions in the nuclear reactors have led to release of radionuclides in sea water and atmosphere. As this paper is written (April 11, 2011), no effective containment of radionuclides and radiation has been achieved. This has raised serious questions on the safety of

Disaster Management and Risk Reduction: Role of Environmental Knowledge; Editors Anil K. Gupta, Sreeja S. Nair, Florian Bemmerlein-Lux and Sandhya Chatterji; Copyright © 2013, Narosa Publishing House, New Delhi

nuclear power reactors especially in the areas known to have seismic activity. Several of Indian's nuclear power reactors are required to be set up along its long coastline (~ 7500 km) where the human population density is also very high. About 30 % people of India's population of 1.20 billion live in the coastal areas and they are highly vulnerable to hydro-meteorological (i.e. cyclones, floods, drought etc.) disasters. Hence, Dr. Swaminathan has written on 15 March 2011 to Shri Jairam Ramesh, Minister of State, Ministry of Environment and Forests, Government of India stating that his concern about the safety of nuclear power plants located along the coast such as Kalpakkam and Kudangulam in Tamil Nadu makes him feel that in addition to necessary technological reinforcements, the bio-shields comprising of mangrove and non-mangrove species in the coastal areas adjoining nuclear power plants should be effectively developed. There would be effective second-line defence. Besides, the risk of sea level rise on account of global warming and the melting of ice and glaciers on high mountain systems and Polar Regions is quite considerable to India and several small developing island states.

2.2 FEED-BACK RELATIONSHIP BETWEEN SOCIAL AND ENVIRONMENTAL FACTORS

At the United Nations Conference on "Human Environment" in Sweden, in 1972, late Mrs. Indira Gandhi the then Prime Minister of India, who led the Indian Delegation declared that 'poverty is the greatest pollutant' and that so long as poverty persists, environmental degradation cannot be prevented. That statement led to an understanding of the vicious spiral between poverty and environmental degradation. Environmental degradation especially in the rural areas (India has about 638,400 villages) accentuates poverty and the poverty in turn results in greater degradation of the environment. The detrimental combination of these two environmental and social dimensions not only contribute to climate change and increased incidence of extreme hydro-meteorological events, but also to enhanced vulnerability to loss of lives, property and livelihood resource base. Kesavan and Swaminathan (2006) have illustrated the vicious feedback relationships as follows:

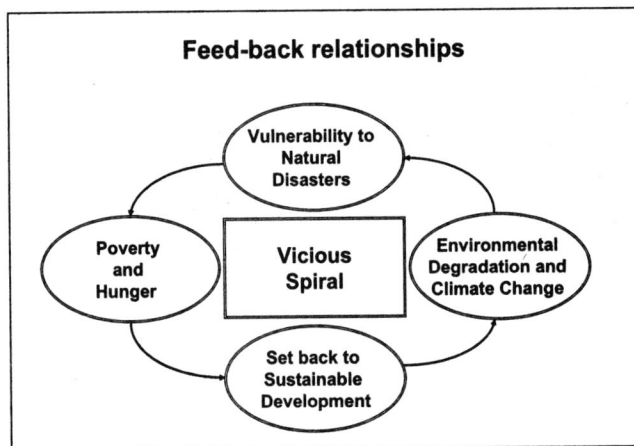

Figure 2.1: Vicious feedback relationships between environmental and social factors and disasters (Kesavan and Swaminathan, 2006)

In a subsequent paper, the authors (Kesavan and Swaminathan 2008) have elaborated how environmental degradation leads to social disintegration through mass exodus of rural families (i.e. 'environmental refugees') to eke out a living in urban areas and the burden of poverty and responsibility of feeding the household members fall on the shoulders of women, often quite young. That is the 'feminization of poverty' (Figure 2.2).

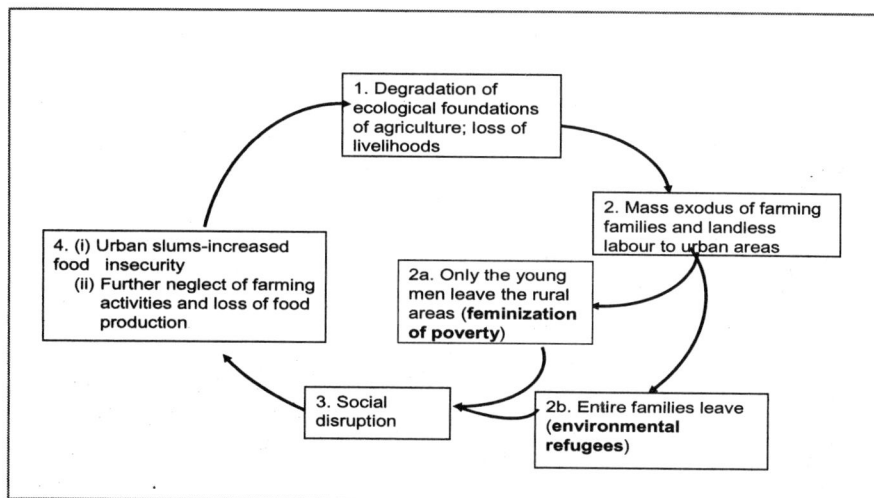

Figure 2.2: Feminization of Poverty' (Environmental degradation, Environmental Refugees) (Kesavan & Swaminathan, 2008)

The terms 'environmental refugees' and 'feminization of poverty' bring out not only the state of abject poverty but also deprivation and helplessness particularly of rural women. When a disaster strikes they are the first ones to perish without being able to offer any resistance and the survivors among them are the ones who have little resilience (i.e. coping capacity) to bounce back to normal or near normal daily life. So, the vulnerability to natural disasters has a strong social and gender dimension. Among the poor, women and children bear the brunt of a natural disaster more severely than men. An example is that in Cuddalore (a coastal town in Tamil Nadu) alone, the December 26, 2004 tsunami resulted in the loss of lives of 391 women compared with that of 146 men. A large number of children clinging to their mothers also perished. Aglionby (2005) has estimated that four times as many women died in tsunami. The rural women, many with frequent pregnancies, high level of malnourishment and lack of benefits of education, skills, economic independence and equal rights in decision-making are particularly the most vulnerable. So, social, economic and gender equities are as essential as technological empowerment of rural communities to achieve disaster risk reduction. Of course, technological development is very important. For instance, the Global Environment Outlook 3: Natural Disasters (http://www.grida.no/geo/geo3/english/448.htm) notes that the level of development is directly proportional to the severity of destruction and number of deaths to a given disaster. It is estimated that the average number of people dying per reported disaster are 22.5, 145.0 and 1052 for countries with high, medium and low level development respectively. Within India, the urban-rural divide in terms of availability of technological support, communication, transport, clean drinking water, sanitation, health care and medical facilities is very wide. Many basic amenities are still lacking in thousands of villages even after 63 years of independence.

It is therefore necessary to integrate the environmental, social and economic dimensions of rural development with disaster preparedness. After assessing the December 26, 2004 tsunami devastation of the Andaman and Nicobar islands, Kesavan and Swaminathan (2007) have emphasized on the need to integrate sustainable development with disaster preparedness for effective risk reduction.

2.3 GROWING CONCERN OVER HYDRO-METEOROLOGICAL DISASTERS

Since poverty and environmental degradation form a vicious spiral, particularly in the rural India, the primary focus must be on sustainable pathways of development to link livelihood and food security of the rural poor with ecological security of the region. A common observation is that degradation of forests directly or indirectly affects the livelihoods of forest-dependent communities and also increases the human vulnerability to natural disasters. Destruction of mangrove forests results in lack of nutrient source in the estuaries for fish, prawns, crabs etc. on the one hand and loss of an effective bio-shield to reduce the velocity of powerful cyclones and tsunami waves on the other. Restoring degraded mangrove forests provides livelihoods for fishermen and also protects them from severe hydro-meteorological disasters.

At this point, it must be mentioned here that among the natural disasters, the hydro-meteorological disasters are increasing both in their frequencies of occurrence and also in their destructive potential. The data published by the UNEP/DEWA/GRID-Europe 2004, suggest that during 1991-2004, the world distribution of disasters by geological, biological and hydro-meteorological origin were 10 %, 14 %and 76 %respectively. With a long coastline of about 7,500 km and thick density of human population (~360 million) India needs to be quite prepared to deal with the risk of cyclones, floods, droughts etc. It is of interest to note that in the 1990s, more than 90 %of those killed in natural disasters lost their lives in hydro-meteorological events (mainly droughts, floods, wind storms etc.,) while earthquakes accounted for 30 %of the estimated risk, they caused just 9 %of all fatalities due to natural disasters. In contrast, hunger caused by famines (due to drought) worldwide killed 42 %of people in the affected regions (IFRC 2001).

The hydro-meteorological disasters are not only increasing in their frequencies of occurrence but their destructive potential is also increasing (Emmanuel 2005). Global warming induced climate change is implicated. Alarming scales of melting and recessions of glaciers in the Himalayas, Poles and Greenland are causing sea level rise. Several small island nations such as Kiribati, Seychelles and Maldives have more than 80 %of land area at less than a metre above the present sea level. Such low-lying islands may have to pay a high price in terms of loss of human lives and property, if sea level rises to the extent forecast by global climate models (UNEP/GRID; A. Rendal 2005). India's Andaman and Nicobar archipelago and Laccadives islands also come under the threat of sea level rise.

The Himalayan glaciers are indeed melting and it is reported that there is an overall reduction from 2,077 km2 in 1962 to 1,628 km2 in 2007. With an eventual disappearance of Himalayan glaciers, the great rivers such as Ganges, Indus, Brahmaputra would no longer be perennial. This initial 'ecological catastrophe' would directly result in "agricultural catastrophe" (Brown, 2008). A similar happening is likely in China for Yangtze and Yellow rivers fed by the Himalayan glaciers. China and India have world's largest human population and are also the two large wheat and rice producing countries (Kesavan and Swaminathan, 2011). Further, data show that with even a slight rise in average night temperature of about 0.5°C, the wheat yield in the north Indian plains decreases by about 450kg/ hectare (Sinha and Swaminathan 1991). Sea level rise in the coastal areas would lead to soil erosion as well as salinization of the soil and fresh water aquifers. These in turn would have adverse implications for food production and food security of India.

2.4 PARADIGM SHIFT IN DISASTER RISK REDUCTION

From the foregoing, it is obvious that resource-poor, poverty-stricken, food insecure, largely illiterate and unskilled rural women and men cannot reduce their vulnerability unless their living conditions improve and they are able to stand on their own feet. It has already been discussed earlier that the famine of rural livelihoods is a major cause of environmental degradation. Rural livelihoods for hundreds of millions of rural women and men cannot be solved by establishing factories, promoting mining activities etc., with a major thrust largely on jobless economic growth. Such activities are invariably akin to "mass production" and "jobless economic growth". In order to link the livelihood security of over 500 million rural people with ecological security of their regions, there is need for development of eco-friendly market oriented rural enterprises by harnessing eco-technologies. The eco-technologies have pro-nature, pro-poor and pro-women orientation resulting from blending of frontier technologies with traditional knowledge and ecological prudence of the rural and tribal communities. The goal is to enable them to manage their local resources in a sustainable manner and develop eco-enterprises using eco-technologies. As described by Kesavan and Swaminathan (2006), knowledge and eco-technological empowerment of the rural women and men is the first step in improving the present helpless state of affairs. Sustainable management of local resources and creation of sustainable livelihoods through technological and knowledge empowerment provide not only means of income generation and food security but also dignity and self-confidence. The largely "subsistence agriculture" should be transformed into dynamic eco-agriculture combining crops (cereal grains, pulses, oilseeds, millets, vegetables, fruits etc.), farm animals for milk and meat and biodiversity. In the coastal villages, culture fisheries, especially ornamental fisheries should be included. In the coastal areas, conservation and enhancement of mangroves enriches harvest of edible marine organisms and also protects lives and livelihoods of the coastal communities against severe cyclonic storms, tsunami etc. Illiteracy or inadequate literacy and lack of skill do not handicap the rural women and men in mastering various eco-technologies for eco-enterprises and making use of modern ICT for tele-conferencing, tele-commerce and for early warning and disaster management purposes. Nearly four decades ago, Swaminathan (1972) had emphasized that rural women and men could master the use of technologies through a pedagogic method of 'learning by doing' and he coined the term 'Techniracy' to describe it. So, the experience at the M.S. Swaminathan Research Foundation is that rural people in the Village Knowledge Centres (VKCs) master computer operations in very short time, say a week, and take to useful eco-technologies like fish to water. Eco-agriculture together with eco-enterprises fights both the famines of food and rural livelihoods. Livelihoods for income generation, agriculture for food production, technological and knowledge empowerment for disaster preparedness and sustainable rural development bring about a "paradigm shift" in the pathway to disaster risk reduction. This 'paradigm shift' initiated by the MSSRF in the 1990s revealed its immense usefulness in (a) forewarning the arrival of tsunami in the fisher village Veerampattinam, in Puducherry on the

26th December 2004 and alerting people to move away from shoreline to higher and safer grounds and (b) in organizing the relief and rehabilitation in a disciplined and organized manner to avoid chaos and to ensure that women, children and old and infirm people received food, water and medical aid.

The two dimensions of "paradigm shift" are the following:

(i) A shift from "post-disaster management" to "pre-disaster preparedness" through technological and knowledge empowerment of the vulnerable rural women and men. Eco-technological empowerment and knowledge empowerment through establishing people-owned, people-managed, Village Knowledge Centres (VKCs) with modern ICT have played a role in the "paradigm shift".

(ii) The second aspect of the "paradigm shift" is that in the VKCs established by the MSSRF, the development of eco-enterprises for livelihood, conservation and enhancement of coastal bio-shield, initiatives for setting up VKCs and disaster preparedness are "bottom-up" and participatory in approach.

After the Indian Ocean tsunami of 26th December 2004 by which time MSSRF had set up several VKCs and also Village Resource Centres (VRCs) with up-and down satellite link in cooperation with the Indian Space Research Organisation (ISRO), the Government of India enacted the Disaster Management Act 2005. It has much in common with the MSSRF's model and its focus is on a paradigm shift from 'post-disaster relief and rehabilitation to improving the pre-disaster preparedness, initiating disaster mitigation projects and strengthening emergency response capacities. It established National Disaster Management Authority (NDMA) as the apex body for disaster management in the country. The State Disaster Management Authorities (SDMAs) chaired by respective Chief Ministers at the state level, and District Disaster Management Authorities (DDMAs) chaired by the respective District Collectors and co-chaired by the elected representative of the "Zilla Parishad" in the respective districts have also been set up. During the last five years of its existence, the NDMA has developed a number of disaster management guidelines for dealing with extreme natural events such as floods, drought, cyclone, earthquake, infestations and other biological disasters, nuclear and radiological emergencies, chemical disasters and disasters caused by terrorist activities, landslides and snow avalanches. It has also developed guidelines for revamping of civil defence, medical preparedness and mass casualty management. It also addresses the technological, social and management dimensions. However, these approaches need to be integrated with basic elements of sustainable rural development, restoration and conservation of ecosystems, as well as eco-technological and knowledge empowerment of resource-poor, small and marginal farming, fishing and landless rural families. These would usher in the much-needed "bottom-up" approach necessary for an effective disaster risk reduction in the rural areas, especially in the remote villages located in disaster-prone regions. Modern ICT tools, particularly the cell phones can play an important role in reaching the last mile and last person. These would greatly help in developing a "bottom-up and participatory" rather than a 'top-down" system of preparedness. These will also enhance the 'resilience' or the "coping capacity" of the disaster-affected people to restore normalcy within a short time. It may be recalled that despite best efforts, the civil societies and the government could not repair extensive damage to boats, fishing nets etc. after December 2004 tsunami and consequently, the fisher families in Tamil Nadu suffered without livelihood and food security for quite a long time. MSSRF facilitated alternate livelihood (poultry, oyster mushroom) for the fisher women in one of the worst-affected coastal village, 'Sadraskuppam', a village near the Kalpakkam nuclear power plants. These equip the rural communities to stand on their feet to face the disaster whenever it could strike all of a sudden. Appropriate protective response would be instantly needed. Hence, the vulnerable people themselves need to be trained, equipped and psychologically prepared.

2.5 SOCIAL, ENVIRONMENTAL, ECONOMIC AND TECHNOLOGICAL FACTORS AT THE GRASSROOTS LEVEL

Since vulnerability has social, environmental, economic and technological dimensions, approaches to achieving disaster risk reduction must also address all of these in an integrated manner. These essentially sum up what M.S. Swaminathan had emphasized in an article, "Beyond tsunami: An agenda for action", in 'The Hindu", (Monday, January 17, 2005). In that article three following initiatives are proposed all along the coast viz., (i) strengthening the ecological foundations of sustainable human security (ii) rehabilitating livelihoods and fostering sustainable livelihood security, and (iii) putting in place knowledge centres in vulnerable coastal villages. Also, these are implemented necessarily in a 'bottom-up', participatory manner. Operationally, a cluster of eco-enterprises based on appropriate eco-technologies (which result when frontier technologies are blended with traditional knowledge and ecological prudence) each with pro-nature, pro-poor, pro-nature and pro-employment is undertaken in a village, it becomes a 'Biovillage' (bios= living). This is indeed transforming a village with conventional subsistence farming and sedentary lifestyle to a more vibrant agri-business unit with a number of eco-enterprises run by enthusiastic self-help groups. Sustainable management of environmental resources for sustainable livelihoods links livelihood security with ecological security. Technological and skill empowerment are through 'techniracy' (i.e. a pedagogic method of learning by doing). Knowledge is power today. Locale-and time-specific, demand-driven information is needed by the resource-poor, rural women and men to overcome the problems encountered in crop and animal husbandry, fisheries, marketing, health care, communication, transport, education etc., and also for awareness of welfare schemes of the state and central governments; when young rural women who have passed 7th or the 8th standard are trained in computer operations, they become the managers of the VKCs. With young women as the Heads of Knowledge Centres, many tangible and intangible benefits occur. Income generation and enhancement of social esteem are clearly tangible. Women connected with VKCs and VRCs gain self-confidence and participate in decision-making processes at home and at grassroots' Panchayat level. Enlightenment of women leads to family planning and better education and better management of food and nutrition security at the household level. And in case of any emergency caused by accidents or natural disasters, they are no longer helpless and incapable of appropriately dealing with the situation. They also master the use of cell phones for contacting police, hospitals, fire service, weather centres etc. These represent preparedness at the grass root level. Women in particular are known for conservation of forest trees, water bodies and biodiversity.

In nutshell, the bio-villages together with VKCs help in integrating sustainable rural development with pre-disaster preparedness. The five Es viz., Economics, Environment, (ecology included), Equity (gender and social), Energy and Employment are essential components of pre-disaster preparedness.

2.6 NEW DIMENSIONS OF DISASTERS

It is a fallacy that technological innovations effectively thwart Malthusian scourge. Every technology exerts a negative impact too. The green revolution of the 1960s and 1970s degenerated into greed revolution and resulted in serious environmental degradation and loss of biodiversity. The modern satellite and computer linked technologies have created electronic pollutants in thespace and on land. So, the real situation is that both humanity and planet Earth are at crossroads. The "ecological footprint" has far exceeded the "bio-capacity" of Earth (Wackernagel et al, 2002). The Copenhagen Accord (2009) does not provide a concrete action plant to contain the emission levels at or below 44 Gt CO_2- eq. The national emissions reduction pledges are insufficient to meet the objective of keeping the global warming to below 2°C. An analysis by Rogelj et al (2010) suggests that the global warming could exceed 3°C by 2020. This means a catastrophe resulting from crossing over the "tipping point". The 'tipping point' with reference to climate change is a point when global climate changes from one stable state to another stable state. Transition to a new state occurs. And the tipping event could be irreversible. Rockstrom et al (2009) show that the planetary boundaries have already been transgressed in a number of parameters viz., concentration of CO_2 in the atmosphere, rate of biodiversity loss, nitrogen cycle, ocean acidification etc. So, tipping point with regard to hydrological cycle will be highly catastrophic.

Notwithstanding controversies whether climate change is natural or man-made, the fact of a climate change is well-established (Climate Change Reconsidered, The Heartland Institute, Chicago, 2010). Predominantly agricultural, densely populated, and very long coastline countries such as India will need to be well-prepared to avert famine of food, drying up of rivers and loss of biodiversity.

India's action plan on climate change is to protect the poor and vulnerable sections of the society through an inclusive and sustainable development strategy, which is sensitive to climate change. The Prime Minister's council on climate change has put emphasis on promoting basic understanding of climate change, adaptation and mitigation, energy efficiency and natural resource conservation. Eight National Missions viz., (i) National Solar Mission (ii) National Mission for Enhanced Energy Efficiency (iii) National Mission on Sustainable Habitat, (iv) National Water Mission (v) National Mission for Sustainable Himalayan Ecosystem (vi) National Mission for a Green India (vii) National Mission for Sustainable Agriculture and (viii) National Mission on Strategic Knowledge for Climate Change with major goals of mitigation, adaptation, and risk management have been identified.

These eight missions have much of their basic tenets and pathways based on the concept of "evergreen revolution" proposed by Swaminathan (1996a; b; 1999; 2000; 2002; 2005). It is particularly suitable to achieve productivity in perpetuity without accompanying ecological harm in millions of resource-poor, small and marginal farms with nearly 2/3 of them located in the rain-fed regions of farming. The National Commission on Farmers (NCF) under the Chairmanship of Professor M.S. Swaminathan submitted its final report in October 2006. It has included a draft National Policy for Farmers incorporating its major recommendations. These include technological, economic, environmental, social and gender aspects which would greatly enhance resilience of Indian agriculture in an era of climate change with increased incidence of hydro-meteorological disasters. There is no indication as yet that these are under implementation now. It would seem that acute and sudden disasters are now receiving some attention, but not the slow and chronic ones especially in the agricultural front. We should remember that drought-related famines have killed far more number of people, than the most serious earthquakes. So, the disaster risk reduction must have appropriate strategies to overcome the impact of a bad monsoon, as well as hydro-meteorological disasters.

References

Aglionby, J. (March 26, 2005). Four times as many women died in tsunami. *The Guardian*. Retrieved from http://www.guardian.co.uk/society/2005/mar/26/.

Brown, L.R. (2008). *Plan B 3.0: Mobilizing to Save Civilization*. New York: W.W. Norton & Company. p398.

Emanuel, K. (2005). Increasing Destructiveness of Tropical Cyclones over the Past 30 Years. *Nature436*, 686-688. doi:10.1038/ Nature03906

IFRC (2001). *World Disasters Report*. See http://www.ifre.org/publicat/wdr 2001/.

Kesavan, P.C., & Swaminathan, M.S. (2006). Managing Extreme Natural Disasters in Coastal Areas. *Philosophical Transactions of the Royal Society, A.,364*,1845,2191-2216.
doi: 10.1098/rsta.2006.1822

Kesavan, P.C., & Swaminathan, M.S. (2007). The 26 December 2004 Tsunami Recalled: Science and Technology for Enhancing Resilience of the Andaman and Nicobar Islands Communities. *Current Science 92, 6*, 743-747.

Kesavan, P.C., & Swaminathan, M.S. (2008). Strategies and Models for agricultural sustainability in developing Asian countries. *Philosophical Transactions of the Royal Society. B, 363*, 1492, 877-891. doi: 10.1098/rstb.2007.2189

Kesavan, P.C., & Swaminathan, M.S. (2011). *Disaster Management in India, Oxford Companion to Economics in India*. New Delhi: Oxford University Press, India (in press).

Nongovernmental International Panel on Climate Change (2010). *Climate Change Reconsidered: The Report of the Nongovernmental International Panel on Climate Change*. Chicago, Illinois, U.S.A.: The Heartland Institute, p708.

Rockstrom, J., Steffen, W., Noone, K., Persson, A., Chapin, F.S., Lambin, E.F., Lenton, T.M., Scheffer, M., Fokle, C., Schellnhuber, H.J., Nykvist, B., Wit, C.A., Hughes, T., Leeuw, S.V.D., Rodhe H., Sorlin,S., Snyder, P.K. Costanza, R., Svedin, U., Falkenmark, M., Karlberg, L., Corell, R.W., Fabry, V.J., Hansen, J., Walker, B., Liverman, D., Richardson, K., Crutzen, P., & Foley, J.A. (2009). A Safe Operating Space for Humanity. *Nature 461*, 472-475.

Rogelj, J., Nabel, J., Chen, C., Hare, W., Markmann, K., Meinshausen, M., Schaeffer, M., Macey, K., & Hohne, N. (2010). Opinion: Copenhagen Accord Pledges are Paltry. *Nature 464*,1126-1128.

Swaminathan, M.S. (1972). *Agricultural Evolution, Productive Employment and Rural Prosperity*. Mysore, India: The Princess Leelavathi Memorial Lecture, University of Mysore.

Swaminathan, M.S. (1991). Deforestation, Climate Change and Sustainable Nutrition Security: A Case Study for India. *Climate Change(19)* 201-209. (doi:10.1007/BF00142227).

Swaminathan, M.S. (1996a). *Sustainable Agriculture: Towards an Evergreen Revolution*. Delhi, India: Konark Publishers, Pvt. Ltd.

Swaminathan, M.S. (1996b). *Sustainable Agriculture: Towards food security*. Delhi, India: Konark Publishers Pvt. Ltd.

Swaminathan, M.S. (2002). *From Rio de Janeiro to Johannesburg: Action Today and not just promises for Tomorrow*. East-West Books (Madras) Pvt. Ltd, p 1-224.

Swaminathan, M.S. (2005, January 17). Beyond Tsunami: An Agenda for Action. *The Hindu*. Retrieved from http://www.hindu.com/2005/01/17/stories/

Swaminathan, M.S. (2000). An Evergreen Revolution. *Biologist 47*, 85-89.

Swaminathan, M.S. (1999). *A Century of Hope:Towards an Era of Harmony with Nature and Freedom from Hunger*. Chennai, India: East West Books (Madras) Pvt. Ltd.

UNEP-GRID (2002). Natural Disasters. *Global Environment Outlook 3: Past, Present and Future Perspectives*. Arendal, Norway: UNEP, GRID. Retrieved from www.grida.no/geo/geo3/english

Wackernagel, M, Schulz, N.B. Deumling, D., Linares A.C., Jenkins, M., Kapos, V., Monfreda, C., Loh J., Myers, N., Norgaard, R. & Randers, J. (2002). Tracking the ecological overshoot of the human economy. *Proceedings of the National Academy of Sciences, USA, 99* (14), 9266-9271. www.pnas.org/cgi.doi10.1073 pnas.142033699

3

Disasters and the Environment:
A Review of Opportunities in Disaster Recovery

Charles Kelly and N.M. Prusty

3.1 INTRODUCTION

The intersection of the environment and disasters has been gaining increasing salience in recent years. Disasters and the subsequent relief and recovery operations resulted in huge environmental damages in the past. Notable examples are Japan earthquake, tsunami and nuclear emergency (March 2011), Haiti earthquake 2010, Sichuan earthquake (May 2008) and South Asian Tsunami (December 2004). The United Nations Environment Program (UNEP) and UN Office for the Coordination of Humanitarian Assistance (OCHA) have a joint environment unit formobilizing and coordinating the international response to environmental emergencies. The American Red Cross and World Wildlife Fund (WWF) US have recently cooperated on developing guidance on environmental sound recovery. A number of other organizations focus on the nexus of the environment and disasters, for instance ProAct Network3 and GroupeUgenceRebilitation and Developpement (Groupe U.R.D.) and so on. Efforts by these organizations have led to the creation of a number of guides and tools to assess the impact of disasters, relief and recovery on the environment. UNEP's Disasters and Conflict Program has assembled a wide range of these guides and tools on disasters and the environment into a single web-based point of access.

Environmental issues are increasingly getting attention following disasters in recent times. For instance both UNEP and the US Agency for International Development (Kelly and Solberg, 2011) conducted rapid disaster impact assessments following the 2010 Haiti earthquake. The Ministry of Environment, WWF Chile / WWF International and Antofagasta Minerals conducted a rapid assessment following the 2010 Chile earthquake and tsunami (Ministry of Environment, 2010). The post-tsunami environment impact assessment carried out in the Sri Lankan coast in Jan. 2005 (Dahdouh-Guebas et al., 2005b). Environment advisors were assigned to the shelter clusters in Haiti following the 2010 earthquake and the 2009 earthquake in Sumatra, Indonesia. UNEP has been active in post conflict and post disasters assessments, including Haiti, Sudan and the Democratic Republic of the Congo, as well as in ongoing conflict in Darfur. Both ProAct Network and Groupe URD have been active in developing reports on environment and disaster

Disaster Management and Risk Reduction: Role of Environmental Knowledge; Editors Anil K. Gupta, Sreeja S. Nair, Florian Bemmerlein-Lux and Sandhya Chatterji; Copyright © 2013, Narosa Publishing House, New Delhi

related issues, or performing field work on specific environment-disaster issues (e.g., improved energy supplies in Darfur).

Figure 3.1: During the emergency phase, shelter and sanitation were among the key Environmental challenges *(Source: UNEP, UNEP in Haiti 2010 Year in Review)*

Figure 3.2: Cutting of young trees and mangroves in shelter and home construction increased dramatically. *(Source: Figures by Allegra Da Silva, Malory Hendrickson and Diego Vallejo, REA Report, Haiti, USAID, 2010)*

Yet, with all this action to date, recognition of the importance of the environment in the post disaster (relief and recovery) context still does not have a high salience in relief or recovery policy making or operations. Existing gaps are (i) while rapid environmental impact assessments are done, the results rarely seem to be used (ii) an environment advisor may be assigned to one sector in one disaster response (e.g. Haiti), another of similar scale disaster will receive limited and late support, at best (e.g., Pakistan flood, 2010) (iii) Relief and recover personnel may raise environment-related questions, finding answers to such questions is neither systematic nor linked to decision making structures.

Figure 3.3: Disaster waste in the aftermath of the Sichuan earthquake, May 2008.
(Source: UNEP, UNEP in China: Building Back Better)

Figure 3.4: Chemical industries damaged due to Sichuan earthquake, lead contamination of debris with hazardous materials to long term environmental damages
(Source: UNEP, UNEP in China: Building Back Better)

For instance, well documented lessons on the provision of fuel efficient stoves have been relearned in several recent disaster responses (iv) While the value of environmental impacts reviews are well established for large scale expenditures such as one finds in disaster recovery, actual recovery plans often avoid such reviews for reasons of expediency, and later result in avoidable environmental damage and a waste of natural resources and recovery funding. For instance, the $11.6 billion recovery plan for the Haiti earthquake received no formal environmental review. As a result, there was no overall idea of the scale of natural resources needed for the recovery or whether there were environmentally more positive options to the plans proposed.

It may seem that the integration of the environment into disaster relief and recovery is a glass half empty. Despite the considerable efforts to bring the environment into the disaster recovery process, real progress and impacts have been limited, with lessons not learned and mistakes repeated. This paper will take a glass half full approach and discuss opportunities for further integrating environmental issues into relief and recovery. The intent is to lay out a set of action points of sorts to identify where further progress can be made towards a fuller integration of environmental issues into the recovery from disaster.

3.2 INTEGRATION OF ENVIRONMENT INTO DISASTER RECOVERY: A PREAMBLE

The integration of the environment into disaster recovery is usually presented in terms of projects to improve environmental conditions, frequently involving tree planting. The following sections focus on managing the impact of recovery assistance on the environment rather than on using recovery funds to impact specific segments (e.g., deforested hillsides) of the environment. While environmental improvement efforts have considerable merit, they can't be pursued in ignorance of the environmental impacts of excessive resource use for housing construction or the sitting of this housing in areas liable to flooding, two common environment-links challenges follows disasters.

3.2.1 Force Results from Assessments

A range of post disaster environmental impact assessment tools exist, including:
   (i)   The Environmental Needs Assessment in Post-Disaster Situations: A Practical Guide for Implementation (UNEP, 2008)
   (ii) Fast Environmental Assessment Tool (Joint Environment Unit, 2009)
   (iii) Technical manual for post-disaster Rapid Environmental Assessment (Eco-engineering Caribbean Limited, 2003), and
   (iv) Guidelines for Rapid Environmental Impact Assessment (Kelly, 2005).

   However, as indicated above, when these tools are used they do not always have a clear and definable impact on recovery plans or process. This failure of impact assessments to have an impact can arise because the assessment results (i) are not relevant to immediate relief and recovery requirements (ii) are not presented in a way which are understandable (iii) point out significant gaps in on-going operations (and thus are embarrassing enough to be ignored) (iv) lack a champion within the upper levels of the recovery planning and management structure.

   Of the four, the most challenging is the last, the lack of a champion for assessment results. Solving the other three issues still will not mean they will be acted on if there is no person or unit which can ensure the results will be used. This need for a champion is particularly acute where procedure or policy allows recovery planning and assistance to proceed without normal environmental reviews (often justified by a lack of time to conduct environmental reviews).The person or unit to champion assessment results may not be evident from an organization chart. Investigation may be needed to identify who within the recovery structure is most amenable to considering environmental issues, and most capable to having the issues addressed. Why a champion may take up the environmental assessment results could lie in a pro-environmental perspective. But it is more likely the reason lies in an interest to avoid problems (and impact assessments normally identify problems), particularly when environment-linked problems have been encountered in the past. Even when a champion is found, interest may only be in one or two issues, which necessitates accepting that one entry point is better than none. Where a champion can't be located it may become necessary to more directly force the assessment results on the recovery process. The most traditional approach to doing this is to engage with local environmental organizations and have them use their capacity for advocacy to promote the assessment results. An alternative, but somewhat more risky, approach is for those involved in the assessment to directly advocate for the assessment results, through the print and electronic media, including blogs and similar communication tools.

The risk, of course, that the recovery institution will simply ignore or pay lip service lead to this type of forcing of the assessment results. At the same time, post disaster environmental impact assessment results which are not acted on are useless so some process for forcing assessment results may need to be pursued if the assessment process is not an exercise in futility.

3.3 ENTRY POINT AND CHALLENGES

Any entry point is better than no entry point at all. This may mean that only a few of the issues identified in an assessment can be effectively pursued or addressed within the recovery structure. When faced with limited entry points it would be idea to select those which match the prioritization of recovery programming, for instance, putting environmentally sound waste management over reforesting hillsides. However, the reality is that entry points, and particularly those created by champions with specific interests, may be of very limited scope and allow for addressing only one or two low priority issues. At the same time, successfully addressing environmental issues at one entry point may lead to greater involvement in other environmental issues across the recovery effort.

Another entry point issue arises as to when environmental issues are to be assessed and considered. In theory, initial assessments should begin as soon as immediate lifesaving needs are being met, and assessment results should be available to decision makers as soon as plans are being developed for the recovery process. In most cases, given that the environment is not seen as a post-disaster priority, the assessment process can begin weeks or even months after the disaster. In this context a rapid assessment of environmental issues may not mean rapidly after a disaster but done rapidly due to limited funding or time. Even if an environmental impact assessment is done months into the recovery period the results can provide guidance on how to redirect on-going assistance or re-direct assistance to address unmet needs. In the same way, if environment-focused assistance, such as training in environmentally sustainable recovery or assistance on specific issues such as waste and debris management begins several months after a disaster then these interventions should be used to establish and expand the scope of environment-focused recovery efforts. In the end, a late entry point is better than no entry point at all.

3.4 BEST POSSIBLE OPTION

As suggested above, it may not be possible for all the issues defined in a post disaster assessment to be address immediately, or over the course of the recovery process at all. As a result, it is necessary to prioritize assessment identified actions by considering (a) what is of greatest need to be done and (b) what will have the greatest impact when it is done. This trade-off process is not one which can be set in stone and should include consultations with the disaster survivors (a core part of any post disaster impact assessment). The process of deciding which assessment results to push forward on, and how to do the pushing, needs to strive not for the perfect fruit, or the lowest hanging fruit, but the best possible fruit which reflects the needs identified by the assessment, the disaster survivors and the resources and institutional support available.

3.5 DEMONSTRATE VALUE

Whether one or more champions have been identified or not, post disaster environmental impact assessments and environment-related technical support need to demonstrate that they can add clear value to relief and recovery operations. The perception is often that environmental issues are not urgent in the post-disaster period and can be addressed later in the recovery effort, if at all. This view often arises from the perception that post disaster environmental interventions are linked to longer term improvements to the environment, through actions such as tree planting or watershed management. Such interventions, while important in the long term, are often not seen as critical to relief and recovery. To gain relevancy in relief and recovery efforts, environment-focused assessments and technical support need to focus on the immediate problems facing these efforts. Such environment-linked interventions include, among others like (i) debris management, including quickly removing and recycling debris (Figure 3.7) (ii) standardized transitional or permanent shelter design to reduce environmentally damaging resource requirements (Figure 3.5) (iii) liquid and solid waste management, particularly establishing waste management systems which improve hygiene and sanitation conditions and limit the spread of diseases (iv) advice on avoiding recovery assistance, which will lead to an overexploitation of natural resources, for instance, provision of boats and equipment which increase fishing capacities above those existing before a disaster, (Figure3.6) and(v) natural hazard assessments to ensure that temporary or permanent settlements are not located in unacceptably hazardous locations.

Figure 3.5: Prototype of the building at Broad Town in Chang Sha, the provincial capital of Hunan, developed by A Broad Air.
(Source: UNEP, UNEP in China: Building Back Better)

Figure 3.6: Cash for work for labour to remove debris.
(Source: Figures by Allegra Da Silva, Malory Hendrickson and Diego Vallejo, REA Report, Haiti, USAID, 2010)

Figure 3.7: Stacked cleaned bricks ready for re-use at the Oxfam GB re-use/recycling yard in Banda Aceh *(Source: ProAct Network)*

This list is not exhaustive. The challenge for environment-focused personnel in the relief or recovery phases is to find ways in which calling attention to environmental issues also generates solutions which can solve problems or make relief and recovery efforts more effective. In other words, assessment and technical support should not just raise environmental problems, but also provide practical solutions which demonstrate the value of integrating an environmental perspective into relief and recovery.

3.6 CHANGE FROM THE FIELD UP

The idea of environmental champion discussed above implies this role is filled by persons or units higher up in the relief and recovery establishment. In fact, significant pro-environmental action can also be initiated at the field level, and particularly through a mid-level management which is aware of environment-base opportunities. For instance, a field manager running camp for relocated disaster survivors can have a considerable influence over waste management, sanitation and general living conditions in the camp. In a similar fashion, the manager of WASH activities for a group of communities can have a significant influence on how latrines are constructed, the supply and use of water or whether waste is dumped in a swamp or composted. It is necessary to reach out to field personnel and provide them with the tools and knowledge needed to implement relief and recovery projects in environmentally wise ways. Such an effort is incorporated into the Global Shelter Cluster's assignment of an environmental advisor to some country-level Clusters.

A broader approach to support the capacities of field personnel is incorporated onto the World Wildlife Fund and American Red Cross Green Recovery and Reconstruction Training Toolkit for Humanitarian Aid and associated training (World Wildlife Fund and the American Red Cross, 2010). Other support is available on such issues as stoves, transitional shelter and refugees, with additional field guidance expected in the near future. The real challenge is to ensuring this wealth of information on environmentally sound relief and recovery reaches the field staffs that need specific information. Making the information available on the web (as it currently the case) is important. But, given the work load pressure which many field personnel face it is unlikely that most will have to time or background to effectively search for and digest information on environmentally sound relief and recovery. As a result, it is likely to be more effective to have an out-reach capacity in the field, through the use of environmental advisors. These advisors do not necessarily need to be experts in relief and recovery (although this will not hurt) but would need to focus on defining what critical environment-related issues are being faced in the field and package information available on addressing these issues in a way which can be used by field personnel in a quick and easy manner. As noted, posting environmental advisors has been done by the Global Shelter Cluster. This effort needs to expand to additional sectors as well as become a standard part of relief and recovery operations rather than being an occasional event.

3.7 KNOWLEDGE AS POWER AND A FOOT IN THE DOOR

The process of furnishing environment-focused information as described above provides the environmental community a degree of power to influence the way in which relief and recovery aid is provided and the impacts which can be expected. Further, many post disaster reconstruction challenges relate to location within the environment (e.g., where to locate disaster displaced, where to rebuild roads, etc.). An environmental and more specifically, an ecosystem services approach to assessing recovery options and impacts can be of considerably utility in the reconstruction process in defining how best to site and provide recovery assistance.

Here again, the focus is not (only) on defining what the environmental issues with reconstruction or recovery efforts are in general, but defining how these process can be sustained by wise use of the environment and how potential problems can be solved in a similar manner. Given the common misperception noted that a focus on the environment is limited to longer term improvement of natural conditions, knowledge about the environment needs to be used to improve the overall recovery process and specifically focus on improvements to specific recovery activities. Providing one package of knowledge which improves one element of the recovery effort also provides a foot in the door for greater access to improve the overall environmental sustainability of recovery efforts.

3.8 CONCLUSIONS

This paper argues that progress in integrating an environmental perspective into disaster relief and recovery has been uneven but also somewhat successful. The paper suggests that many of the tools and guidance needed for a better integration of environmental perspective into relief and recovery are already available. Under these circumstances, moving the integration process forward requires efforts in six areas viz. (i) effectively using assessment results (ii) Finding entry points into the relief and recovery process (iii) focusing on the best possible match of reducing negative environmental impacts and recovery programming (iv) demonstrating the clear value of an environmental perspective on recovery (v) ensuring field personnel know about how to effectively integrate an environmental perspective into their work, and (vi) using knowledge about the environment to influence and shape the recovery agenda.

The prospect is for environmental considerations, and an awareness of environmental issues, to be increasingly accepted as an integral part of relief and recovery. However, this process can occur through demonstrated failures of relief and recovery due to a failure to consider environmental impacts and options or this process can occur though a more proactive integration and problem avoidance, as suggested above. The argument for this latter course is that it will lead to better relief and recovery, and less suffering and hardship for the disaster survivors. And anything which will reduce the suffering of disaster survivors should be pursued with vigour.

References

Dahdouh, G.F., Jayatissa, L.P., Nitto D., Di Bosire, J.O., Seen D. Lo, and Koedam, N., (2005). How effective were mangroves as a defence against the recent tsunami? *Current Biology 15*(12): R443-446.

Dahdouh-Guebas Ecoengineering Caribbean Limited (2003). Technical Manual for Post-Disaster Rapid Environmental Assessment, Volumes 1 and 2, Environment & Sustainable Development Unit, Organization of the Eastern Caribbean States, St. Augustine http://www.caribank.org/titanweb/cdb/webcms.nsf/AllDoc/9E2B73C29C5CB1A1042573D100546360/$File/OECSManualVolume_1_Final.pdf.

IASC (2006). Protecting persons affected by natural disasters: IASC Operational Guidelines on Human Rights and Natural Disasters. Washington DC: Brookings-Bern Project on Internal Displacement. http://www.humanitarianinfo.org/iasc/pageloader.aspx

IASC (2012).*Multi Cluster/Sector Initial RapidAssessment (MIRA)- Provisional Version.*Washington DC: Inter Agency Standing Committee. http://www.humanitarianinfo.org/iasc/pageloader.aspx

Joint Environment Unit (2009).*Fast Environmental Assessment Tool (FEAT).*United Nations Environment Program and Office for the Coordination of Humanitarian Assistance.Geneva and New York.http://ochanet.unocha.org/p/Documents/FEAT_Version_1.1.pdf.

Kelly, C. (2005) *Guidelines for Rapid Environmental Impact Assessment in Disasters.*Version 4.04. London: Benfield Hazard Research Centre. http://www.benfieldhrc.org/rea_index.htm

Kelly, C., and Solberg, S. (2011). *Rapid Environmental Impact Assessment: Haiti Earthquake January 12, 2010.* Sun Mountain International, CHF International and US Agency for International Development. http://chfinternational.org/files/PNADS052.pdf.

Ministry of Environment, Government of Chile, WWF Chile/WWF International and Antofagasta Minerals (2011). Final Report, *"Rapid Environmental Impact Assessment, Chile Earthquake and Tsunami"*, February 27, 2010.
http://www.worldwildlife.org/what/partners/humanitarian/WWFBinaryitem19130.pdf

Shelter Centre, ProACT Network & Disaster Water Recovery (n.d). Planning Centralised Building Waste Management Programmes in Response to Large Disasters.
http://www.proactnetwork.org/proactwebsite/media/download/BriefTecnicalGuides/Waste_Guide.pdf

UNEP (2008).*The Environmental Needs Assessment in Post-Disaster Situations: A Practical Guide for Implementation.*http://oneresponse.info/Disasters/Haiti/Environment/publicdocuments.

UNEP(2010). UNEP in China: Building Back Better UNEP (2010). UNEP in Haiti: 2010 Year in Review. Retrieved from http://postconflict.unep.ch/publications/.

USAID (2010). *Rapid EnvironmentalImpact Assessment: Haiti Earthquake - January 12, 2010.*

UNOCHA (n.d). Environmental Emergencies: Resources. http://www.unocha.org/what-we-do/coordination-tools/environmental-emergencies/resources.

URD (2010, March 9). Guide to reducing the environmental impact of humanitarian action - Work underway. http://www.urd.org/spip.php?page=imprimir_articulo&id_article=322

World Wildlife Fund and the American Red Cross (2010).*Green Recovery and Reconstruction Training Toolkit for Humanitarian Aid.* Washington. http://green-recovery.org.

4

Comprehensive Drought Hazard Analysis using Geospatial Tools: A Study of Bundelkhand Region, India

Anjali Singh, Sreeja S. Nair, Anil K. Gupta, P.K. Joshi and V.K. Sehgal

4.1 INTRODUCTION

A natural disaster is the consequence of a hazard condition or event (e.g., drought, flood, volcanic eruption, earthquake, or a landslide) that affects living beings and ecosystems, and causes environmental, social or financial disruption beyond the immediate coping capacity of the system. The resulting impacts inversely depend on the capacity of the population or the system to resist a disaster through hazard control, avoidance, and tolerance mechanisms, whereas 'resilience' is understood as the ability to bounce back after the stress is over.

Drought is an insidious hazard of nature and it ranks first among natural disasters throughout the world in terms of the number of persons directly affected (Hagman, 1984; Hewitt, 1997). It is considered to be one of the most complexes and least understood among all natural disasters due to wider range of environmental inputs and influences over the occurrences, impacts and mitigation strategies – near impossible for a single discipline or two to be capable of complete understanding. It is interesting to mention that contrary to other natural hazards like floods, earthquakes, and hurricanes, the impacts caused by drought are usually non-structural but spread over larger geographical areas. Its effect accumulates slowly over a considerable period of time and lingers on for years.

Drought means differently to different people depending on their specific interests. To the farmer drought means a shortage of moisture in the root zone of his crops. To the hydrologist, it suggests below average water levels in the streams, lakes, reservoirs, and the like. To the economist, it means a shortage which affects the established economy. To an ecologists, drought draw interdisciplinary concerns of meteorological, land and water relations for natural resources and human development. Drought, therefore, is known as creeping phenomenon, that is a normal part of climate for virtually all regions of the world; it results in serious environmental, social and economic impacts often complex to understand and much more difficult to anticipate (Wilhite, 2000).

Disaster Management and Risk Reduction: Role of Environmental Knowledge; Editors Anil K. Gupta, Sreeja S. Nair, Florian Bemmerlein-Lux and Sandhya Chatterji; Copyright © 2013, Narosa Publishing House, New Delhi

4.2 CLIMATE CHANGE AND DROUGHT

The impact of global climate-change on the incidence of natural disasters particularly hydro-meteorological ones, for e.g., floods, drought, cyclone, forest fire, pest infestations, etc. having been recurred will be in the recent 4th Assessment Report (2007) of Intergovernmental Panel on Climate Change (IPCC), along a range of national and international literature. Although the implications of global environmental change have implications for all natural disasters but has much less impact on occurrence of geophysical disasters. Historical evidence envisaged that droughts and floods will become more common and more severe (IADB, 2009).

A major challenge for drought research is to develop suitable methods and techniques for forecasting the onset and termination points of a drought. Another but equally challenging task is the dissemination of drought research results for practical usage and wider applications in planning drought-proofing and preparedness (Panu & Sharma, 2002). Since drought studies require huge amount of continuous data and because of its peculiar characteristics there has been a lack of progress in drought proofing, preparedness and management throughout the world in general and in particular in India. Despite of significant drought research, studies that dealt with the global picture of drought patterns and impacts are limited. Fewer studies dealt with the global mapping of drought-related indicators. Peel et al., (2004, 2005), conducted an analysis of precipitation and runoff periods of consecutive years below the median for 3,863 precipitation and 1,236 runoff stations worldwide. Fleig et al., (2006), carried out similar research using daily flow time-series data from 16 selected river basins worldwide. The above studies were conducted using observed data, which is useful in examining geographical differences in the statistical nature of droughts but are constrained by limited observation points. Dai et al., (2004) had developed a global monthly dataset of the well-known Palmer Drought Severity Index (PDSI) for 1870-2002 using historical data on precipitation and temperature on a 2.5° × 2.5° grid and established that, globally, very dry areas had more than doubled since the 1970s. Below et al., (2007) have undertaken a comprehensive review of 807 drought and 76 famine entries from 1900 to 2004 in the EM-DAT database (Emergency Events Database: www.emdat.be/) and revised estimates of global drought-related deaths. The Natural Disaster Hotspots project of the World Bank has assessed the global risks of two disaster-related outcomes – mortality and economic losses - on a 2.5'×2.5' grid by considering physical exposure and historical loss rates (Dilley et al., 2005).

Various studies on drought have been carried out by using different drought indices. The studies reflected selection of different indices depending on the topography, geography, seasonality and availability of data (Palmer, 1965; Hayes, 2006). Recently an attempt has been made to assess spatio-temporal variability of drought in the United States central plains by using Standardized Precipitation Index. It helped in identifying isolated drying regions which overlap regions of heavy water use and irrigation, suggesting possible detrimental effects on agricultural production and aquifer levels if the trend continues (Logana, 2010). An attempt was made to characterize the spatial and temporal variability's of precipitation in Sonora and to conduct a meteorological drought intensity–duration–frequency analysis based on annual and warm season precipitation records (Alegria and Watkins, 2007). Drought indices are essential components of an efficient drought watching system and it aimed at providing an abstract picture of drought conditions to help the decisions. Groundwater Resource Index (GRI) can be used as a reliable tool useful in a multi-analysis approach for monitoring and forecasting drought conditions in Mediterranean climate (Mendicino et al., 2008). A study was taken up to find the relationship between the satellite-based Vegetation Condition Index (VCI) and a number of frequently used meteorological drought indices was evaluated using data from all 254 Texas counties during 18 growing-seasons (March to August, 1982–1999). These results demonstrate that care must be taken when using the VCI for monitoring drought because it is not highly correlated with station-based meteorological drought indices and it is strongly influenced by spatially varying environmental factors (Quiring and Ganesh, 2010).

The Bundelkhand region is one of the most poverty stricken regions in India and is currently experiencing recurring drought. The government of a developing country like India and in this case the states of Uttar Pradesh (U.P.) and Madhya Pradesh (M.P.) are facing the dilemma of linking disaster mitigation and environment protection for sustainable development. In the last four to five years, Bundelkhand has been in top news stories for the drought conditions and distress that has plagued it. There has been news of mass migration, starvation deaths, farmer suicides and even the 'mortgaging' of women over the years. According to the J. S. Samra committee report, historically, the Bundelkhand region of U.P. and M.P. had a drought every 16 years in Eighteenth and nineteenth centuries, which increased by three times during the period 1968 to 1992 (Samra, 2008). The most recent and continued period of poor rainfall recorded in Bundelkhand was in 2004-07, when below average and erratic rain was reported in most part of the region in all the years. There are several manifestations of drought like late arrival of rains, early withdrawal, long break in between, lack of sufficient water in reservoirs and drying up of wells leading to crop failure and even unsowing of the crops which ultimately curtail livelihood and may lead to migration. Certain efforts to study various aspects of natural resources, agroforestry, water management, and perspective planning for the region has been reported in publications by Indian Grassland and Fodder Research Institute (IGFRI), The National Research Centre for Agroforestry (NRCAF), U.P. Council of Agricultural Research (UPCAR), etc. However, a systematic scientific research on drought hazard identification and vulnerability assessment in Bundelkhand has been lacking despite of the region being known for droughts. Government have poured 'huge amount of money' in all possible mitigation measures, expected outcomes could not achieve a satisfaction level in any part of the region. It indicates that there is gap between what is being done for the drought mitigation in different district of Bundelkhand and what is actually required. This shows efforts done so far are not prioritized according to the need of the situation.

Present study is an attempt to scientifically identify and assess the hazards and vulnerability on spatial scales, by using remote sensing and GIS techniques. The aim of the study was to develop a drought risk map for the region that can serve as a major basis for risk management planning and programmes including effective warning and monitoring framework.

The present work piloted a new methodology of comprehensive drought analysis covering three major types, i.e., meteorological, hydrological and agricultural with the use of limited and freely available data. The objectives of this research were following:

(a) To select suitable drought indices for assessment of meteorological, hydrological and agricultural drought.
(b) To identify districts exposed to different type of drought hazards.
(c) To prepare composite drought hazard map (all three types of drought) for Bundelkhand region.

4.3 STUDY AREA

The Bundelkhand region lies at the heart of India located below the Indo-Gangetic plain to the north with the undulating Vindhyan mountain range spread across the northwest to the south. The region span across thirteen districts: seven in Uttar Pradesh - Jhansi, Jalaun, Lalitpur, Hamirpur, Mahoba, Banda and Chitrakut, and six in Madhya Pradesh - Datia, Tikamgarh, Chattarpur, Damoh, Sagar and Panna. It covers an area of 7.08 million hectares (mha) and is located between 23020' and 26020' N latitude and 78020' and 81040'E longitude (figure 4.1).

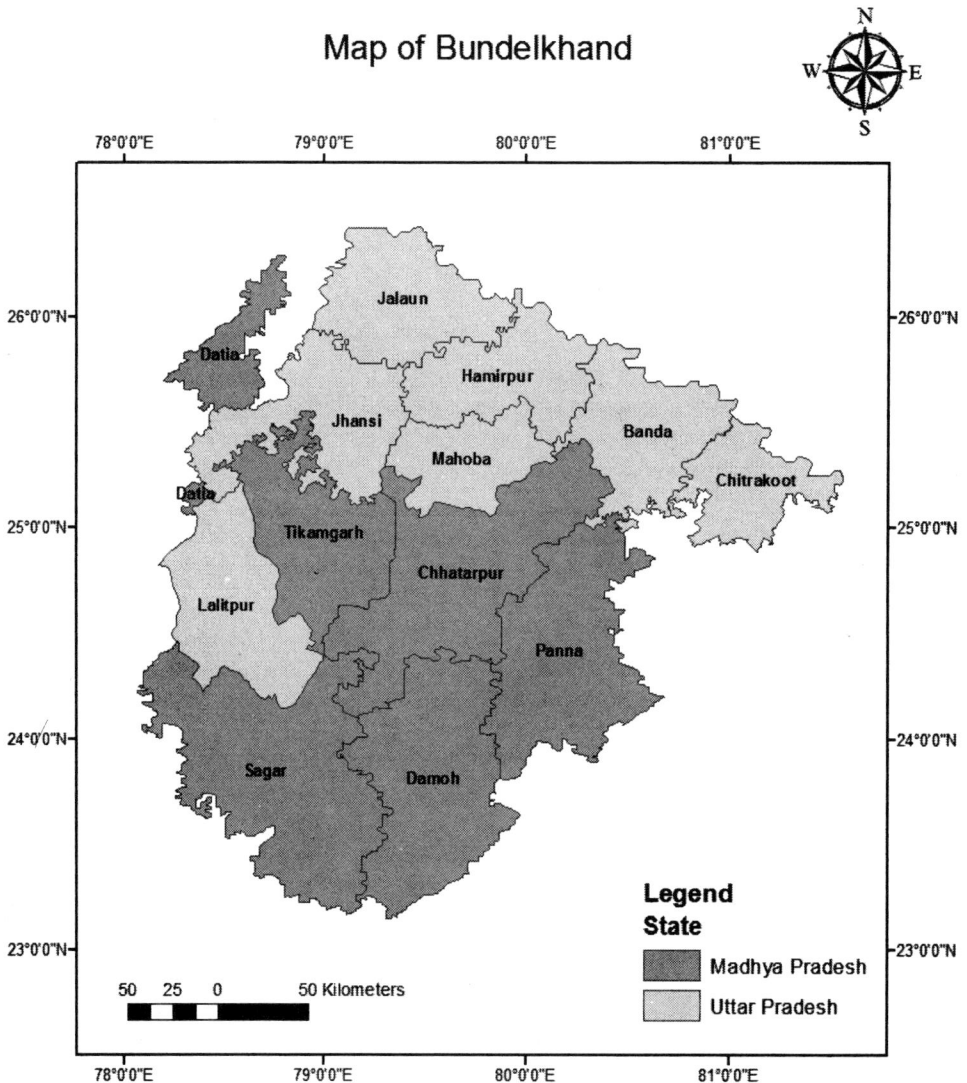

Figure 4.1: Map showing the study Area showing Bundelkhand districts

About 7.08 Million hectares (M ha) Bundelkhand region is ravenous, undulating and hillocks are bounded by Vindhyan Plateau in south, river Yamuna in north, river Ken in east and rivers Betwa and Pahuj in west. While the Yamuna flows from west to east, its first order tributaries viz., Betwa, Ken, Pahuj, Baghain, and Paisuni flow from south to north. Second order tributaries of the Yamuna namely, Dhasan, Jamni, Birma, Sonar, Katne, Bewas, Kopra etc., also drain the area.

Rainfall is the major source of surface and ground water resources for raising biomass and other utilities. The average annual rainfall of Bundelkhand in Uttar Pradesh is 876.1 mm with a range of 786.6 to 945.5 mm. In Madhya Pradesh portion the average rainfall is 990.9 mm with a range of 767.8 to 1086.7 mm and is 13 % more than the Uttar Pradesh part. About 90 % of the rainfall is received in the monsoon season of July to September in about 30-35 events or spells. Delayed on set of rains, early withdrawal or long dry spells in between also lead to drought like situation.

4.4 METHODOLOGY

The following flow chart depicts the methodology followed in this study.

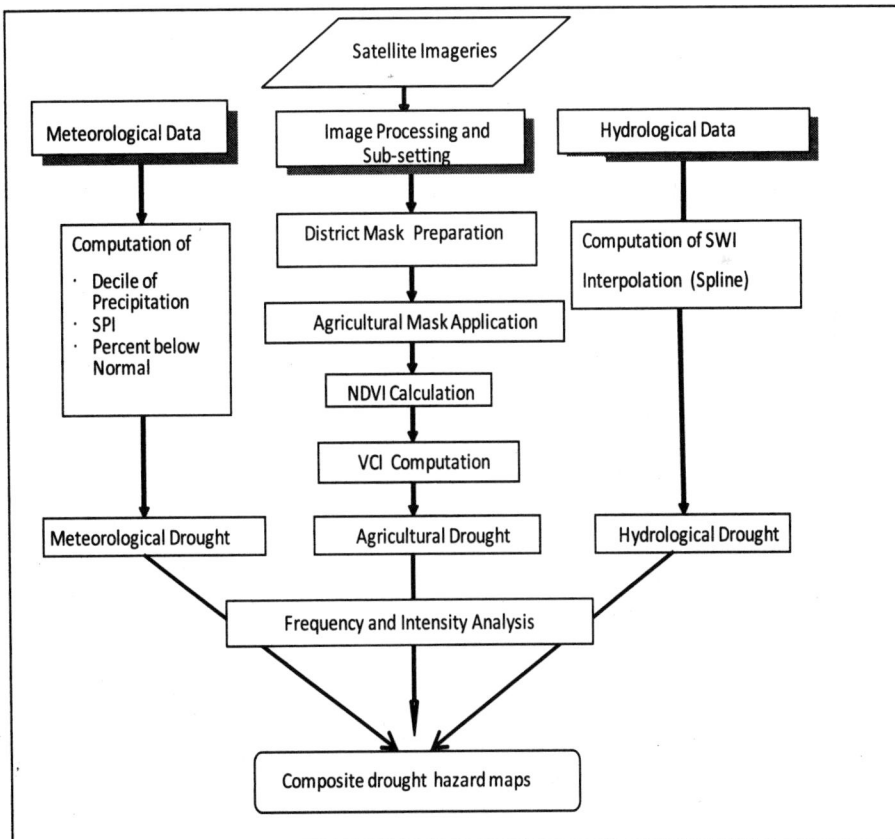

Figure 4.2: Methodology flowchart

4.4.1 Drought Indices

Drought indices assimilate thousands of bits of environmental data related with rainfall, stream flow and other water supply indicators, vegetation vigour, etc. to present a concise picture of drought risk scenario. A drought index value is typically a single number, far more useful than raw data for decision making as it allows comparisons on temporal and spatial scales helping prioritization possible for the planners and making it easier to communicate the information to diverse users (Wilhite et al., 2000). Different indices were selected for the present study on the basis of five preferred criteria, these were: index must suit the topography and geography of the region, it should be flexible and able to detect short-term droughts, it should match with the data availability, it should be easy to compute and easy to understand. Meteorological drought indices chosen for the study were 'decile of precipitation', 'Standard Precipitation Index (SPI)' and 'per cent by normal'. For hydrological drought, 'Standardized Water level Index (SWI)' and for agricultural drought, Normalised Difference Vegetation Index (NDVI) and Vegetation Condition Index (VCI) were selected.

Failure of monsoon is the prime cause of drought and it becomes very important to know the rainfall pattern of historical times of this region. To study the frequency and severity of meteorological drought 'decile of precipitation' was computed using 102 years rainfall dataset. Decile is a non-parametric index and useful to understand overall pattern and it is used only in the areas with undulating topography. For more concrete results SPI was computed for years from 1970 to 2002.

4.4.2 Analysis of Datasets

Monthly rainfall data from year 1998 to 2009 for the 13 districts of Bundelkhand was procured from Indian Meteorological Department (IMD), New Delhi. Monthly and annual rainfall data from 1901 to 2002 was acquired from India Water Portal website. Monthly data of groundwater for years 1998 to 2010 for 264 stations distributed over thirteen districts of Bundelkhand was obtained from Central Ground Water Board, Faridabad. Satellite scenes of South-East Asia of 11th September from 1998 to 2009 were downloaded from SPOT Vegetation data available freely at website (http://free.vgt.vito.be/). The analysis has been done in three steps.

Step I dealt with data processing and computation of the indices for all three types of drought. Step II included data analysis which included frequency and intensity calculations over the 12 years, combining frequency and intensity to produce hazard and vulnerability maps and lastly identifying the chronology of the drought occurrence. In the Step III, hazard and vulnerability were assigned with appropriate ranks to their categories and weightage to arrive at the final composite drought risk map.

Step I - The data was pre-processed for index computation. Bundelkhand receives maximum amount of rainfall during June to September. Percent by departure of these 4 months was calculated to analyse the severity of meteorological drought for 12 years. According to IMD's criteria, negative deviation up to 19 % is considered 'normal', from 19 to 59 % is said to be 'moderate' and beyond 60 % is considered as 'severe' meteorological drought. Result was generated into pictorial format for better representation of spatial and temporal changes over the period.

The present study was based on administrative boundary of the districts, and therefore, surface water analysis could not be accomplished. Besides there is no well-established methodology existing for hydrological drought assessment using reservoir water levels and water bodies such as rivers and lakes. 26 % of agriculture in Bundelkhand is using groundwater for irrigation, 10 % is from reservoirs and rest remains rainfed (Samra, 2008). Selection of hydrological drought index, therefore, included aspects, viz., it should be easy to understand, carry physical meaning, sensitive to wide range of drought conditions, independent of area of application, reveal drought with short lag after its occurrence and be based on the data which are readily available. Thus, Surface Water level Index (SWI) was selected for the assessment of hydrological drought. It is one of the most preferred indexes for drought detection (Bhuiyan, 2006). For SWI computation pre and post monsoon months were selected to understand the role of monsoon on water resources, *i.e.*, May and November, respectively. Standardized Water Level Index has been computed to scale ground water recharge deficit. The SWI expression is given by

$$SWI = (W_{ij} - W_{im})/\sigma$$

W_{ij} is the seasonal water level for the ith well and jth observation. W_{im} is the long term seasonal mean and σ is standard deviation.

Table 4.1: Standardized Water Level Index

SWI Values	
SWI > 2.0	Extreme drought
SWI > 1.5	Severe drought
SWI > 1.0	Moderate drought
SWI > 0.0	Mild drought
SWI < 0.0	No-drought

(Source: Bhuyian, 2006)

Data was processed and SWI was computed for all 264 ground water monitoring stations in the Bundelkhand. Interpolation was done using Inverse Distance Mean method; however, its results were not satisfactory because the surface layers generated were not smooth. Then a Regularized Spline method was adopted for creation of surface layers. This method gives a smooth surface and satisfactory results. This spline is a general purpose interpolation method which fits a minimum-curvature surface through the input points. This method is best for gradually varying surfaces such as water-table depths.

NDVI reflects vegetation vigour, density and biomass. It varies in a range of -1 to +1. Among all the vegetation indices that are available, it is a universally acceptable index for operational drought assessment because of its simplicity in calculation, easy to interpret and its ability to partially compensate for the effects of atmosphere, illumination geometry (http://dsc.nrsc.gov.in).

NDVI is derived as under

$$NDVI = (NIR - Red) / (NIR + Red)$$

Where, near Infra-Red and Red are the reflected radiations in these two spectral bands. For agricultural drought assessment "agriculture mask" was generated to mask the forests and other non-agricultural area. District-wise masks were prepared by using district shape files and agricultural mask was applied on that respective district mask. NDVI is indicator of environmental stress but it is difficult to categories drought severity on the basis of NDVI. VCI was selected to overcome this problem. VCI was designed to signal out the impact of meteorological condition on the vegetation of that area. It shows how close the NDVI of the current month is to the minimum.

$$VCIj = (NDVIj - NDVI\ min\ /\ NDVImax - NDVImin)$$

Where NDVI min and·max are calculated from long-term record for that month and 'j' is the index of current month. VCI values around 0.5 reflect fair vegetation condition. VCI between 0.5 to 1 indicates optimal to normal condition. Different drought severity is indicated by VCI values below 0.5. Kogan (1995) illustrated that the VCI threshold of 0.35 may be used to reflect extreme drought condition. Low VCI values over years served consecutive time interval points to drought development (http://dms.iwmi.org/). Agricultural drought is closely related with weather impacts. In NDVI, strong ecological component subdues the weather component. On the other hand, VCI separates the short-term weather-related NDVI fluctuations from the long-term ecosystem changes (Kogan, 1990, 1995). Therefore, while NDVI shows seasonal vegetation dynamics, VCI varies in the range 0 and 100 % reflect relative changes in the vegetation condition from extremely bad to optimal (Kogan, 1995; Kogan et al., 2003).

Step II- Results got from Step-I were further analysed to calculate drought frequency and intensity. No. of drought occurrences in 12 years were then categorised into different classes i.e. extremely high, very high, high, moderate and low and results were shown in map form separately for each type of drought. From the sum of deviations from reference level gives drought severity. Drought intensity was calculated for each district over 12 years of study period for each type of drought. Finally, drought intensity was classified into five classes i.e. extreme, severe, high, moderate and mild. Meteorological hydrological and agricultural drought analysis, frequency and intensity values were multiplied and then again classified into 5 classes. Correlation was carried out for each district over 12 years between meteorological drought and hydrological drought and between meteorological and agricultural drought with and without lag Step.

Step III - For preparation of composite drought hazard map, three maps *i.e.* meteorological drought, agricultural drought and hydrological drought hazard maps were used. Since all three are classified in same categories, each category was assigned a rank. Similarly, each drought variable i.e. rainfall, agriculture and groundwater was given weightage based on expert knowledge. Afterwards, these three layers were integrated to get one composite value for each district. Finally composite drought risk map was again classified into four classes i.e. Severe, high, moderate and mild risk.

Table 4.2: Weightages to categories

Category	Rank
Extreme	5
Severe	4
High	3
Moderate	2
Mild	1

Table 4.3: Weight age (influence values) assigned to drought variables

Drought Variables	Weight age
Meteorological (Rainfall)	0.35
Hydrology (Groundwater)	0.20
Agricultural (Agriculture)	0.45

4.5 RESULTS AND DISCUSSIONS

Observation and interpretations of the results on different aspects of meteorological, hydrological and agricultural drought in Bundelkhand region of central India are presented in the following sections:

4.5.1 Drought Hazard Mapping

4.5.1.1 Meteorological Drought

Meteorological drought is usually defined on the basis of the degree of dryness (in comparison to some "normal" or average amount) and the duration of the dry period. Definitions of meteor-ological drought must be considered as region specific since the atmospheric conditions that result in deficiencies of precipitation are highly variable from region to region (NDMC). According to the Inter-ministerial team survey in Bundelkhand, more than 19 % deficit in the normal rainfall, late onset of monsoons, early withdrawal of rains, long breaks during growing season and permutations or combinations of these factors are manifestations of triggering drought. The cumulative build up of meteorological droughts has rippled into hydrological drought with a complex set of highly differentiated adverse impacts and trade-offs.

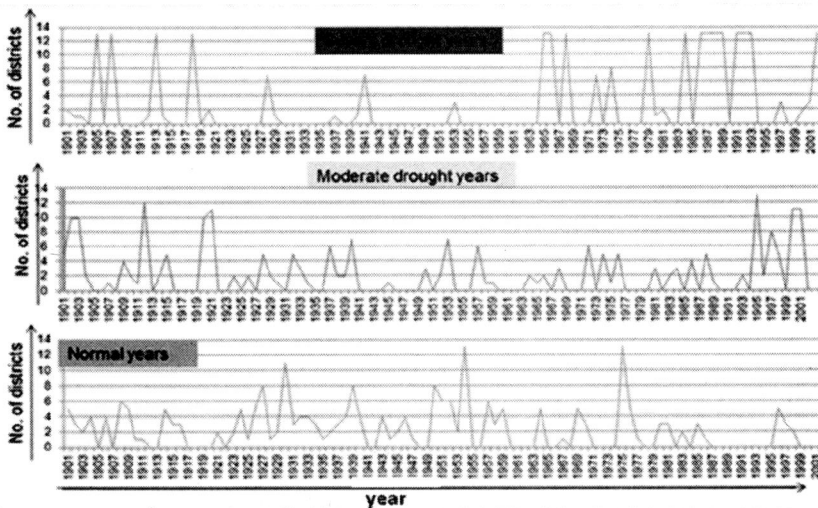

Figure 4.3: Deciles of Precipitation

5.1.1.2   Deciles of Precipitation

In figure 4.3 horizontal axis represents time period while the vertical axis represents the number of districts in Bundelkhand region experiencing drought. The graph provides a picture of twentieth century. It clearly indicates that the number of normal years has decreased and the moderate and severe drought years have sharply increased in last 30 years.

5.1.1.3   Percent by Normal

The results obtained from the 'percent by normal' method have been shown in the form of thematic maps (Figure 4.4) to reveal the patterns of meteorological drought in Bundelkhand region. Severe to moderate drought was observed during 2004 to 2009 in different districts of this region. Hamirpur, Mahoba and Jhansi districts experienced consecutive drought from 2004 to 2009. However, the year 1999 and 2003 witnessed above to normal rainfall during the period of study. During the last 6 years, U.P. Bundelkhand has faced more consecutive drought as compared to the M.P. Bundelkhand districts. It has been found that the U.P. part of the region experienced rainfall deficit of 35 % in 2004-05, 26 % in 2005-06, 41 % in the year 2006-07, which went up to 54 % in 2007-08, 39 % in 2008-09 and 42 % in 2009-10. All the districts experienced meteorological drought in 2009. These results are verified with the Central Ministerial Report, 2008.

In Madhya Pradesh, the rainfall was almost normal during 2004-05 and 2005-06 except in Tikamgarh and Datia which experienced drought in 4 consecutive years. In 2006-07 the region experienced overall shortfall of 41 %, with 6 Bundelkhand districts receiving deficit rainfall ranging from 45 % to 82 %. Severe drought occurred in Tikamgarh and Panna. The overall shortfall in precipitation went up to 41 % during 2007-08. Except Panna, all districts received normal rainfall in 2008-2009 and in the following year, once again the entire districts experienced moderate drought with overall deficiency of 32 %. In Madhya Pradesh, the rainfall was almost normal during 2004-05 and 2005-06 except in Tikamgarh and Datia which experienced drought in 4 consecutive years. In 2006-07 the region experienced overall shortfall of 41 %, with 6 Bundelkhand districts receiving deficit rainfall ranging from 45 % to 82 %. Severe drought occurred in Tikamgarh and Panna. The overall shortfall in precipitation went up to 41 % during 2007-08. Except Panna, all districts received normal rainfall in 2008-2009 and in the following year, once again all the districts experienced moderate drought with overall deficiency of 32 %.

4.5.2. Hydrological drought

Hydrological drought is associated with the effects of periods of precipitation shortfalls on surface or subsurface water supply (i.e., stream flow, reservoir and lake levels, and groundwater). The frequency and severity of hydrological drought is often defined on a watershed or river basin scale. Although all droughts originate with a deficiency of precipitation, hydrologists are more concerned with how this deficiency plays out through the hydrologic system.

Hydrological droughts are usually out of phase with or lag behind the occurrence of meteorological and agricultural droughts. It takes longer for precipitation deficiencies to show up in components of the hydrological system such as soil moisture, stream flow, and groundwater as well as reservoir levels. As a result, these impacts are out of phase with impacts on other economic sectors (Wilhite et al., 1985)

4.5.2.1. Standardized Water level Index (SWI)

The entire region has been classified into 5 categories based on interpolated maps of SWI as shown in figure 5a and 5b. Bhuyian et al., (2006) used SWI and analysed dynamic water stress in the Aravali region in India for mapping of SWI with SPI and VHI. The result of SWI from 1998 to 2009 indicated that water level declined over the period in various districts both during pre and post monsoon months. During 1998 to 2000, most parts of the region were free from water stress due to normal rainfall. Lalitpur, Panna and Chattarpur districts faced extreme to moderate drought in 2001, and it is worth mention that meteorological drought was recorded in previous year 2000 in these districts as well. From 2004 onwards a remarkable decline in water level has been noticed which aggravated year after year by the monsoon failure. Year 2006 and 2007 were hydrological drought years for the entire Bundelkhand region. Datia, Tikamgarh, Chitrakoot and Jhansi districts faced extreme drought for 3 consecutive years.

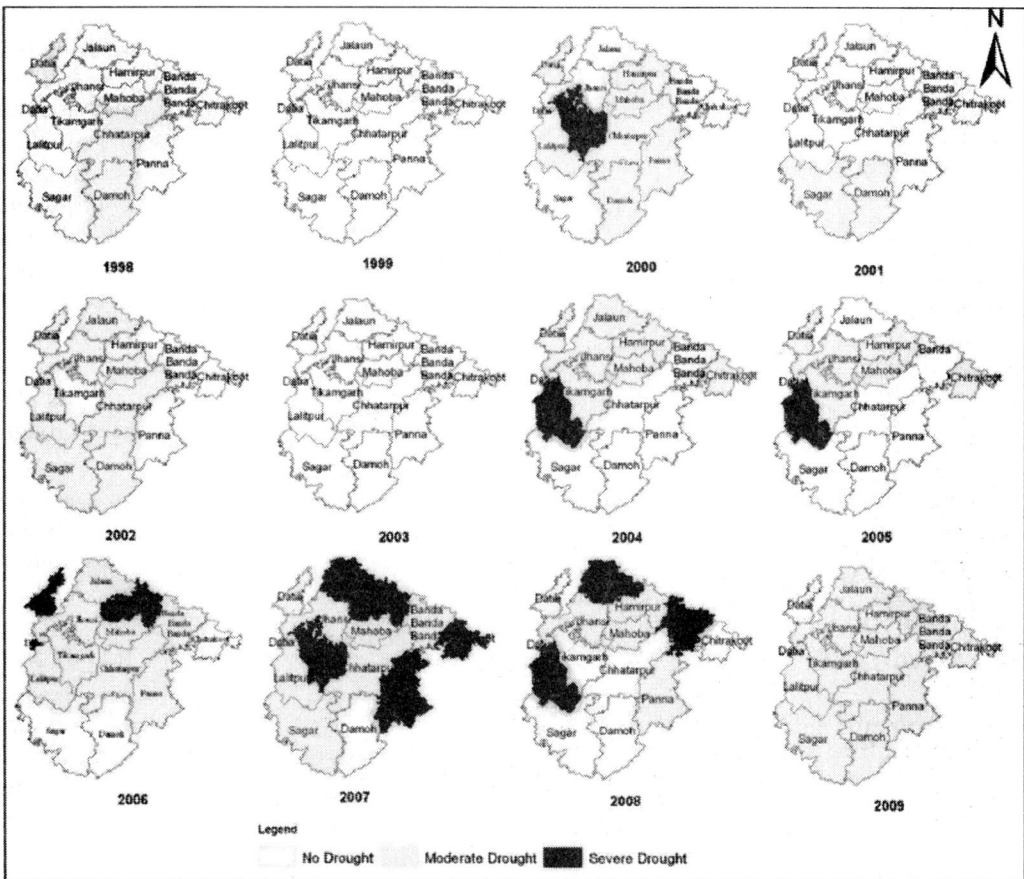

Figure 4.4: Thematic map representing meteorological drought during the period 1998 to 2009

SWI for Datia in 2006 was as high as 4.35, whereas extreme drought in 2007 the SWI was up to 5.50 in Tikamgarh district, 5.01 in Damoh district and 4.93 for Jhansi district. It is important to mention that 2006 and 2007 were also meteorological drought years for Bundelkhand region. 3 out of 13 districts became free from drought in 2008 while other still remained under extreme to moderate drought. The continuous decline of water level during the period of 2004 to 2007 may be attributed to the reduced rainfall runoff pouring into reservoirs. Only 17 % of the total capacity of reservoirs was filled in 2008. The situation was even more alarming in the Madhya Pradesh (MP) portion of Bundelkhand region. As against the total storage capacity of 950 MCM in 19 reservoirs, the actual filling of water progressively reduced from 52 % in 2004 to 10 % in 2007 (Samra, 2008).

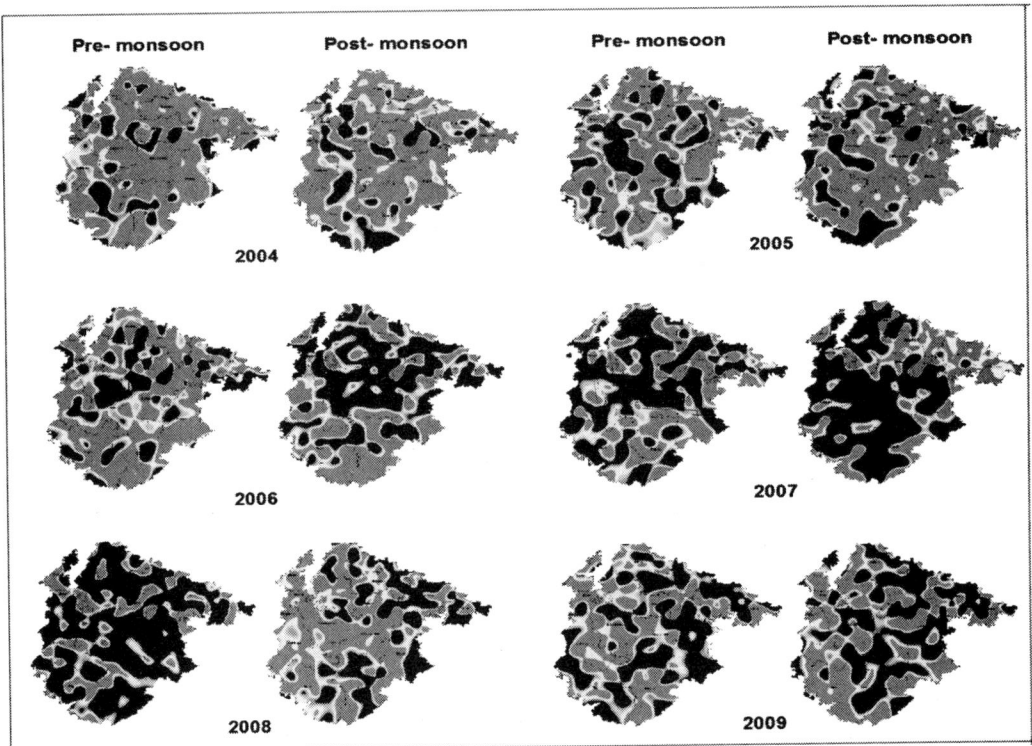

Figure 4.5a: Hydrological drought maps from year 2004 to 2009

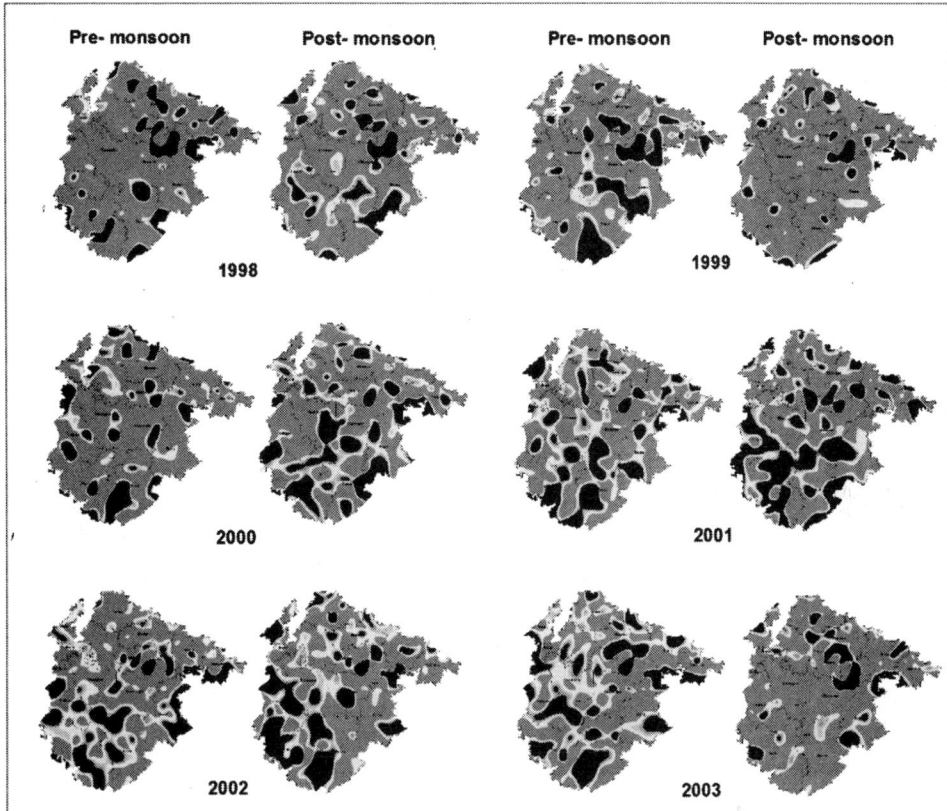

Figure 4.5 b. Hydrological drought maps from year 1998 to 2003

Legend

	Normal
	Mild
	Moderate
	Severe
	Extreme

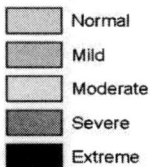

4.5.3 Agricultural Drought

Agricultural drought links various characteristics of meteorological and hydrological drought to agricultural impacts, differences between actual and potential evapo-transpiration, soil water deficits, reduced groundwater or reservoir levels, and so forth. Demand for water from plants depends on prevailing weather conditions, biological characteristics of the specific plant, its Step of growth, and the physical and biological properties of the soil. A good definition of agricultural drought should be able to account for the variable susceptibility of crops during different Steps of crop development, from emergence to maturity (Whilite et.al., 1985).

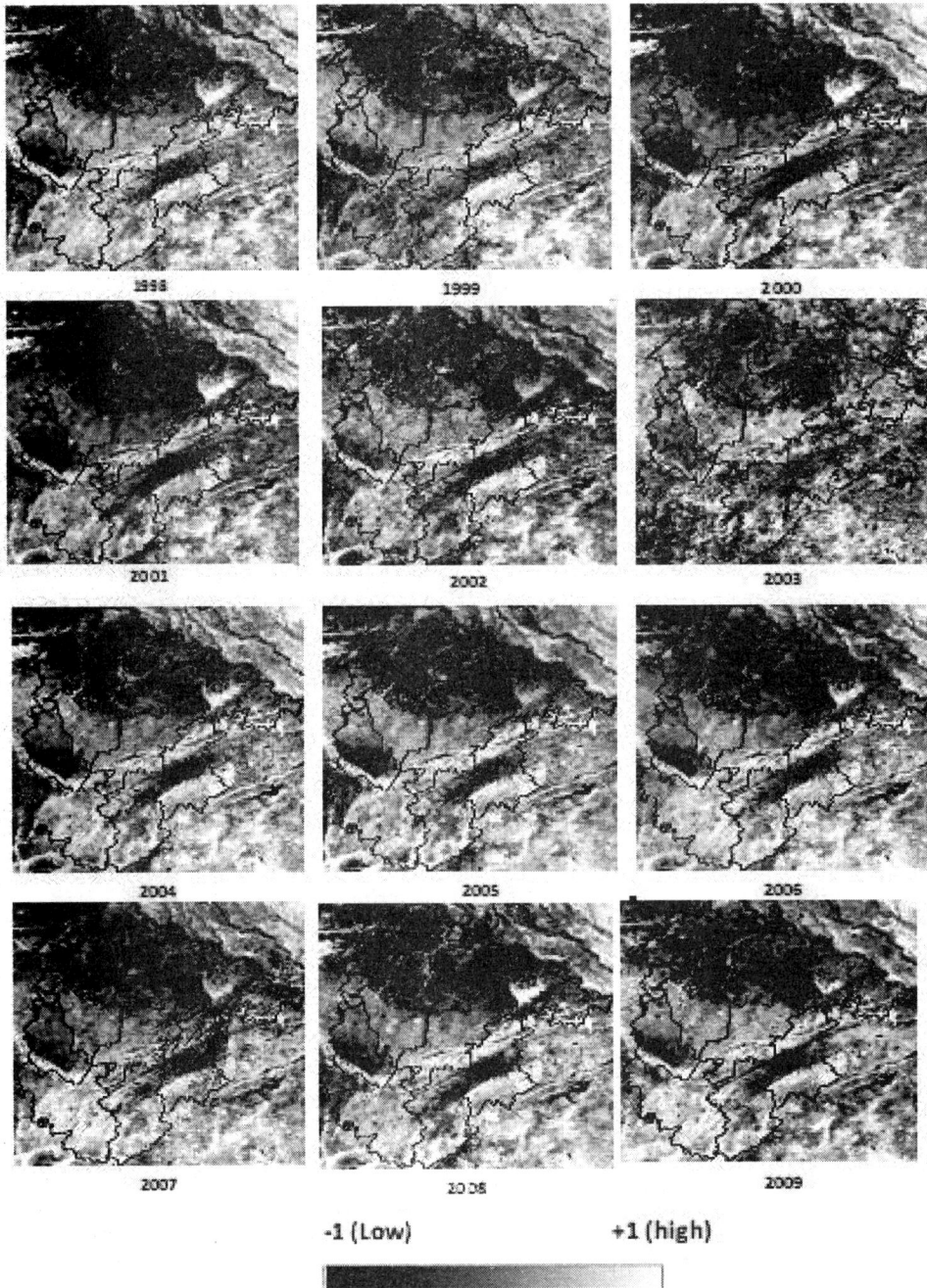

Figure 4.6: NDVI Images of Bundelkhand region from 1998 to 2009

4.5.3.1 Normalized Difference Vegetation Index

The NDVI images reflect vegetation vigour, density and biomass. NDVI images corresponding to the mid-September period of Bundelkhand region has been analysed. Vegetation under stress has been found in the northern portion of the region as compared to the southern part, where a healthier vegetation cover was observed throughout the study period. Doi (2001) examined the serial relationship between NDVI and rainfall occurrences at different stations in the Thar Desert region. NDVI has been used successfully to identify stressed and damaged crops and pastures, but suffered with interpretive problems when the results were extrapolated over non-homogeneous areas. Kogan (1990) introduced the method of "geographic filtering of NDVI" and enumerated the concept of NDVI stratification and applications of the modified NDVI for more accurate monitoring of vegetation in non-homogeneous areas.

4.5.3.2 Vegetation Condition Index

As this study is focused on agricultural drought, and not on vegetation vigour, an agricultural mask was prepared to exclude all the pixels falling in non-agricultural area from the analysis. The smoothed NDVI data is normalized and VCI values were calculated for each year for all the thirteen districts of Bundelkhand region and the results are shown in figure 4.7. The region showed normal vegetation (95% < VCI> 51%) in year 1998. During 1999, all the districts except Mahoba and Chattarpur were observed to be under severe to extreme agricultural drought. Vegetation under severe stress was evident from the VCI value dropped to 5 % in Damoh district.

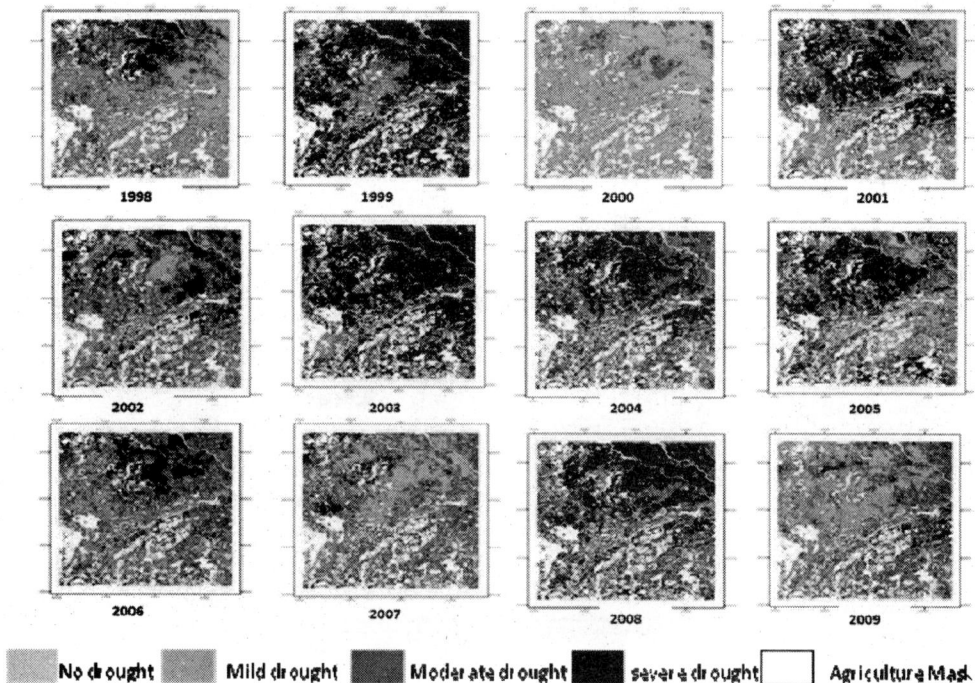

Figure 4.7: Agriculture drought interpretations based on VCI for Bundelkhand region

However, there was a remarkable improvement in vegetation condition in the following year and VCI ranged between 84 % and 95 %. 2000 was the best year for vegetation health among the 12 year period analysed in the study. Vegetation condition remained fluctuating due to the variation in rainfall in the following two years. Year 2002, was a normal year in terms of agriculture drought whereas 2003 was a complete drought year for the entire region. The worst situation of drought impact prevailed in this year. Drought continued again for the subsequent two years in most districts. The region was almost drought free in the years of 2007 and 2009 whereas in 2008 various districts were affected by agricultural drought of different degrees.

4.5.4 Drought Frequency and Intensity Analysis

Frequency and intensity (severity) maps obtained from the analysis of meteorological and agricultural drought were categorized into five classes of drought i.e. Extremely High, Very high, High, Moderate and Mild; whereas the hydrological drought frequency map was categorized into three classes i.e. Severe, High and Moderate. These maps are shown in figure 4.8 and 4.9. Mahoba, Sagar and Damoh districts faced the highest number of hydrological drought i.e., 5 to 6 drought years out of 12 observed years.

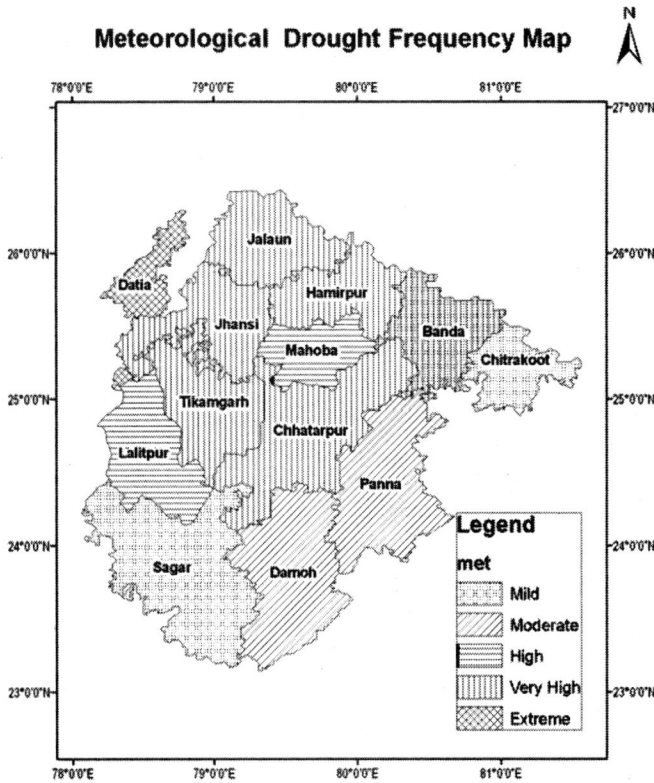

Agricultural Drought Frequency Map

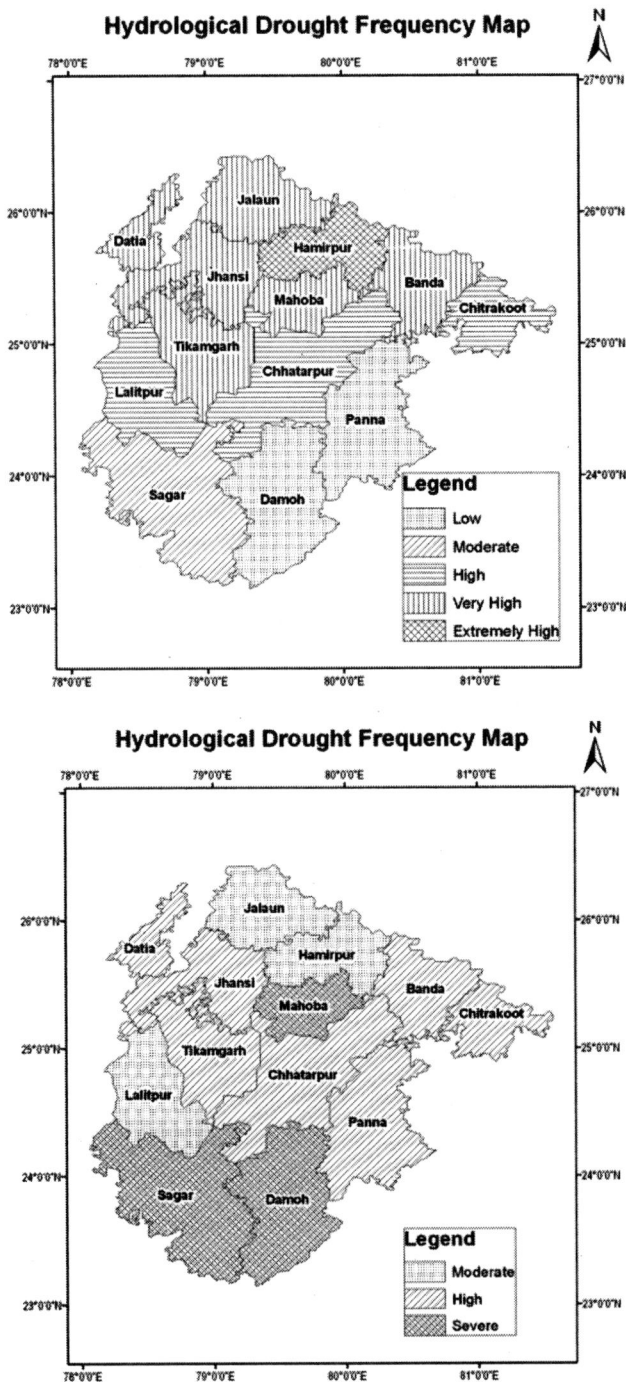

Figure 4.8: Frequency maps of meteorological, agricultural and hydrological drought

Meteorological Drought Intensity

Agricultural Drought Intensity

Hydrological Drought Intensity

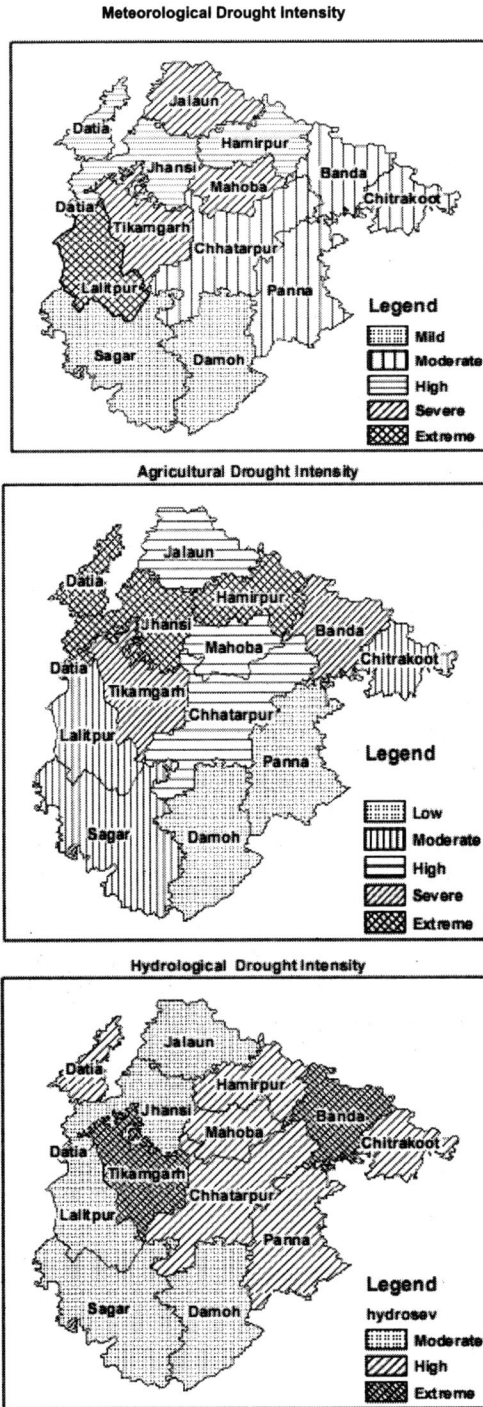

Figure 4.9: Intensity maps of meteorological, agricultural and hydrological drought

As evident in the meteorological drought hazard frequency map, Datia district faced the highest number of drought years (9 times) followed by Lalitpur and Mahoba in the range of 6 to 8 times in 12 observed years. Tikamgarh, Jhansi, Jalaun, Hamirpur and Chhatarpur districts faced 5 to 6 droughts in 12 observed years. The agricultural drought frequency was the highest in Hamirpur i.e. 8 out of 12 observed years followed by Datia, Tikamgarh, Jhansi, Jalaun, Mahoba and Banda facing 4 to 6 drought years in 12 observed years.

Specific maps representing intensity (severity) of meteorological, hydrological and agricultural droughts were prepared and drought intensities were categorized (Figure 4.11). Meteorological drought incidences were extremely severe in Lalitpur district; agricultural droughts in Datia, Jhansi and Hamirpur, and hydrological droughts were more severe in Tikamgarh and Banda districts of Bundelkhand region. The results of the correlation analysis between meteorological and agricultural drought for all the 13 districts from 1998 to 2009 are given in the table below.

Table 4.4: Correlation between meteorological and agricultural drought

District	With zero lag	With 1 year lag
Banda	-0.33483	-0.07646
Chitrakoot	-0.90221	-0.18436
Hamirpur	-0.46048	0.473797
Jalaun	-0.17115	0.168141
Jhansi	-0.25909	0.282094
Lalitpur	-0.14638	-0.02247
Mahoba	-0.33692	0.395351
Chhatarpur	-0.81569	0.174062
Damoh	-0.37758	0.416816
Datia	-0.51005	0.303156
Sagar	-0.71041	0.462438
Panna	-0.67833	0.202472
Tikamgarh	-0.50788	0.37793

It is important to mention that meteorological drought has been represented in decreasing values of percent by normal, while hydrological drought is represented with increasing SWI values. Therefore, negative values of the coefficient represent a higher correlation. It is evident from these results that hydrological drought follows meteorological drought without a time lag, whereas agricultural drought follows meteorological drought with a time lag of a year.

Table 4.5: Correlation between meteorological and hydrological drought

District	With Zero year lag	With 1year lag	With 2 year lag	With 3 year lag
Banda	-0.89664	-0.58658	-0.53985	-0.32887
Chitrakoot	-0.42107	-0.38818	-0.79538	-0.36141
Hamirpur	-0.22709	-0.38316	-0.89637	-0.51286
Jalaun	-0.59294	-0.65567	-0.58307	-0.19532
Jhansi	-0.61299	-0.03042	-0.18485	-0.18926
Lalitpur	-0.11162	-0.18708	-0.33495	0.10719
Mahoba	-0.79556	-0.38071	-0.58882	-0.10781
Chhatarpur	-0.86521	-0.08559	-0.17298	0.32488
Damoh	-0.03784	0.26590	0.79486	0.44961
Datia	-0.72710	-0.39377	-0.15281	0.43852
Sagar	-0.76011	-0.49113	-0.05289	0.13221
Panna	-0.65246	-0.21415	-0.37806	0.02635
Tikamgarh	-0.72815	0.13632	0.58308	0.38472

4.5.5 Hazard Analysis (Frequency and Intensity)

Drought hazard map for meteorological, hydrological and agriculture were prepared based on drought frequency and intensity. The study area is classified into five classes of drought severity *i.e.*, extreme, severe, high, moderate and mild. The outcome maps are given in the Figure 4.10.

It has been interpreted that Datia district experienced extremely high frequency of drought - 9 out of 12 years. Almost every year there was meteorological drought in Datia, whereas severely high drought occurred in Lalitpur and Mahoba districts - 8 in 12 years. However, drought intensity was observed to be moderate in Datia. In Lalitpur district droughts were of very intensive type but the frequency came under the moderate category. Thus, it can be concluded that drought hazard was overall of extreme level in Lalitpur followed by severe drought hazard in Datia and Mahoba districts of Bundelkhand region.

The response of the environmental system (i.e. in terms of hydrological and agricultural system) to meteorological drought has also been analyzed. The frequency of hydrological drought increased after 2004 in almost all the districts of Bundelkhand and relatively higher frequency was recorded for Sagar, Damoh and Mahoba districts and very severe drought occurrences observed in Tikamgarh and Banda districts. Tikamgarh, Banda and Mahoba were assessed to be the hydrologically the most vulnerable districts of Bundelkhand region.

In case of agricultural drought, results showed that Hamirpur experienced very high frequency of drought (8 out of the 12 years). High drought intensity was reported for Datia, Jhansi and Hamirpur districts. Overall, the agricultural vulnerability was found to be extreme in Datia, Jhansi and Hamirpur followed by severe in Tikamgarh and Banda districts in Madhya Pradesh.

The hazard map indicated very high probability of meteorological drought in Lalitpur, whereas Datia, Jhansi and Hamirpur are indicated as most vulnerable to agricultural drought and Tikamgarh and Banda as most vulnerable to hydrological drought.

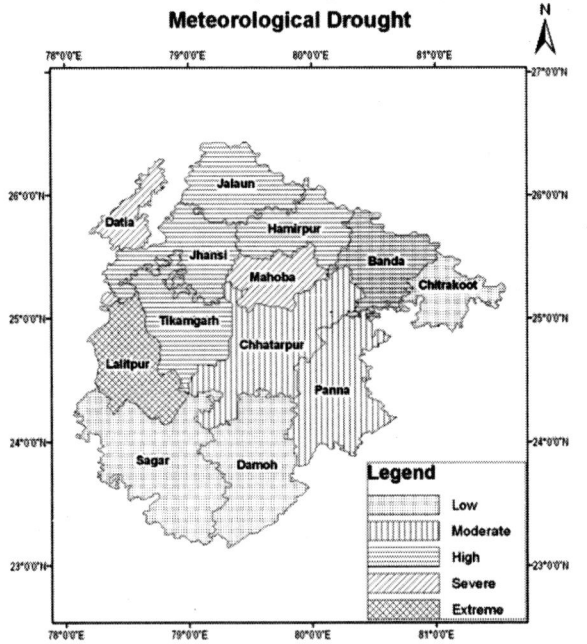

Meteorological Drought

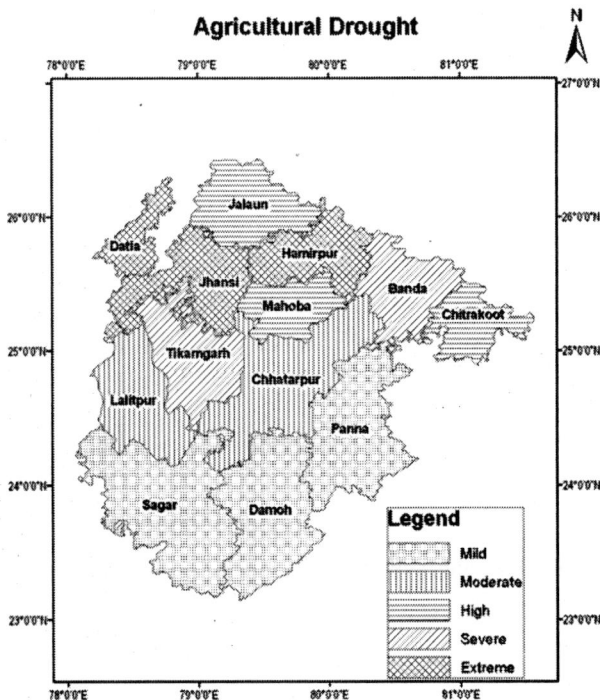

Agricultural Drought

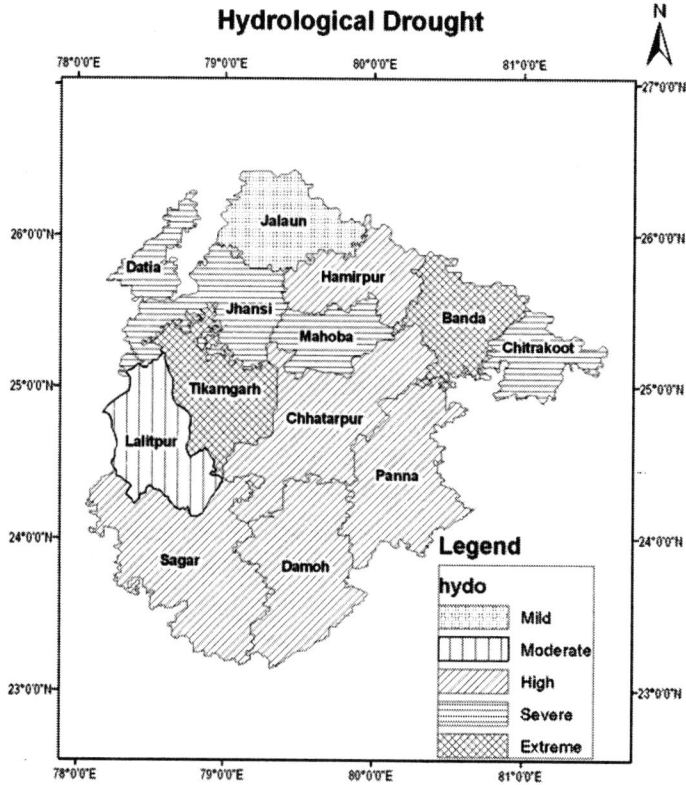

Figure 4.10: Maps of meteorological, agricultural and hydrological drought

4.5.6 Composite Drought Hazard Mapping

Aim of the present study was to develop a 'Composite Drought Hazard Map' for the Bundelkhand region by integrating the hazard frequency and intensity maps for Meteorological, Hydrological and Agricultural drought. This composite vulnerability map has been prepared combining effect of all the three droughts – meteorological, hydrological, and agricultural, for individual districts in the region. The composite was prepared after assigning weightages for all the three categories of drought. The results showed that the districts Datia, Tikamgarh, Jhansi, Mahoba and Hamirpur districts were facing severe drought vulnerability followed by Lalitpur, Jalaun and Banda. In Datia the contribution of agricultural drought was more visible than meteorological and hydrological drought. In Tikamgarh the hydrological drought contributed more for the high vulnerability. In Jhansi and Hamirpur districts the agricultural drought was more evident than the other two types, and pushing these districts into the severe vulnerability category. Both meteorological and hydrological drought equally contributed to the high risk in Mahoba district.

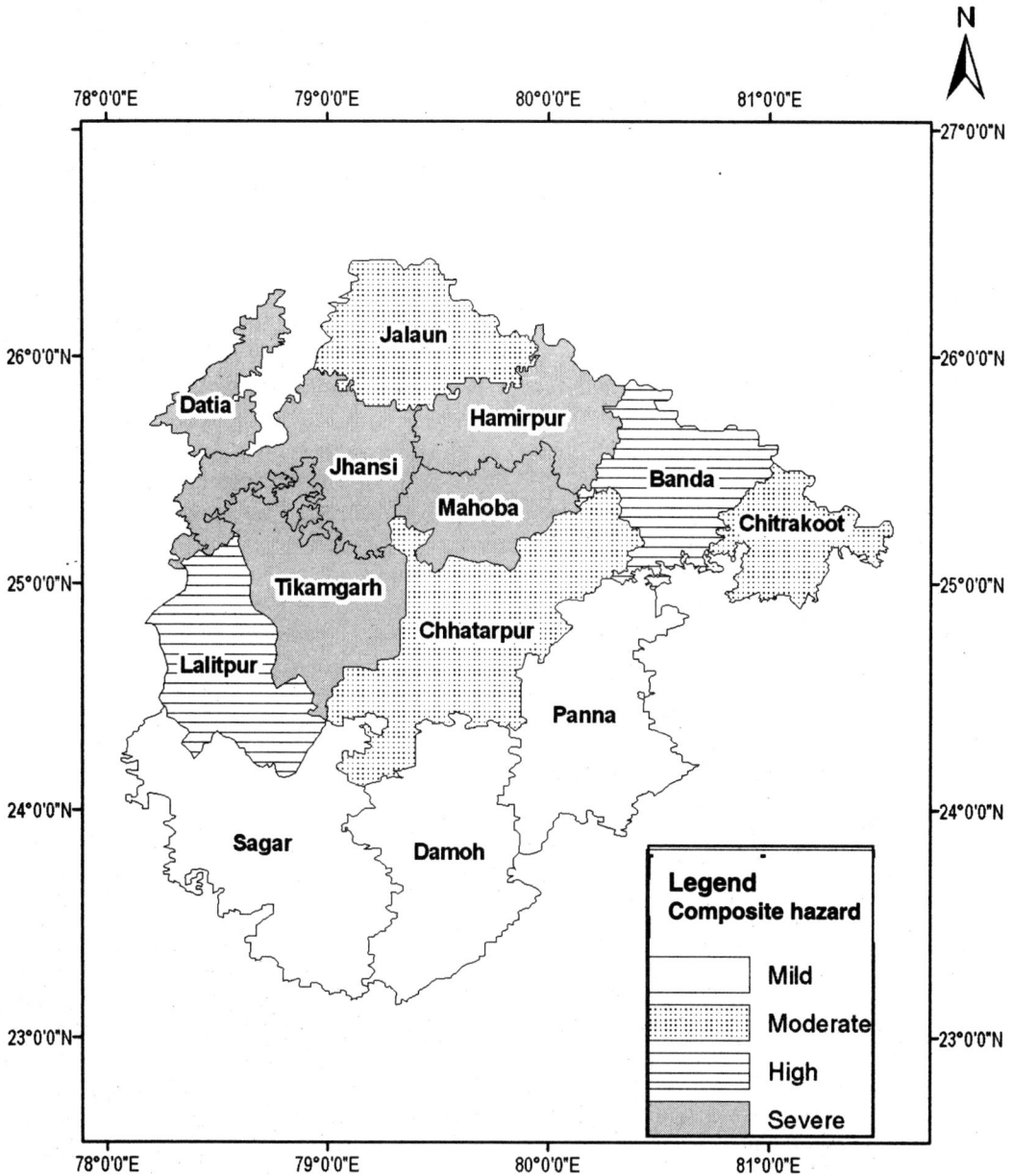

Figure 4.11: Composite drought hazard map of Bundelkhand

4.6 CONCLUSION

This study presented a systematic analysis of frequency of occurrence and its severity of drought hazards, using meteorological (rainfall), hydrological (ground water) and agricultural (using SPOT vegetation) data for the Bundelkhand region. It attempted to unlock the word "drought" itself and redefined the role of meteorological, hydrological, and agricultural droughts. Typological, spatial and temporal analysis followed by comprehensive analysis was performed. It was an attempt of applying six different indices into one assessment. Caution has been used to avoid any overlapping of data. Interestingly, the occurrence of a meteorological drought did not necessarily result in a hydrological or an agricultural drought at district level. This has given an understanding on the significance of site specific interventions that relate to short or long term mitigation of drought impacts, besides adaptability of people's actions in particular of water and agriculture management practices. In addition, this has also indicated the role of ecological system's resilience and people's attitude towards dealing with a calamitous situation. A clear explanation of these terms will enhance the understanding and better equip the policy makers and mitigation analysts engaged in drought mitigation program in the region. An interesting observation regarding the results of three different droughts is the fact that almost all results have been validated with the findings of Inter-Ministerial Team of Govt. of India 2008. The conclusions drawn in respect of Bundelkhand in this study are strategically important from holistic management of droughts by policy makers and implementing agencies. This study has implications for drawing better plans for early warnings of drought, drought proofing mechanisms and implementation of drought preparedness and decision support systems for better prioritization of actions at spatial and temporal extents.

References

Below, R., Grover-Kopec, E., and Dilley, M. (2007). Documenting Drought-Related Disasters. *The Journal of Environment & Development 16(3):* 328-344.

Bhuiyan, C. (2004). Various drought indices for monitoring drought condition in Aravalli terrain of India, In: Proceedings of the XXth ISPRS Conference, Int. Soc. Photogrammetry and Remote Sensing, Istanbul.

Bhuiyan, C., Singha, R.P., and Kogan, F.N. (2006). Monitoring drought dynamics in the Aravalli region (India) using different indices based on ground and remote sensing data. *International Journal of Applied Earth Observation and Geoinformation,* 8(4),289-302.

Cutter, S.L. (2003). *Hazards Vulnerability and Environmental Justice.* London: Earth Scan.

Department of Agricultural Cooperation (2009). Manual for Drought Management. New Delhi: Department of Agriculture and Cooperation, Ministry of Agriculture.

Dilley, M., Chen, R.S., Deichmann, U., Lerner-Lam, A., and Arnold, M. (2005). *Natural disaster hotspots: A global risk analysis. Washington. DC: World Bank, Hazard Management Unit.*

Inter-American Development Bank (IDB), Working Papaer Series IDB-WP-124. The Economics of Natural Disasters, Eduardo Cavallo Ilan Noy 2009.

Doi, R.D. (2001). Vegetation response of rainfall in Rajasthan using AVHRR imagery. *Journal of the Indian Society of Remote Sensing 29* (4), 213-224. DOI: 10.1007/BF02995726.

Eriyagama, N., Smakhtin, V., and Gamage, N. (2009). Mapping Drought Patterns and Impacts: A global perspective, IWMI Research Report 133. Colombo, Sri Lanka: International Water Management Institute.

Fleig, A.K., Tallaksen, L.M., Hisdal, H., and Demuth, S. (2006). A global evaluation of stream flow drought characteristics. *Hydrology and Earth System Sciences Discussions 10 (4),* 535-552.

Hagman, G. (1984). *Prevention better than cure. Report on Human and Environmental Disasters in the Third World.* Prepared for the Swedish Red Cross. Stockholm.

IADB (2009). *Economic Survey of Natural Disasters.* Inter-American Development Bank.

Kogan, F.N. (1995). Drought of the late 1980s in the United States as derived from NOAA polar-orbiting satellite data. *Bulletin of the American Meteorological Society, 76,* 655–668.

Logana, K.E., Brunsell, N.A., Jonesa, A.R., and Feddemaa, J.J. (2010). Assessing spatiotemporal variability of drought in the U.S. central plains. *Journal of Arid Environments 74(2)* 247-255.

Mendicino, G., Senatorea., A., and Versacea, P. (2008). A Groundwater Resource Index (GRI) for drought monitoring and forecasting in a Mediterranean climate. *Journal of Hydrology 357 (3-4)* 15 August. pp 282-302.

Nagarajan, R (2003). *Drought: Assesment, Monitoring Management.* New Delhi: Capital Publishing Company.

Nair, S.S., and Gupta, A.K. (2010). Climate Change and Disaster Management: Data Needs for Risk Analysis, Decisions and Planning. In: Proc. Volume of National Workshop on Climate Change: Data Requirement and Availability, Bangalore, April 16-17, 2009, ISEC, ISO & Ministry of Statistics, India. Pp 89-114

Panu, U.S., and Sharma, T.C. (2002). Challenges in drought research: Some perspectives and future directions. *Hydrological Sciences-Journal—des Sciences Hydrologiques. 47(S),* Special Issue: Towards Integrated Water Resources Management for Sustainable Development.

Samra, J.S. (2004). *Review and Analysis of Drought Monitoring, Declaration and Management in India.* Working Paper 84. Colombo, Srilanka: International Water Management Institute.

Samra, J.S. (2008). *Report on Drought Mitigation Strategy for Bundelkhand Region of Uttar Pradesh and Madhya Pradesh.* New Delhi : Inter ministerial Team.

Steven, M.Q., and Srinivasan, G. (2010). Evaluating the utility of the Vegetation Condition Index (VCI) for monitoring meteorological drought in Texas, *Agricultural and Forest Meteorology 150(3),* 330-339

Wilhite D.A., (2000). Drought as a Natural Hazard: Concepts and Definitions (Chapter 1, pp. 3-18). In: D.A. Wilhite (Ed.): *Drought: A Global Assessment* (1), London, U.K: Routledge Publishers.

Wilhite, D.A., Hayes, M. J., and Svoboda, M.D. (2000). Drought Monitoring and Assessment: Status and trends in the United States. In J. V. Vogt, and F. Somma (Eds.), *Drought and drought mitigation in Europe* (pp. 149–160). Dordrecht: Kluwer Academic Publishers.

Wilhite, D.A., Hayes, M. J., Knutson, C., and Smith, K.H., (2004). The Basics of Drought Planning: A 10-Step Process. http://www.drought.unl.edu/plan/handbook/10step.pdf.

Wilhite, D.A. (1993). *Drought Assessment, Management and Planning Theory and course* Study. Boston : Kluwer Academic.

Wilhite, D.A., Glantz, M.H. (1985). Understanding the Drought Phenomenon: The Role of Definitions. *Water International 10* (3),111–120.

Flood Warning in Bavaria, Germany

Alfons Vogelbacher

5.1 INTRODUCTION

Timely warning of flood hazard is an essential part of precautious flood protection. Prolongation of the forecast lead time enables more substantial protection measures like the building up of mobile walls. The effectiveness of the warnings mainly depends on the awareness and preparedness of the recipients. Today the enhancement in information media i.e. the internet has led to a much better reach ability of flood information and warning for everyone.

In Germany, the responsibility for a flood information service is assigned to the federal states. Until 2000, the flood information service in Bavaria primarily served as a reporting service that collected water gauge records and forwarded these – supplied with a trend comment – to those concerned. The amount of quantitative flood forecasts was limited. Simple methods were used, e.g. regression functions for gauges.In the last 12 years, Bavaria has suffered under a series of extreme floods. There was a large flood at the Whitsuntide in the year 1999 in Southern Bavaria and the Alps. The centres of damage have been located at the Danube River and the southern tributaries Iller, Lech and Isar. Three years later the heavy rain falls in August, 2002 caused extreme floods of the Danube River and its northern tributary Regen in East Bavaria. In January 2003, the northern parts of Bavaria, especially the Basin of the River Fränkische Saale, a tributary to the Main in the Rhine Basin, suffered a large flood. Recently, the flood in August, 2005 hit the same region as 1999. Partly the flood peaks even surpassed the peaks of 1999.

Because of the increasing damage and the increasing number of floods - in particularly the Whitsuntide Flood in 1999 - the Bavarian State government established an Action Program for sustainable flood protection led by a long term flood prevention and protection strategy until 2020. The Action Program is part of the Integrated River Basin Management. Actions have been taken at the same time in the fields of natural retention, structural flood measures and flood precaution including new challenges like Climate Change. According to the resolution of the Bavarian State Government in May, 2001 there will be an investment of 2.3 billion Euros until 2020 in the fields of improvement of natural detention, improvement and construction of structural flood measures and flood precaution.

Part of this program was the development of flood forecast models as well as the implementation of an automatic online rain gauge network and an optimization of the existing

Disaster Management and Risk Reduction: Role of Environmental Knowledge; Editors Anil K. Gupta, Sreeja S. Nair, Florian Bemmerlein-Lux and Sandhya Chatterji; Copyright © 2013, Narosa Publishing House, New Delhi

river gauge network. The dissemination of flood information based on means of modern communication and its reliability was improved. Today, hydrological forecasts have become an important part of the flood information service in Bavaria, since hydrological models cover almost the total area of Bavaria (70 000 km²). The hydrological forecasts are calculated daily on business days and more frequently during flooding periods in five regional flood forecast centres. The decision-makers within the Bavarian Water Management Authority can access the results, i.e. forecasts for around 600 gauge stations over whole of Bavaria. Additionally, the forecasts of about 100 selected gauging stations are published in the web site (http://www.hnd.bayern.de) for a horizon of 6–24 h depending on the catchment area.

5.2 ORGANISATION OF FLOOD WARNING IN BAVARIA

Flood warning systems in river basins and on lakes are the responsibility of the German federal states. For this purpose, Bavaria has set up a Flood Warning Service that collects data on water levels, run-off and precipitation, analyses this information, draws up flood alert plans and warns people affected.

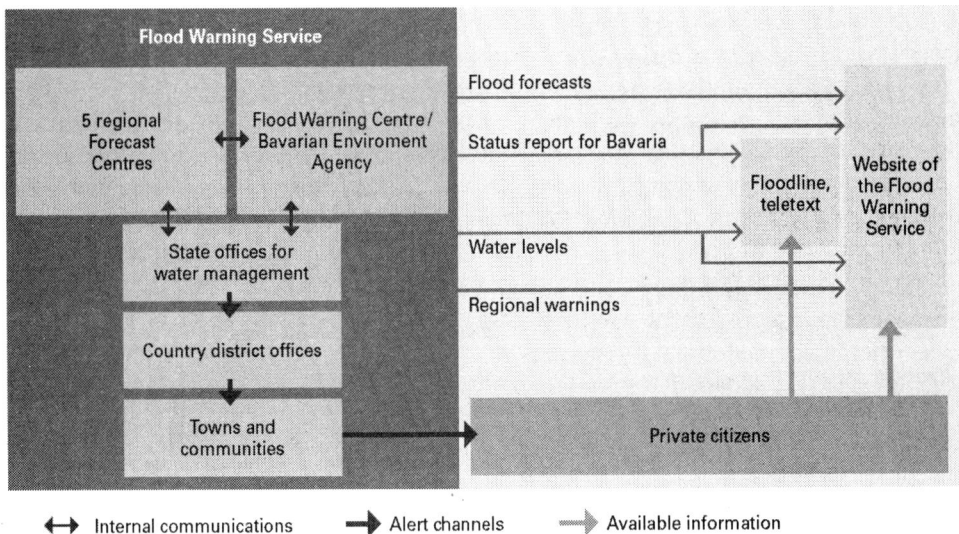

Figure 5.1: Reporting and information scheme of the Bavarian Flood Information Service

Connected to the Flood Warning Service are the state offices for water management, the county district offices, towns and communities. The coordinating unit is the Flood Warning Centre in the Bavarian Environment Agency (Figure 5.1). Five regional Flood Forecast Centres are responsible for calculating and preparing flood predictions for the basins of the rivers Main, Danube, Isar, Iller-Lech and Inn respectively. The respective centres are put into action as soon as rivers or lakes rise above defined threshold values. The water levels at the gauging stations are then read on an hourly basis and flood predictions are updated continuously. The state offices for water management put out regional alerts, while the Flood Warning Centre issues a flood status report for all Bavaria.

Alert plans make sure that this information is passed on through the county district offices to the affected towns and communities. Towns and communities are the last link in the alert channel and play a particularly important role. The alert plans specify who is to be warned, when and how, and what measures are to be taken at which gauge levels. For this case, the local authorities have plans of endangered areas or buildings and plans for the organization of flood defencesystems. The flood warnings have to be actively transmitted to the concerned recipients. Once warned, the recipient has the responsibility to inform himself about the threatening flood. For this purpose, the state offices for water management, the regional flood forecast centres and the flood information centre provide updated data, forecasts and flood reports via the website as well as flood news via a telephone service to the public and authorities. Actual water levels of all gauging stations are provided via TV-text and telephone service.

5.3 ROLE AND OPERATION OF THE FLOOD FORECAST CENTRES

In Bavaria there are five flood forecast centres corresponding to the main river catchments (Danube, Inn and Main) and Danube tributary catchments where large reservoirs have to be operated (Iller-Lech and Isar). They are responsible for operational flood forecast. The decentralised and river catchment-related organisation of the flood forecast allows the use of experiences with local knowledge and the operation of large reservoirs *in-situ*. The model-based flood forecast service supports the services of the regional water authorities and the flood information centre. In the case of the flood forecast centres Iller-Lech and Isar, respectively, the flood forecast service additionally assists in the management of reservoirs. The state offices for water management are responsible for local flood forecasts and the dissemination of flood information in the regional flood information service, the application of models for reservoir operation and for the data service (collecting and providing data of the river gauge network). The regional flood forecast centre informs the main reporting offices within the forecast area and the flood information centre, if due to the flood forecast a lasting overrun of the warning limits has to be expected. After receiving this information, the main reporting offices have to provide the river gauge data on an hourly basis.

In case of a flood event the flood forecasts are calculated three or four times per day at defined points of time. If necessary, the flood forecasts are updated every hour. The forecast centres calculate forecasts for all river gauges implemented in the flood forecast models. The complete flood forecast simulation results are made available for the regional water authorities, regional governments and ministries by a Java-Client application within the data network of the Bavarian water management administration. In the intranet and internet, only forecasts of selected river gauges and for reduced forecast horizons are published. The regional water authorities have access to the flood forecasts for all river gauges and can use these results for their own forecasts.

5.4 WATER LEVEL AND FLOOD RISK

The basic Information in flood warning is the water level at the gauge site – measured or forecasted. For interpretation in respect to the flood risk and to protective measures more information is needed. First, there are 4 alert levels that deliver plain and simple information on the extent of flooding. At each individual gauging station it is determined which water levels correspond to the respective flood alert levels (Figure 5.2).

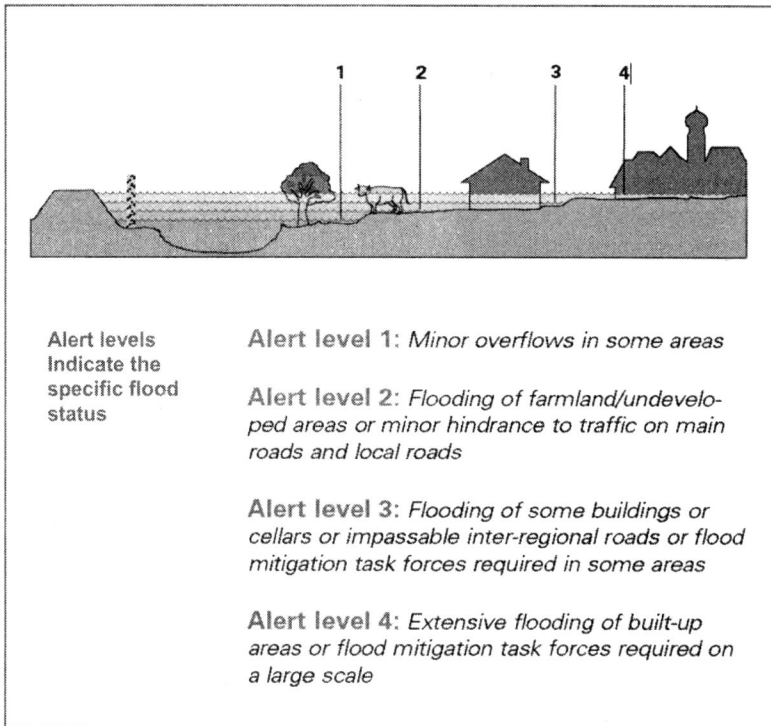

Alert levels
Indicate the
specific flood
status

Alert level 1: *Minor overflows in some areas*

Alert level 2: *Flooding of farmland/undevelo-
ped areas or minor hindrance to traffic on main
roads and local roads*

Alert level 3: *Flooding of some buildings or
cellars or impassable inter-regional roads or flood
mitigation task forces required in some areas*

Alert level 4: *Extensive flooding of built-up
areas or flood mitigation task forces required on
a large scale*

Figure 5.2: Flood Alert Levels

Second, there are tables for each gauging station connecting water level to a description of the flood extent, protective measures and flood hazard (Table 5.1). The communities are responsible for the maintenance of plans linking water level to flood risk and protective response. They keep plans of endangered areas or buildings and plans for the organisation of flood defence systems. Protective responses in Bavaria are: manageable flood polders, major storage dams in headwater, flood protection walls, mobile closures, superstructures, sandbags and evacuation.

5.5 INFORMATION ABOUT FLOOD RISK, VULNERABILITY AND HAZARD

Online Maps (www.iug.bayern.de) of the designated flood areas and preliminary assured flood areas can be reached by the public. The Information Service Flood Affected Areas (IÜG) gives an overview of the status of the investigation and determination of flood risk in Bavaria. Flooded areas for frequent, medium and rare flood events and water-sensitive areas are displayed. Where available, also the water depths in flooded areas and observed floods are shown.

The EU Floods Directive on the Assessment and Management of Flood Risks went into force in 2007. This Directive now requires Member States to assess if all water courses and coast lines are at risk from flooding, to map the flood extent and assets and humans at risk in these areas and to take adequate and coordinated measures to reduce this flood risk. This Directive also reinforces the rights of the public to access this information and to have a right of codetermination in the planning process.

Table 5.1: Flood hazards and protective measures in relation to the water level at gauging Station Passau Donau (cut version)

W cm	Place	Type of measure or hazard
630	Passau	Advance warning to the police by the regulatory agency
720	Passau	Flooding of the upper Zollände in Passau and the harbour area Racklau.
720	Passau	Flooding of the Fritz-Schäffer-Promenade at the tavern - Tiroler-, removal of the motor vehicles.
740	Passau	Flooding of the Fritz-Schäffer-Promenade, traffic stoppage.
750	Passau	Evacuation of the parking lots at Schanzl.
750	Passau	Traffic blocking at the upper Donaulände
770	Passau	Flooding of the high road ST 2132 at Löwmühle
770	Passau	Evacuation of garages and souvenir shops at the lower Donaulände. flooding of the harbour area Racklau
780	Passau	Highest water lever for water navigation (HSW).
790	Passau	Flooding of the approach to the Nagelschmied- and Höllgasse
800	Passau	Flooding of the posterior Donaulände downstream of the Schanzl bridge.
800	Passau	Warning by loudspeaker for the upper Donaulände, Regensburgerstreet, Rindermarkt and Ort.
810	Passau	'Ort' partly flooded.

The Directive's aim is to reduce and manage the risks that floods pose to human health, the environment, cultural heritage and economic activity. The Directive requires Member States to first carry out a preliminary assessment by 2011 to identify the river basins and associated coastal areas at risk of flooding. For such zones they have to draw up flood risk maps by 2013 and establish flood risk management plans focused on prevention, protection and preparedness by 2015 (http://ec.europa.eu/environment/water/flood_risk/index.htm).

5.6 ONLINE FLOOD INFORMATION SERVICE

The website of the Flood Warning Service (Figure 5.3) gives access to detailed background information and to the latest water level readings and recorded measurements. Maps and Charts give a quick overview of the current status 24 hours a day. These sites give access to water level and run-off data from river and lake monitoring stations as well as to measurements provided by precipitation and snow depth monitoring stations. In a flood event the Flood Bulletin (status report) gives an overview of the flood situation and a forecast of the expected further development. The bulletin is updated several times a day. In their Warnings the state offices for water management provide a detailed description of current and predicted flooding for each one of their county districts that are threatened. In addition reports on past flood Events as well as Links, addresses and phone numbers of other contact partners can be found on the website (www.hnd.bayern.de).

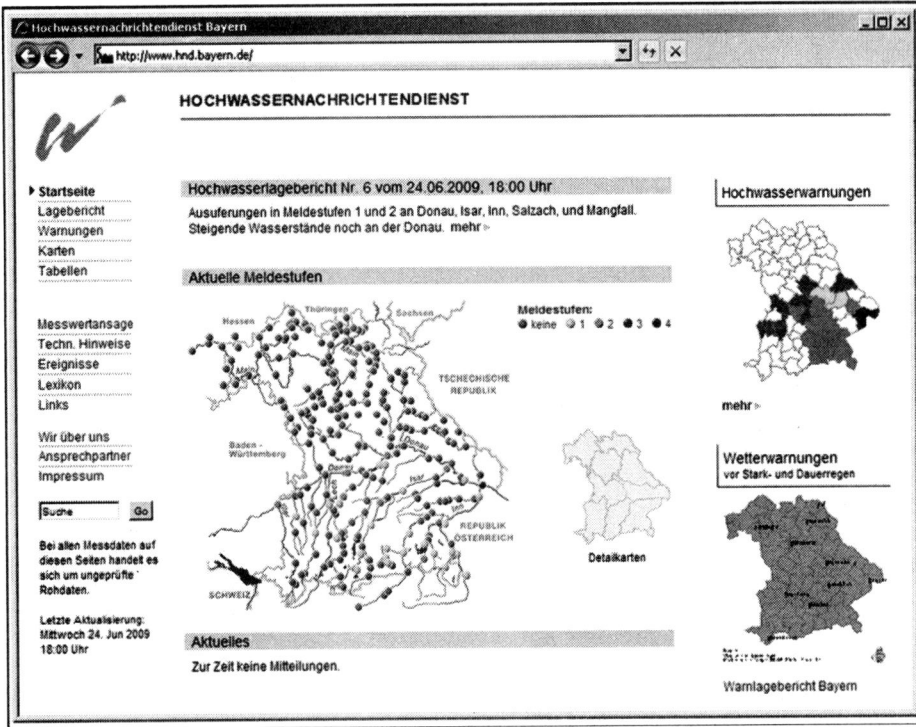

Figure 5.3: Website of the Flood Information Service

5.7 REAL-TIME DATA COLLECTION

The flood warning system uses all reachable data of the different measurement networks in hydrology and meteorology, weather radar products and numerical weather forecasts from the weather services and data of the hydro power plants.
Transmission of the data is done by:

- Polled Systems (sampling of data via telephone landline, digital mobile as GSM (GPRS).
- Push Systems (active transmission from station to server via ISDN or GSM).
- NFS file transfer in the computer network.
- Ftp transfer and http-request between computer in different networks and from neighbouring countries.

5.8 WATER LEVEL AND DISCHARGE DATA

The river gauge network in Bavaria consists of about 600 stations, 560 of which are equipped with telemetric data transmission. 320 river gauges are so-called "report river gauges" for the flood information service. The extreme flood events in the Danube river catchment in May, 1999 and August, 2002 as well as in January, 2003 in the Main river catchment revealed that equipment and data transfer were not sufficient.

Figure 5.4: Gauging Station at the river Isar in flood stage

Failures and breakdowns of the measurement installation and data supply occurred. To ensure the data supply, most of the river gauges have been equipped with redundant measurement devices and telecommunication channels. For telemetric data transfer, the conventional telephone network or mobile telephone systems GPRS are used. The remaining river gauges in the flood information service have at least redundant measurement devices. Furthermore, there were numerous river gauges where the extreme floodwater stages could not be assessed. Constructional measures will be taken within the next years.For the flood forecast models, discharge is the observation, and thus reliable stage-discharge-relationships are needed. However, often discharge is not measured during flood situations and additionally it is associated with large uncertainties. To check and improve the rating curves especially in the high-flow extrapolation range, a project has been started at the Bavarian Environment Agency where hydraulic simulations for about half of the Bavarian river gauges shall be conducted.

There are also water level data available from the Hydrographical Services of Austria, from the adjacent Federal States of Baden-Wuerttemberg, Thuringia, Hesse as well as from the Federal Waterway and Navigation Administration for the national waterways. For the flood forecast models, discharge data at the power stations are required. These data are obtained in a data exchange with different operators of hydro power stations. The data of most of the external partners are imported per ftp or http-request via the internet into the main database of the flood information service.

5.9 HYDRO METEOROLOGICAL INPUTS

Most of the precipitation data are provided by a joint automatic monitoring network, operated by the German Weather Service and the Bavarian Environment Agency with about 285 stations. Additionally, data from the monitoring networks of the private company Meteomedia, the Bavarian Agency of Agriculture and the Bavarian Avalanche Warning Service are used. Because of transboundary tributaries, also precipitation data of the Central Institute of Meteorology and Geodynamics, Vienna, and the Austrian hydrographical services as well as of the federal states of

Baden-Wuerttemberg and Hesse are collected. Overall, precipitation data of 700 stations are available online.

Besides precipitation, snow height, snow water equivalent, air temperature, wind speed and radiation are the most important meteorological observations for snow melt calculation. However, these quantities are measured in a lower spatial resolution. The snow measurement network has been considerably increased within the last years. At present, the snow water equivalent is measured at approx. 120 stations.

The spatial assessment of precipitation can be improved by using weather radar measurements. The adjustment of radar signals to the measurement data of the ground precipitation network allows a spatially high-resolution assessment of rainfall events. The operational products of the German Weather Service project RADOLAN (radar online-adjustment) are available since 2005. The composite data of the radar echoes and the products adjusted to the ground monitoring network are received each hour via FTP from the German Weather Service. Since 2008, the RADOLAN-data can be used directly as input data for the operational flood forecasts.

5.10 NUMERICAL WEATHER PREDICTION (NWP)

Several weather forecasts are available for operational flood forecasting in Bavaria. Usually, the results of the numerical weather forecast models of the German Weather service are used. COSMO-EU is the main product of the German Weather service for the forecast horizon of 3 days (www.cosmo-model.org). The most recent product is COSMO-DE, which has been used for operational flood forecasting in Bavaria since 2008. The high horizontal resolution of 2.8 km allows direct simulation of thunderstorm cells of the size of only a few kilometres. Forecast runs are started every 3 hours. Within the model runs, updated radar measurements are integrated.

For Southern Bavaria, the results of the ALADIN-Austria model operated by the Central Institute of Meteorology and Geodynamics in Vienna are available additionally (Vogelbacher 2007). For the purpose of comparison, forecast products such as the American GFS (Global Forecast System) model or the ECWMF (European Centre for Medium-Range Weather Forecasts) model are used at times. Since December 2007, the model output of COSMO-LEPS (Consortium for Small-Scale Modelling – Limited-area Ensemble Prediction System) of 16 ensemble members is available and used in flood forecasts.To spatially assess the water release from snowmelt and rainfall, simulation results of the snowmelt model SNOW4 of the German Weather Service are used. Water yield and water equivalent of the snow cover are given in a spatial resolution of 1×1 km in an hourly time step.

Weather and storm warnings of the German Weather Service are sent via FAX, e-mail or are made available via the internet. Further information can be obtained by telephone contact with the meteorologist-in-charge at the regional office of the German Weather Service in Munich. For defined geographical regions in Bavaria, the meteorologist gives an evaluation for areal precipitation amounts.

5.11 ON-LINE DATA BASE

All of the data are continuously fed into a set of MySQL relational data base tables to form the online master data base. The master data base is replicated continuously on up to 15 other web servers with a Linux-Apache-MySQL system. These servers are located at 3 regional floods forecast centres and in the computing centre of the state of Bavaria for internet access. If the master data base fails the other servers can take over the function of a master. The automated import procedures can be run on each of the servers in the network.

Metadata for gauging-stations, meteo-stations, precipitation-stations, data transfer, statistics, level-discharge curves, flood stages, warning levels, flood-risk plans etc. Alert and dissemination plan with 3991 addresses and 1425 messages, reports, like the flood bulletins and warnings.

The data base contains the following (Table 5.2)

Table 5.2: Data and number of stations

Data	Number of stations
Water level & Discharge	1077 with Forecasts 184
Precipitation	919
Snowfall level	627
Water Equivalent	377
Atmospheric pressure	71
Global Radiation	191
Relative Air Humidity	406
Sunshine Duration	92
Air Temperature	511
Dew Point Temperature	148
Wind Velocity	365
Snowmelt and Rain from SNOW4-model	1km×1km grid-data
Water Equivalent of the Snow Cover from SNOW4-model	1km×1km grid-data

5.12 FLOOD FORECAST MODEL SYSTEM

Forecast models completely cover the area of Bavaria (Figure 5.5). The rainfall–runoff model predominantly used is LARSIM (Ludwig and Bremicker, 2006). Because of its robust and relative simple components, the model is suitable for operational forecast (Gerlinger et al., 2000). LARSIM can be applied as an event-based flood model as well as a water balance model for continuous simulations. For the time being, the operational LARSIM flood forecast models are implemented in event-based mode. The water balance model mode is currently being introduced. For the Iller catchment the water balance mode of LARSIM has been implemented, and is now in an operational pilot phase.

The rainfall–runoff models used are deterministic models, which calculate discharge rates as response to precipitation input. In the simplest case, the effective precipitation is determined with a runoff-coefficient depending on the antecedent moisture content of the area. According to non-linear relations between precipitation and resulting discharge, the runoff-producing precipitation is adjusted by superposition calculations. The catchment parameters can be subdivided either in gridded or irregular sub-catchments. Each sub-catchment is characterized by coordinates, elevation, length of watercourses and schematic river cross-section with roughness coefficients. On a grid or sub-catchment basis, the relevant runoff processes are calculated using specific hydrological methods (Laurent et al., 2010).

Snow accumulation and snowmelt can be considered as well as artificial influences (e.g. storage basins, diversions, water transfer between different catchments). The model calculations are based on hourly data of precipitation, discharge, precipitation forecasts, as well as on snowmelt calculations and forecasts. The forecasts for the tributaries based on rainfall–runoff models are linked to hydrodynamic models.The basis for the flood routing model at the Main river is the hydrodynamic model WAVOS of the German Federal Institute of Hydrology. The hydrodynamic model FLORIS 2000 (Reichel, 2001) is run for the Danube, the Lech and for the Inn River. The standard operating regulations of the hydropower stations are included in these models.

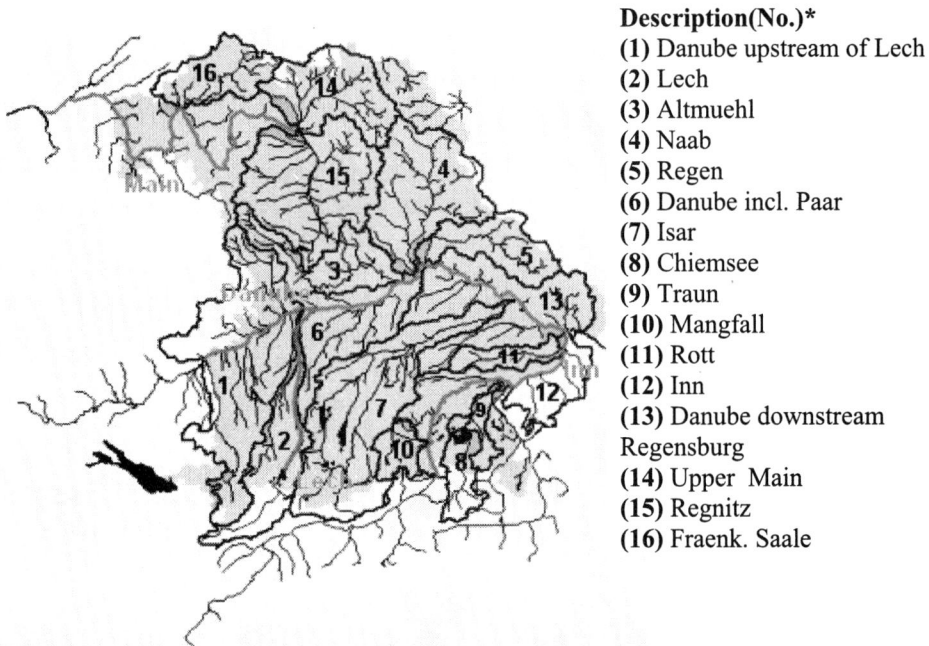

Description(No.)*
(1) Danube upstream of Lech
(2) Lech
(3) Altmuehl
(4) Naab
(5) Regen
(6) Danube incl. Paar
(7) Isar
(8) Chiemsee
(9) Traun
(10) Mangfall
(11) Rott
(12) Inn
(13) Danube downstream Regensburg
(14) Upper Main
(15) Regnitz
(16) Fraenk. Saale

Figure 5.5: Overview of the hydrodynamic and rainfall-runoff models used for operational flood forecast in Bavaria

The operational flood forecast model system works in a so-called served operation. First, the rainfall–runoff models have to be run to the connecting gauge for the hydrodynamic model.

Besides forecasts of the rainfall–runoff models for the Bavarian region, external flood forecasts for tributaries are needed and implemented in the system.For example, the Danube forecasts upstream of the Iller tributary are produced by the Federal State of Baden-Wuerttemberg. For the Inn catchment, flood forecasts are needed for the Inn upstream of Kufstein which are supplied by the federal state of Tyrol (Austria) and for the Inn tributary Salzach, supplied by the federal state of Salzburg (Austria). The Salzach River is within the responsibility of the federal government of Salzburg where they use HYDRIS (Hydrological information system for flood forecast).Using rainfall–runoff models, the forecast period is limited by the precipitation forecast. To adjust to the COSMO-EU model of the German Weather Service, 72 hours forecasts are produced by default. However, forecasts for shorter periods are published on the internet. At present, the hydrodynamic models are run for 48 hours forecasts (Vogelbacher, 2007). Since December 2007, ensemble precipitation forecasts have been used operationally for the catchments of the Rivers Regen, Fraänkische Saale, Upper Main, Regnitz and Mangfall and are currently being evaluated.

5.13 UNCERTAINTY OF FORECAST

Experiences with published forecasts during former flood events have shown the need for communicating the uncertainty associated with hydrological forecasts to the public and the responsible persons in civil protection. Expectations on the reliability of flood forecasts are high. Publishing only one forecast as a quasi-deterministic forecast with one value at a given time even raises these expectations by pretending to be exact. Therefore, one aim of computing uncertainty is to publish it along with the corresponding forecast. The resulting illustration should make the numeric forecast less absolute for the average user and communicate the probability of a certain water level to be reached.

The uncertainty in the flood forecast is the result of uncertainties in input data, in model simplifications, in the estimation of the model parameters and in operational practice of forecast generation. Each of these components bears a particular uncertainty, which affects the uncertainty of the simulation output. In many cases, especially in headwaters, meteorological forecasts are the most dominant source of input data uncertainty. Often there are large differences in rainfall amounts between forecasts originating from different meteorological models and between the different model runs of the same model. Not only the amount, but also the spatial and temporal distribution of precipitation can vary between different forecasts and the measured precipitation affecting the output of the hydrological model, especially in smaller catchments. Therefore, ensemble forecasts should be included in the calculation of total output uncertainty to account for the dynamic uncertainty of the meteorological forecast. If the forecast horizon lies within the travel time of a flood wave observed upstream, the runoff forecast is expected to be more accurate as it only involves the single process of flood routing.

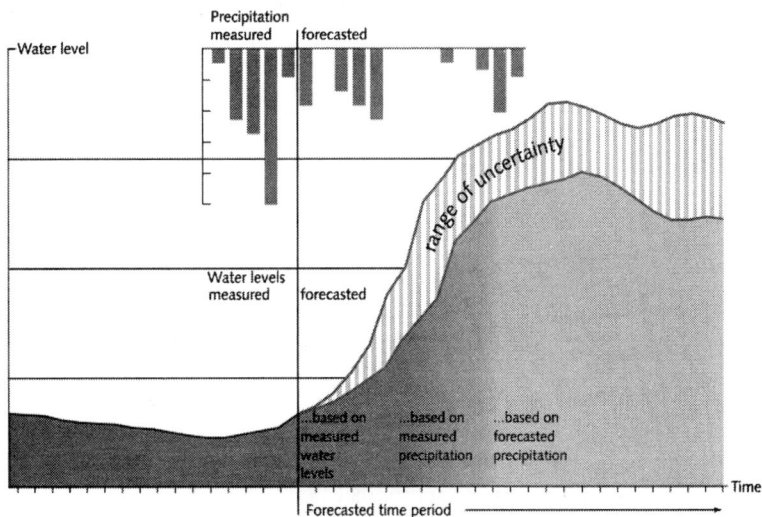

Figure 5.6: Relation between uncertainty and the lead time.

The relative influence of the different sources of uncertainty depends on different factors like forecast horizon, meteorological conditions and up or downstream location of the catchment. With increasing forecast horizon, the influence of the uncertainty of the meteorological forecast increases (Figure 5.6).

Considering all the sources of uncertainty, the optimal approach would be to do multiple forecast realizations by randomly varying all of the sources of uncertainty within their range (Monte Carlo simulation). Because of operational constraints, a relatively straightforward approach for the calculation of forecast uncertainty has been applied in Bavaria. Long time series of former, archived flood forecasts have been compared to gauge observations calculating the relative error for each time step within the forecast horizon. From the error distribution obtained for each gauge, the relative error on the 10 and 90 per cent exceedance probability level is used to illustrate the uncertainty on each new forecast (Figure 5.7). In addition to these "static" uncertainties the results of ensemble predictions (Figure5.8) are analysed and a combination of the "static" uncertainty with the "dynamic" uncertainty is calculated. Especially in headwaters, where the precipitation forecast is the dominating source of uncertainty, this is seen as an advisable procedure. The procedure is explained in detail in Laurent et al., (2010).

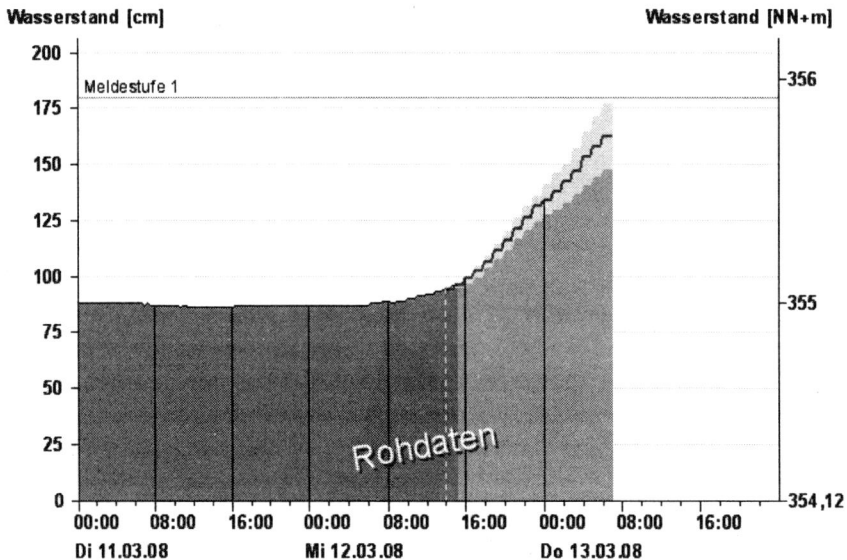

Figure 5.7: Uncertainty range in deterministic lead forecast

Communicating uncertainty to the users of the hydrological forecast is a very important task. As dealing with probabilistic numbers is not common to most users, descriptions and explanations should also be understandable to all. Adequate illustrations and descriptions should be evaluated in relation to the normal and advanced user. Additionally, experiences gained during future flood events might help in further adaptation of methods of communication.

For advanced users such as decision-makers in the water management authority, the published uncertainty should furthermore serve as a tool for better risk assessment. For example, the person in charge of operating a flood control basin can then base his decisions on probabilistic numbers in addition to the deterministic forecast instead of only on the latter. By judging, what probability threshold value should be taken into account the responsibility for operator increases. Therefore, it is very important to teach users how to interpret the published forecasts, especially the probabilistic part.

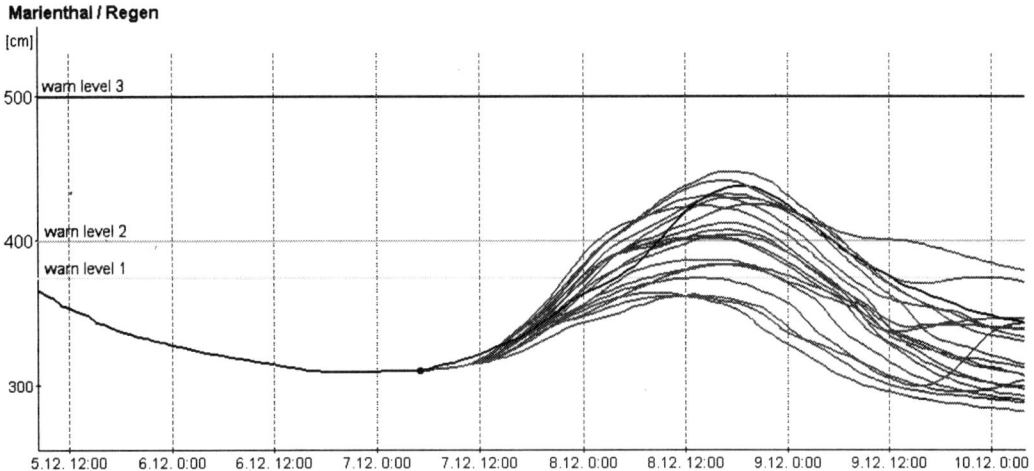

Figure 5.8: Using all available meteorological forecasts as input into a rainfall–runoff model for a medium-sized catchment in the Bavarian Forest shows the broad variety in the resulting forecasts of water level (green lines). The observed water level is marked blue.

Understanding the sources of uncertainty and their meaning also helps improving the correct usage of published prognostic data. This task is especially important since, compared to the normal weather forecasts; flood forecasts rarely gain the same every-day importance for the average user. Therefore, users are mostly lacking personal experience in evaluating the reliability of hydrological forecasts.

Overall, analysis and experiences with forecasts over the last years show that the accuracy achieved so far could be improved upon in many cases. Hence, calculating and communicating uncertainty is one goal. A major goal, though, remains reducing the existing uncertainty in the data, the model and its operation. Appropriate operations and projects remain a permanent task.

References

Gerlinger, K., and Demuth, N. (2000). Operational Flood Forecasting for the Moselle River Basin in *PIK-Report No. 65 (1/2): Proceedings of the European Conference on Advances in Flood Research 2000*, Potsdam Institute for Climate Impact Research (pp. 546 – 558).

Laurent, S., Hangen-Brodersen, Ch., Ehret , U., Meyer, I., Moritz, K., Vogelbacher, A., and Holle, F. K. (2010). Forecast Uncertainties in the Operational Flood Forecasting of the Bavarian Danube Catchment. In M. Brilly (Ed.) *Hydrological Processes of the Danube River Basin*. London: Springer.

Ludwig, K. and Bremicker, M. (2006). The Water Balance Model LARSIM - Design, Content and Applications, *Freiburger Schriften zur Hydrologie, Band 22*.

Reichel, G. (2001). FluxDSS und FLORIS2000 - ein leistungsfähiges Paket zur Modellierung der Fließvorgänge in komplexen Systemen, *Österreichische Wasserwirtschaft, Jahrgang 53, Heft 5/6*.

Vogelbacher, A. (2007). Unsicherheiten bei der Abflussvorhersage. (Uncertainties of runoff forecasts) In "Flutpolder - Hochwasserrückhaltebecken im Nebenschluss, *Bericht Nr. 113 des Lehrstuhls und der Versuchsanstalt für Wasserbau und Wasserwirtschaft der Technischen Universität München.*

Vogelbacher, A., Daamen, K., Holle, F. K., Meyer, I., and Roser, S. (August, 2006). *Hydrological aspects of the flood in August, 2005 in the Bavarian Danube Catchment.* Paper presented at 23rd Conference of the Danube Countries on the Hydrological Forecasting and Hydrological Bases of Water Management. Belgrade, Republic of Serbia.

Hot Spots Analysis of Tornado for Bangladesh:
A GIS and Remote Sensing Approach

Hasem Roney

6.1 INTRODUCTION

A fully developed tornado is one of the most terrifying weather events. The destruction caused by tornadoes is due mainly to the violence of the winds. Bangladesh experiences mostly the violent storm tornados every year and it faces the deadliest tornados comparatively than other countries of the world. Tornados kill an average of 179 people per year (USA 80) in Bangladesh (Geo-fact sheet, 2006), from 1967-1996 around 5373 tornado related deaths were observed and were most severe in the world (Paul and Rejuan, 2004). The 1989's tornado killed at least 1300 people and almost half of the world (Paul, 1998).

"Due to climate change, the nature's disastrous phenomena tornado and Norwester's are frequently occurring though short lived tornados are very hard to be tied directly to climate change due to limited climatology" (Science Daily, 2008). Global warming has a close relation with the formation of tornado as an increase in the sea surface temperature increases atmospheric moisture content and it causes severe weather and tornado activity particularly in cool season (Trapp et al., 2007). Occurrences of tornado most commonly defended on the time of day because of solar heating and it occurs in between 3 p.m. - 7 p.m. when temperatures ranges in Bangladesh between 33.3°C - 11.7°C (Kelly, Schefer et al., 2009 & Banglapedia, 2008) and average wind speeds prevail at 220-270 km/h (Finch & Dewan, 2000) whereas it stretches about 2-3 km across and stay on the ground for 3 miles (Paul, 2004).

Most people of the Bangladesh lives in fragile houses and majority of the populations are poverty stricken (Grazulis, 2001). Fatalities due to tornado remain high in Bangladesh mostly the rural central part experiences 75 per cent of strong tornados (Finch & Dewan, 2000) and the denser is exacerbated due to high population density- 809 per sq.km, poor quality of construction- 55.6 per cent kutcha road at central part and lack of tornado safety knowledge (Pual, 2001; BBS, 2001; Banglapedia, 2008 and Stewart, 2010). Its worse impact is that it has thrown away the houses and animals to the long distances and the flying debris such as corrugated iron caused decapitation, lacerations or loss of limbs and diarrheal diseases (Finch & Dewan, 2000). In

Disaster Management and Risk Reduction: Role of Environmental Knowledge; Editors Anil K. Gupta, Sreeja S. Nair, Florian Bemmerlein-Lux and Sandhya Chatterji; Copyright © 2013, Narosa Publishing House, New Delhi

Bangladesh, women and children are more vulnerable than men to natural hazards like tornado (Ikeda, 1995 and Canon et al., 2005). The majority of tornadoes occur in agricultural areas as Crops need moisture to grow and the temperature variation associated with changing seasons (Stewart, 2010) and agriculture is the primary economic activities of 60 per cent people of Bangladesh (BBS, 2001) and main occupation (43.6%) of identified 21 districts of Bangladesh (Banglapedia, 2008). Agricultural loss due to tornado was reported as 5,000-12,000 taka per acre and agricultural production will decline erratically by 2050 due to heavier and more erratic rainfall (Bangladesh Climate Change Strategy and Action Plan, 2008).

Though the occurrences of tornado and norwester's cannot be prevented but the magnitude of impact can be prevented by developing apposite tornado countermeasures. The construction of tornado shelters and strong wall made housing are the popular means of tornado management strategy for Bangladesh in general. It has been observed that government has given more emphasis on cyclone and flood hazard management and paid less attention to the adverse impact of tornado (Paul, 2001) though many environmental threats were already reported (Rasid and Mallik, 1996). In order to lessen negative consequences of tornado, identified hotspot area must be considered and proper countermeasures should be adopted.

Remote Sensing and Geographic Information System has been widely used to map and model different types of hazards (Werner, 2001; Aziz et al., 2003). As tornados are short lived events so it is very difficult to identify and provide instantaneous and synoptic view necessary for the estimation of tornado vulnerability assessment. The developed countries are using NOAA data for reporting tornado and Bangladesh is not using such modern technologies though fujita scale 5 (F5) levels of tornado mostly visit here (Yamane et al., 2009). There are only four national networks of Doppler Radar Station (Geo fact sheet, 2006 and JICA, 2010) to cover and collect data of tornado hazard. To mitigate this problem Bangladesh government is heartily trying (Bangladesh climate change strategy and action plan, 2008) but it is very difficult to keep update their database due to the lack of resources (Dewan et al., 2006). The study mainly focuses to identify the most devastating places which are also known as hotspot zones due to tornado in Bangladesh in memorable history in terms of deaths, frequency, injury and economical loss etc.

6.2 STUDY AREA

The study area was the whole Bangladesh which lies in between 20°34′ and 26° 38′ N latitudes and 88° 01′ and 92° 41′ E latitudes. This area is mainly located on flat low laying areas and a broad deltaic plain largely formed by Ganges, the Brahmaputra and the Meghna. The topography is extremely flat and most elevations are less than 10 meters above sea level (Rasheed, 2008). It lies in the subtropical monsoon zone and experiences a humid climatic condition (Dewan et al., 2006) for which it is the breeding ground of most natural hazards specially tornado in Bangladesh (Stewart, 2010). Further the focused study area experiences annual rainfall range from 1200 mm to 5000 mm (Rasheed, 2008) and average annual temperature range from 11.7°C to 33.3°C (Banglapedia, 2008). The more focused study area i.e. middle part of Bangladesh remains dry and experiences less rainfall than other parts of Bangladesh though three mighty rivers passes through.

6.3 DATA PREPARATION

Tornados data preparation is very difficult as data are not well documented (Ono et al., 2010) and mostly we have to rely on newspaper reports, Bangladesh Meteorological Department and

Bangladesh Space Research and Remote Sensing Organization (SPARSO) etc. Another hardship is that peoples are mostly confused with identifying tornado and cyclone (Yamane et al., 2009). So both primary and secondary data were used for the study. The primary data include house types, health impact and media access etc. It is not possible to measure the proximity of hazards source as the probability of tornado point source occurring is .0363 (Stewart, 2010).

Frequency of tornado and their death toll over 40 years were collected from different sources. The Geo coded values were mostly encoded from the Bangladesh and East India tornado prediction website .Besides more values were compiled from different sources to obtain a hotspot zone for the Bangladesh.

6.4 CONSIDERATION OF HOTSPOT ZONE

Geographic Information Systems were widely used to analyze the features within the context of neighbouring features. A statistically significant hotspot may not contain highest values of features. It is very difficult to create a generally accepted hotspot zone due to the lacking of Geo-coded data of overall Bangladesh and as tornado has the less probability of spatially occurrences. So a simple procedure is adopted to identify the spatially related tornado events. The most severe and well documented 84 occurrences were used to identify hotspot zone (Finch, 2007) from 1875 to 2010 for Bangladesh. The values were further corrected to identify hotspot area.

Getis-Ord Gi* algorithm is mostly used to calculate and analyze the spatially co-related feature values. To find the optimum locations of tornado occurrences, geo-coded points are used (Finch, 2007) and the points are integrated within GIS environment. For the understanding of tornado events the geo-coded values are integrated in a number of geo-processing tools. To obtain the statistically significant hotspot the following algorithms are used. Firstly, the geo-coded values are integrated and a threshold distance determines the general geographic locations. For obtaining a weighted point feature, collect events tools converts the integrated values indicating the number of events occurring in those locations. By running the hotspot analysis with rendering tools it will ensure the centre of large hotspot cluster with Euclidean distance band and 4125.35 default neighbourhood. The low to high events of occurrences are shown in Figure (with graduated colour). Users have little control on the analysis using rendering tools by generating spatial weight matrix to assess different conceptualization of relationships of features.

Intensity of occurrences are derived for the conceptualization of spatial relationships from the Getis-Ord Gi* tool. To consider the threshold distances in between features like 'Zone of indifference' work well to combine the inverse and fixed distance band. Here the 'Manhattan distance method' measured the distance by summing the differences between point coordinates and Clusters tolerance are found identical. Then Z-scores (Gi-Z score) were calculated to apply a cold to hot graduated colour rendering to the z-score field values. The Gi-Z scores were obtained at a standard deviation from < -2.58 to > 2.58 and the Gi-Z scores of rendered hotspot results were obtain within < -2.0 to > 2.0.

A widely varying feature density was calculated and rendered hotspot results were generated. Interpolation tools used to create a rendered surface of Gi-Z score values by compressing the cell size (.0128). The graduated colour red to bluish grey indicates the high to low value of occurrences and red circle identified is the place of most occurrences of tornados. This surface area extent from 91.6074 to 88.9842 and 25.6058 to 22.6942 longitude and latitude respectively with 0.79 standard deviation.

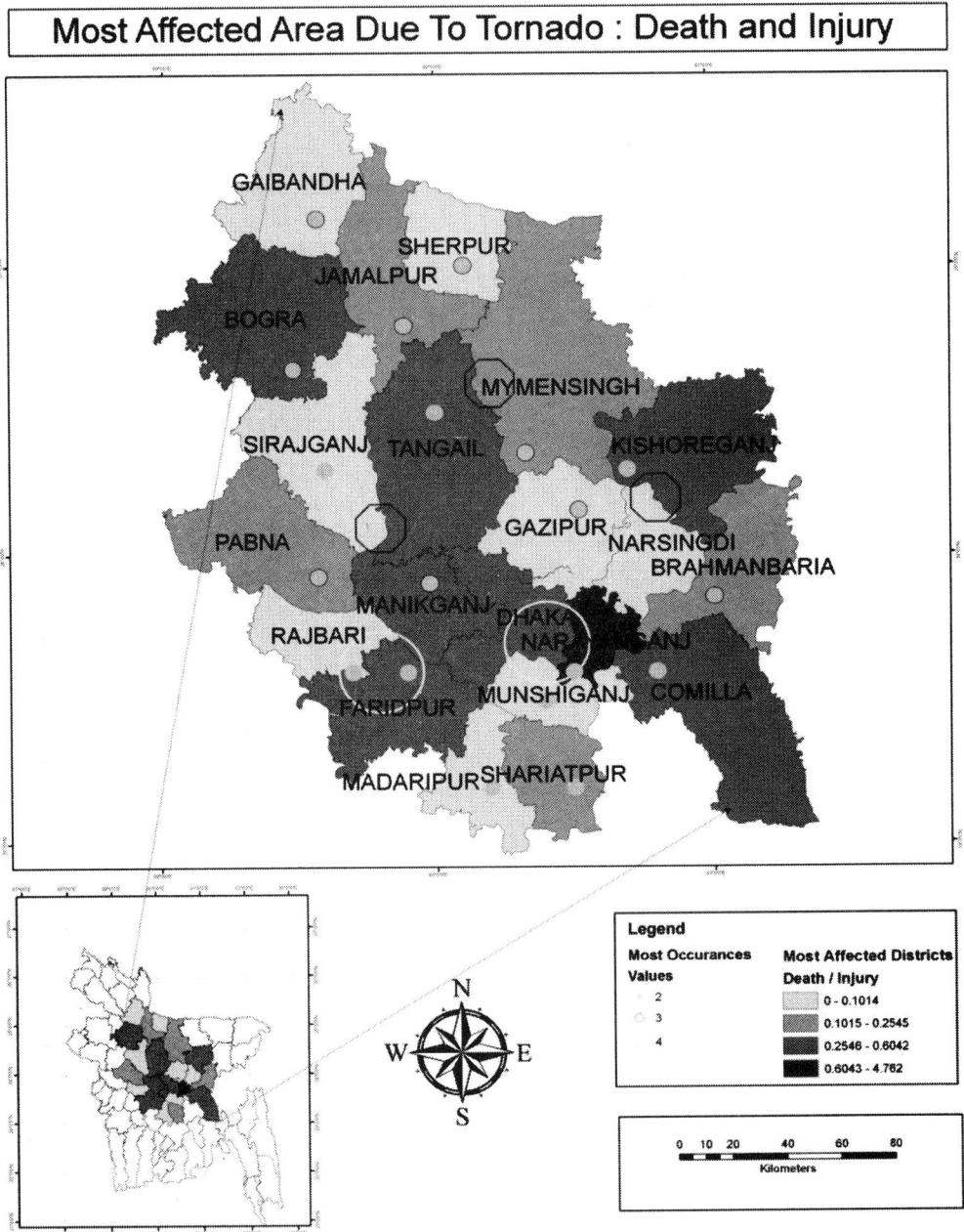

Figure 6.1: Most affected areas of Bangladesh due to tornado are calculated by evaluating death and injury ratio for 21 vulnerable districts of Bangladesh.

Table 6.1: Possible results of hotspot zone identification with standard deviation, geo-coded values and rendered hotspot values.

Scores	GiZ score (Std. Devia.)	Some geo-coded values		Rendered HS results
		Latitudes	Longitudes	(Gi-Z score)
1	‹-2.58	23.7	90.07	‹ -2.0
2	-2.58 t0 1.65	23.6	89.8	-2.0 to -1.0
3	1.65 to 1.96	24.3	90.7	-1.0 to1.0
4	1.96 to 2.58	23.6	90.06	1.0 to 2.0
5	›2.58	23.8	88.6	›2.0

*Std. Dev = Standard Deviation; Geo-coded values = Lat/Long values; HS = hot spot

The extension surface encircling by different colour represents the 21 districts of central-north Bengal with about 8000 square miles (Figure 6.1).So with the help of Getis Ord Gi-Z score a hotspot was determined to analyze the neglected climatic phenomenon.

6.5 ANALYSIS ON HOTSPOT ZONE

Climate has a significant role to the formulation of tornado as it is related to global warming and Bangladesh lies in between a triangle where it is bounded by Indian great hills on two sides and Bay of Bengal on the other side. So the country's geography weakens its capacity to withstand natural disasters and to adapt to climate change (DFID, 2008).The poor people, children and women are vulnerable as they are unable to recover themselves from a disaster promptly (resilience). The hotspot zone is an area of about 41809.53 sq.km (Figure 6.3) with a population of 43, 013, 917 and 51.3 per cent women; where dominant economic activity is agriculture of more than 43.6 per cent people of the embedded 21vulnerable districts.(BBS, 2001).House types are traditional kutcha house which is very defenceless to strong winds.

Table 6.2: The number of occurrences in each district block identified by hotspot analysis

Sl.no.	Occurrences	No of District Block	Events summary
1	2	16	32
2	3	3	9
3	4	2	18
Total			59

Tornado mostly occurred in relatively small areas of central north part of Bangladesh, where vast agricultural and open land exists. The moist and potentially unstable air has a tendency to rise and to become violent wind (GeoFact sheet, 2006).The dry line mostly passes through the Central-west part of Bangladesh (Finch, 2007). Out of 84 occurrences, it was found on an average 59 occurred in this vast central-northern part of Bangladesh, and suffered most fatalities than any other parts of the country (Finch, 2007 and BMO, 2010).

The death and injury records were collected and calculated to obtain the best analytical results in support of hotspot zone. The following (Table 6.3) shows the situation of death and injury history of the most affected 21 districts. Then it is classified for statistical analysis. After statistical manipulations, 21 vulnerable districts are found and mapped with the help of Geographical Information Science tools.

Table 6.3: Death and Injury ratio calculations for vulnerable districts (1960-2010)

Sl.no	District name	Death	Injury	Death/injury
1	Tangail	1450	1560	0.929487
2	Sirajganj	1131	2276	0.496924
3	Pabna	272	1796	0.151448
4	Rajbari	1770	1690	1.047337
5	Faridpur	1285	1415	0.908127
6	Madaripur	1521	1420	1.071127
7	Shariatpur	1456	1045	1.393301
8	Comilla	370	870	0.425287
9	Narayanganj	200	42	4.761905
10	Dhaka	2027	3975	0.509937
11	Narshingdi	1118	225	4.968889
12	B. Baria	124	165	0.751515
13	Gazipur	948	2560	0.370313
14	Kishoreganj	1426	1115	1.278924
15	Munshiganj	1253	2760	0.453986
16	Jamalpur	1244	362	3.436464
17	Sherpur	17	123	0.138211
18	Gaibanda	600	927	0.647249
19	Bogra	816	192	4.25
20	Manikganj	1648	1285	1.28249
21	Mymensingh	1369	3238	0.422792

Table 6.4: Obtained classes and area for vulnerable districts

Sl.no	Class	Districts	Area (%)
1	0 - 0.1014	Gaibanda, Sherpur, Sirajgonj, Rajbari, Gazipur, Narshingdi, Munshigonj, Madaripur.	8.22
2	0.1015 - 0.2545	Brahmonbaria, Mymensingh, Jamalpur, Pabna, Shariatpur.	7.04
3	0.2545 - 0.6042	Dhaka, Comilla, Faridpur, Manikgonj, Tangail.	6.72
4	0.6043 - 4.762	Narayangonj, Kishoregonj, Bogra.	6.33

Sources: Meteorological Department of Bangladesh and Bangladesh Space Research & Remote Sensing Organization (SPARSO).

6.6 CONCLUSION

Tornado is a neglected climatic event in Bangladesh. It has a quite few surface synoptic observatories to obtain thermodynamic data. Tornado forecasting is in its infancy with 28 active weather stations and 4 Doppler Radars which is very few for collecting data of short lived tornado. In this study tornado hotspot areas were identified to provide a means for the potential vulnerability analysis, which used to mitigate future economic, agricultural and human loss. The vulnerable area is classified into four zones. The calculations of areas under different classes revealed that the 28.31 per cent of total Bangladesh is, in vulnerable area with capital Dhaka. All possible combination of hotspot map was prepared based on death and injury ratio. It is evident that flimsy houses made with fragile traditional materials and tin sheets are very crucial things in times of tornado occurrences. More resilient shelters should be installed to reduce the vulnerability of tornado hazards. To cope with devastating climatic short lived event-tornado, awareness and mitigation measures should be taken to lessen the fatalities of the people of hotspot areas.

6.7 ACKNOWLEDGEMENTS

It's my alacrity to address that this paper I fully owe to my friend Istiak Ahmed Bhuyan (Student of Geography & Environment, University of Dhaka) and also to Dr. Ashraf Mahmood Dewan (Curtin University of Technology, Perth, Australia and Professor of Geography & Environment; University of Dhaka, Dhaka; Bangladesh) who encourages me to do this research.

References

Bangladesh Meteorological Department (BMO), Accessed from www.bmo.org/Bangladesh

Dewan, A., Kumamato, M., T., and Nishigaki. M. (2006). Flood Hazard Delineation in Greater Dhaka, Bangladesh Using an Integrated GIS and Remote Sensing Approach. *Geocarto International, 21(2)*, 33-38.

Finch, J.D. (2007). Bangladesh and East India Tornado Prediction Site: Background and Information. Retrieved from http://bangladeshtornadoes.org/bengaltornadoes.html

Finch, J.D., and Dewan, A.M. (2000). Tornados in Bangladesh and East India. Bangladesh and East India Tornado Prediction Site. Retrieved from http://bangladeshtornadoes.org /bengaltornadoes.html

Grazuli, T.P. (1991). *Significant Tornadoes, 1880-1989.* Vol. 1: Discussion and Analysis. St. Johnsbury, VT: Environmental Films.

Grazulis, T.P. (1993). *Significant Tornados-1680-1991.* Environmental Films. St. Johnsbury, VT.

Grazulis, T.P. (2001). *The tornado: Natures ultimate wind storm.* USA: University of Oklahoma Press.

Kelly, D.L., Schaefer, J. T., McNulty, R. P. Doswell, C. A., and Abbey, R. F. (1978). An Augmented Tornado Climatology. *Monthly Weather Revenue.106* (8), 1172–1183. doi: http://dx.doi.org/10.1175/1520-0493(1978)106<1172:AATC>2.0.CO;2

Ono, Y. (1997). Climatology of Tornados in Bangladesh, 1990-1994. *J. Met. 22*(222), 325-340.

Ono, Y. (1997). *Risk Factors in Mortality from tornados- A case study of the May 13, 1996 tornado in Bangladesh.* A thesis submitted to Kent State University (unpublished).

Ono, Y. (2001). *Design and Adoption of Household Tornado Shelters to Mitigate the Tornado Hazard in Bangladesh.* An unpublished Ph.D. Dissertation, Kent State University, Kent, Ohio.

Ono, Y., and Schmidlin, W.T. (2010). *Design and Adoption of Household Tornado Shelters for Bangladesh.* Springer: Science + Business Media.

Paul, B. K. (1998). Coping with the 1996 Tornado in Tangail, Bangladesh: An Analysis of Field Data. *The Professional Geographer,* 50 (3), 287-301. doi: 10.1111/0033-0124.00121.

Paul, B. K., and Bhuiyan, R.H. (2004). The April 2004 Tornado in North-Central Bangladesh: A Case for Introducing Tornado Forecasting and Warning Systems. *Quick Response Research Report 169.* University of Colorado.

Rasheed, K.B.S. (2008). *Bangladesh Resource and Environmental Profile* (p:4-10). New Market, Dhaka: A H Development Publishing House.

Spilbury, L., and Spilbury, R. (2009). *Natural Disaster in Action: Hurricane and Tornadoes in Action.* The Rosen Publishing Group Inc, 29 east 21st. Street, New York , NY 10010.

Stewart, C. (2010). *Tornado's The Most Violent Storm.* The Oklahoma Institute of Disaster and Emergency Medicine. University of Oklahoma, Tulsa, Japan.

Trapp, R.J., Diffenbaugh, N.S., Brooks, H.E., Baldwin, M.E., Robinson, E.D., and Pal, J.S. (2007). Changes in severe thunderstorm environment frequency during the 21st century caused by anthropogenically enhanced global radiative forcing. *Proceedings of the National Academy of Sciemce. U. S. A., 104*(50), 19719 –19723 (doi:10.1073/pnas.0705494104).

Yamane, Y., Hayashi, T., Dewan, A.M., and Akter, F. (2010). Severe Local Convective Storms in Bangladesh: Part-2. Environmental Conditions. *Atmospheric Research. 95*, 407-418.

7

Probabilistic Risk Assessment
A Study of South Andaman Island, India

Shrikant Maury and S. Balaji

7.1 INTRODUCTION

Andaman Island faces a number of natural disasters as the island is having a fragile geographic and environmental setting. There is an urgent need to understand the environmental complexity and morphological diversity of this island. Historically, there is not much information available on disasters and their sites but during last decades many natural disasters happened. Amongst many disasters, Asian tsunami 2004 was proved disastrous to this island. Tsunami 2004 not only devastated the infrastructure but also affected geographical environment and coastal system infringed of social life of this island. The post-math tsunami impacts still trigger imbalance on various consequences, driving major environmental and climatic changes of this Island. In particular, low-lying areas are strongly affected by coastal flooding or by active processes of shoreline erosion and sedimentation pose the most serious consequences for local communities (Claudio Szlafsztein and Horst Sterr, 2007). Due to tsunami inundation, and even adjacent coastal areas behave in quite different ways, varying according to social, economic and environmental conditions, prompting efforts to classify the coastline and subdivide it into (relatively homogeneous) units (Inham and Nordstrom, 1971). Since the proven work have been done to minimize the risk but it is not much worth for mitigation activities. The ingression of sea water damaging the environment and agricultural lands resulted in the deep salinization of soils and adversely affected all kinds of vegetation (Velmurugan et al., 2006). So, there is a need for integrated study, approaches and methodologies to understand the complexity of island system. Experience with several methodologies has yielded not only successful results, but their application also revealed many problems and deficiencies (Klein and Maciver, 1999).

The Probabilistic Risk Assessment of the South Andaman Islands provides estimation with much reliability to determine the frequencies and probabilities of various events modelled in a PRA. The PRA is to provide methods for estimating the parameters used in PRA models and for quantifying the uncertainties in the estimates. PRA is a matured technology and it has emerged as an advanced tool especially during the last decade, widely applicable in assessment of risk and their life cycle. PRA risk assessments are useful to identify many risks inherent in the

Disaster Management and Risk Reduction: Role of Environmental Knowledge; Editors Anil K. Gupta, Sreeja S. Nair, Florian Bemmerlein-Lux and Sandhya Chatterji; Copyright © 2013, Narosa Publishing House, New Delhi

developmental projects and policies. The assessment results can be used prior to implementing the development plans. Risk can be understood as detrimental outcomes which are directly or indirectly pose and accrued on the subjected environment which becomes vulnerable than degradable and the complex mechanisms evoke a disaster. PRA associated risk assessment is characterized by two aspects; first is severity of adverse consequence and second the likelihood of occurrence of the given adverse consequences. Probability risk assessment usually explains the quantitative answer of initiating events, severity of the potential detriments, and probability of initiating events. However, the qualitative nature of these assessments can lead to inconsistency and imprecision in risk characterization that make risk prioritization difficult and risk aggregation impossible (Teri et al., 2009). Successful PRA studies are helpful in risk reduction before it poses a disaster and also inform recommendations for improving the design and operational risk controls.GIS based approach has also been adopted for this study. This encompasses the quantitative risk assessment, vulnerability mapping and safety related issues for disaster risk reduction. This is a cost effective and reliable technique for the assessment of disaster risks. The present study focuses to identify and mitigate the risks, improve the community resilience, preparedness, and recovery of the affected destination and also to provide tangible evidence of the safety, security arrangements for residents and agro-industrial development.

7.2 PROBABILITY MODELS

7.2.1 Poisson Process Model

Poisson Process Model models are good approximation when the probability is small and the number of trials is large. PRA models will be most reliable in the foundational material and it involves the development of models that delineate the response of systems and operators to accident initiating events (Atwood et al., 2003). The probability that an event will occur in any specified short exposure time period is approximately proportional to the length of the time period. In other words, there is a rate $\lambda > 0$, such that for any interval with short exposure time Δt the probability of an occurrence in the interval is approximately as follows (Thompson, 2003):

(1) $\lambda \times \Delta t.$

It is always conventional to select the simple model with much reliability. In addition to support this equation if an event initiating at time t and its fails during a short time period from t to $t + \Delta t$ is approximately proportional to the length of the exposure period Δt then the equation (1) will be as below.

(2) $\lambda \times \Delta t = Pr(t < T \le t + \Delta t \,/\, T > t)$

Where λ= frequency of occurrence; μ= the mean number of occurrences; t=exposure time; X numbers of failures; T is random time of failure; Pr damage rating. This model is much simple and sophisticated and much suitable to study PRA for my study areas.

7.2.2 Regression Data Equation

Hald's regression data equation provides better information on regression analysis of multiple factors related to vulnerability assessment and risk analysis. The equation is as follows (Ghosh et al., 2006).

(3) $y_i = \beta_0 + \beta_1 x_{1\,i} + \beta_2 x_{2\,i} \ldots\ldots + \beta_p x_{pi} + \varepsilon_i, \; i = 1,\ldots,n,$

Where $_p$ is the number of repressor variables in the model, β_0, β_1,...β_p are unknown parameters, and $_{\varepsilon i}$'s are independent errors having a N $(0, \sigma^2)$ distribution.

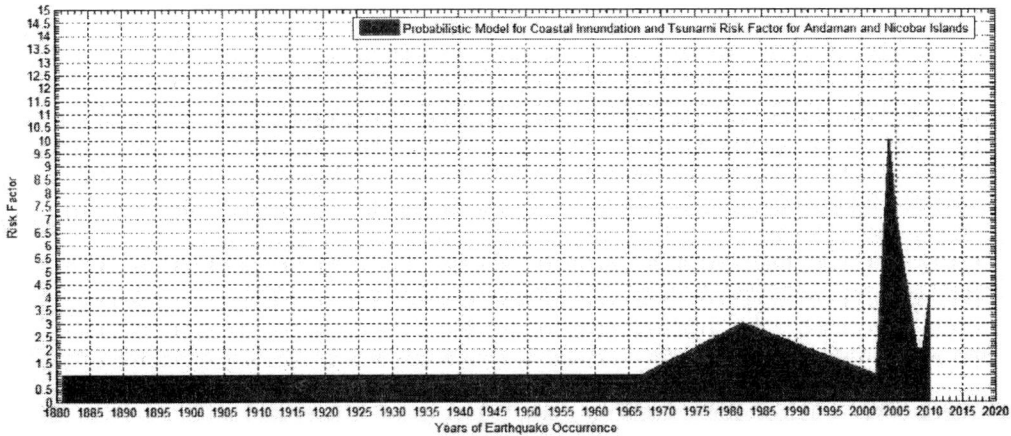

Figure 7.1: Probability based model of tsunami and coastal inundation occurrence due earthquakes

7.3 STUDY AREA

The South Andaman Island lies in the Bay of Bengal in N-S direction and administratively ambit between 10°00' to 12°12' North latitudes and 92° to 94° East longitudes (Figure-7.2) selected for this study. Including all 115 islands the area of this island is about 1445 km^2 (area of creeks not included) and bounded by Andaman sea and having 196, 992 population (2001 census). Major landscape of this study area is divided into low to moderately high and steep hills, intermountain narrow valleys and gradually sloping coastal tracts including swamps. The Ground elevation in South Andaman Islands ranges from the sea level up to 459 metres, the maximum peak in South Andaman. The group of Islands receives rainfall to a tune about 3000 mm. annually representing tropical humid climate. The South Andaman Island is one of the most natural disaster prone zones, very frequent to earthquakes which are often most destructive and also inherently poses various vulnerable natural hazards such as catastrophic tsunamis, coastal floods, coastal land subsidence and landslides etc. The Mw 9.3 tsunami genic Sumatra-Andaman earthquake of December 26, 2004 affected the island and its environment widely. Post-tsunami impacts still triggering risks to the island environment by their devastating effects. These impacts have been noted at several places.

The post-tsunami major impacts were coastal flooding, depletion of land resources due to ingression of sea water and change of land use pattern. This tsunami catastrophe also led intrusion of sea water in shallow fractured aquifers and permanent submergence of coastal lands into seawater. The effects of seawater intrusion have been observed at several places indicating alteration in groundwater quality parameters (Laxminarayana and Kar, 2010). Moreover, there is no proper conservation, regular monitoring and lack of management skills which is also another cause for potential risk. Consequently, the above practices, severe impacts of natural hazards and climate changes affect the agriculture and rural development of South Andaman Island (Deirdre Shurland and Pieter de Jong, 2008). The culmination of these consequences can become a nestling ground for another disaster.

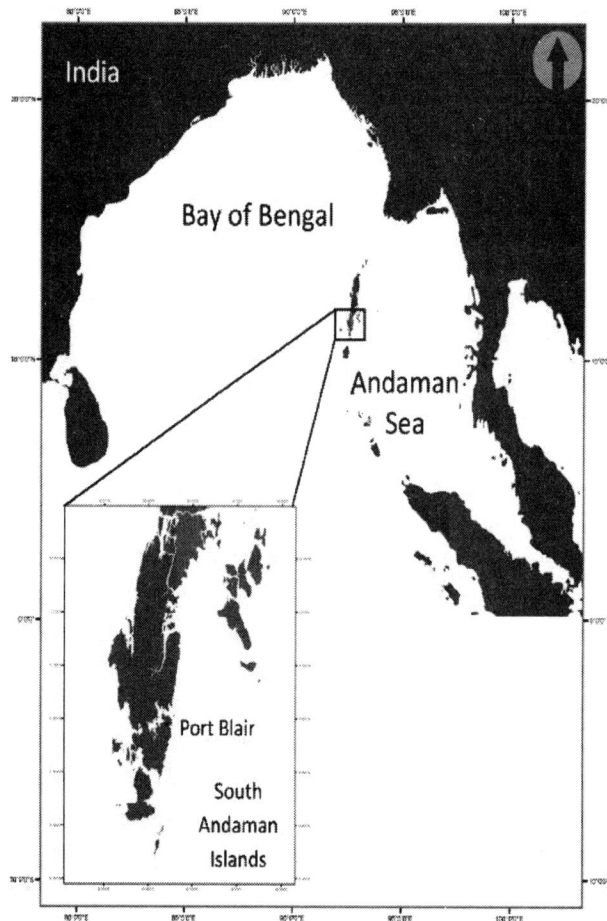

Figure 7.2: Local map of the study area

7.4. MATERIALS AND METHODS

7.4.1 Field Surveys

The field surveys were carried out at different parts of South Andaman and various other islands covering Rutland Island, Chidiya Tapu, Port Blair, Wandoor, Tirur, Ferrergunj, Mithakhari, Bamboob Flat, Havelock Island, Neil Island and Baratang islands using Ground Penetrating Radar (GPR) with multiple antennas. The satellite imagery interpretations were validated in the field with limited field check-up. The strike directions of the profile, erosion and accretion phenomenon were noted using the Brunton compass, Total station and Garmin GPS for positioning.

7.4.2 Remote Sensing

Remote sensing, because of the repetitive and synoptic nature, has proved to be an excellent tool to monitor and study the inundated area and change in land use, land cover and morphology before and after tsunami.

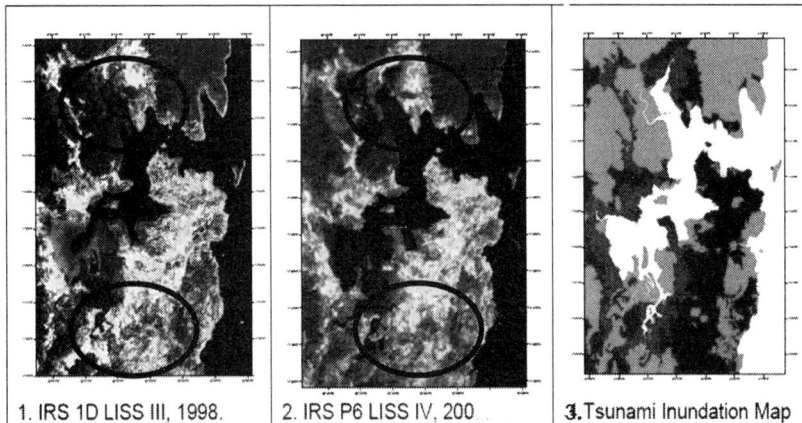

Figure 7.3: Satellite imagery showing the tsunami inundated areas of Sippighat, Chauldhari, Mithakhari Bambooflat and Port Blair Area, South Andaman.

The IRS 1D LISS - III, 1998 and IRS P6 LISS IV 2007 satellite based on the tonal, textural and associated factors. The SOI top sheet 1958 and Google Earth Pro were also used as an ancillary supporting data (Figure 7.2).

7.4.3 Geographical Information System

The ArcGIS software was used for the digitizing and layers used for analysis. Different layers like land use, tsunami inundated area and contour lines for generation of Digital Elevation Model (DEM) were prepared. DEM are useful for understanding the risks due to tsunamis and coastal flooding to make emergency evacuation and allocate population to the safer place also for development planning.

7.4.4 Geophysical Studies

Ground Penetrating Radar (GPR) is used to unravel the subsurface details like folds, faults, litho units, soil types and buried objects. Different antennas like 80 MHz and 200 MHz were used to extract the subsurface information. The GPR raw data was analysed and processed in Radan 5.1 software and the information could bring out the subsurface details in time section which is converted into depth section. The GPR study will be much helpful in identifying vulnerable zones due to landslide and seismic activities. GPR can also be much useful in demarcation of depth cover of soil of agriculture land and which in turn is helpful in mitigation of the agriculture hazards. The GPR profile has enabled to delineate subsurface ocean faults at depth and subsurface structures like folds and dipping structures (Figure 7.4).

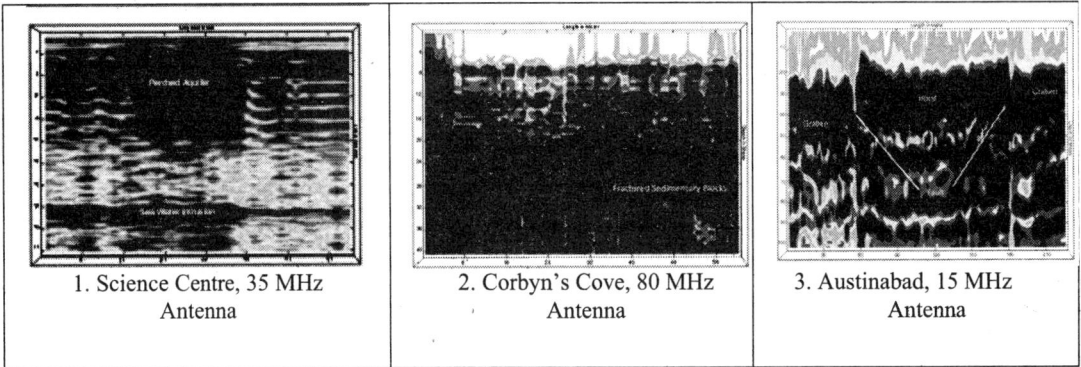

| 1. Science Centre, 35 MHz Antenna | 2. Corbyn's Cove, 80 MHz Antenna | 3. Austinabad, 15 MHz Antenna |

Figure 7.4: Subsurface structures and ocean fault of South Andaman Islands.

7.5 NATURAL HAZARDS AFFECTED THE ANDAMAN ISLANDS

Natural hazards of this island like tsunamis, earthquake, land subsidence, land submergence, landslides and sea water intrusion are the real threat to the coastal community and infrastructure. The visual inspections of the coastline and tsunami hazardous prone zone of South Andaman Islands were made during the field visits. In Corbyn's Cove, there is simultaneous erosion and accretion is taking place. Because of the excessive erosion, the beach cliff rocks were eroded severely and finally led to the formation of wave cut platform near the Science centre, Port Blair. In the eroded coasts, the trees and vegetation were uprooted and soil cover is lost. Trees were destroyed completely because of higher salinity and destroyed the coastal properties (Figure 7.5).

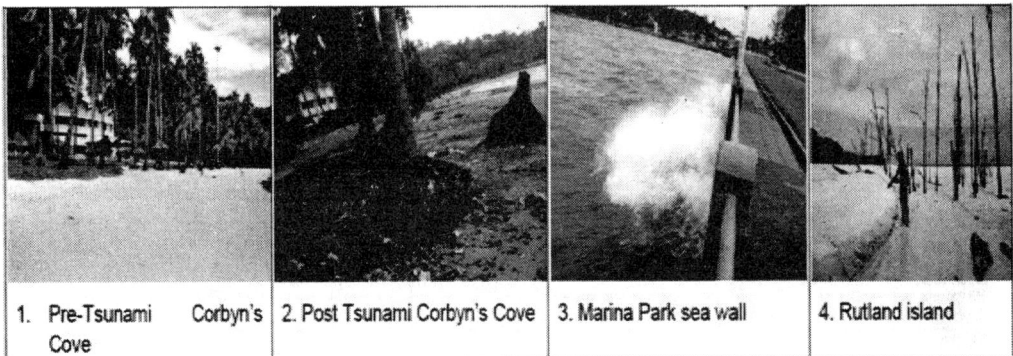

| 1. Pre-Tsunami Corbyn's Cove | 2. Post Tsunami Corbyn's Cove | 3. Marina Park sea wall | 4. Rutland island |

Figure 7.5: Destruction caused by tsunami in South Andaman Islands.

Before the 2004 tsunami, the landslide was not so frequent but after the tsunami, the landslides are much frequent and severe as it can be seen near the hilly area of North Cinque Island and hilly residential area of Brookshabad during the monsoon period (Figure 7.6). The mud eruption of Baratang area is a big havoc (Mishra et al., 2007) for local residential people and the subsurface faults are still active (Figures 7.6 and 7.4).

Figure 7.6: landslides, mudslides and volcanic eruption after 2004 tsunami.

Coastal areas, near Carbyn's Cove, there are sea water inundations taken place after the tsunami 2004.The water quality was good before the tsunami and it has deteriorated after the in as evident from the water sampling of wells near the Carbyn's cove. Rutland Island is completely inundated during tsunami which resulted in severe damage to coastal ecosystems in the island (Figure 7.7).

Figure 7.7: Damaged ecosystems due to increased salinity as a result of tsunami inundation.

In South Andaman Islands, coastal lands are uplifted and submerged at several places as a sequential to the 2004 Great Andaman-Sumatra earthquake (Figure 7.8).

| 1. Collinpur Beach | 2. Neil Island | 3. Havelock Island | 4. Kodiyaghat |

Figure 7.8: Post tsunami morphological changes like uplift and subsidence.

7.6 VULNERABILITY AND PROBABILISTIC RISK ASSESSMENT

The environment of the South Andaman is fragile and much variable in aspect of vulnerability thus every place seems to be erratic and specific in nature of risk. Probabilistic risk assessment is very essential for the south Andaman Islands to assess the capabilities and immediately respond to the natural disasters like tsunamis, earthquakes and further to prevent the sea water logging into limited land resources. It has been observed that the ingression of sea water towards lands continued and the low lying land is becoming unfertile and hard to establish any infrastructure. Vulnerability and risk assessment depends much more on logic and high integrity of different technologies for different consequences of the environment and nature. The special distribution of affected land use pattern shows that sea water majorly entered through the creeks and inundated large portions of land use area (Velmurugan et al., 2006). This will be much useful to establish community-based disaster preparedness and prevention. As we can see from maps, the eastern coast is highly vulnerable to the tsunami impact (Figures 7.6, 7.7, and 7.8).

7.7 DISCUSSION

Remote sensing data has clearly revealed the loss of land area after tsunami (Figure 7.2). The DEM model also depicts the area prone to inundation and hence can be used for disaster management planning(Figure 7.3). The GPR depth image clearly brought the subsurface ocean faults and its impact on sea water intrusion and coastal erosion (Figure 7.4). The landslides, land subsidence, coastal erosion, sea water intrusion, uplifted coral beds all revealed the ongoing neo-tectonic activities and coastal hazard impact.

The major tsunami affected sites in terms of land use area were identified at Sippighat, Chaudhari, Gopalnagar, Mithakhari, Bamboo flat, Hazaribagh, northern region of Rutland Island and SW region of Baratang. While at Havelock Island, land use was very less affected and at Neil Island not showing any tsunami inundation in terms of land use (Table 7.1). The tsunami inundation results provided an understanding that the ecosystem of the island is highly fragile towards tsunami. It has been observed that some area do not have tsunami inundated water and logged land but represents higher salinity in soil.

Table 7.1: Tsunami impact on land use of South Andaman Island

S. No.	Island Name	Year	Land use area in (km²)	Year	Land use Area in (km²)	Change in land use area (km²)	Landuse under tsunami water (%)
1	Andaman Main	1998	113.88	2007	89.76	24.12	21.18
2	Baratang Island	1998	14.14	2007	11	3.14	22.20
3	Havelock Island	1998	10.44	2007	10.23	0.21	2.01
4	Neil Island	1998	04.00	2007	04.00	0.00	0.00
5	Rutland Island	1998	2.72	2007	1.56	1.16	42.65
6	Total Tsunami water submerged Area	1998	141.18	2007	112.55	28.63	20.28

The GPR study reveals that most of the part of the island having highly neo tectonic structures like highly active faults and fractures and new geomorphologic structures leading to new hazards. These areas should not be used for any developmental planning and related activities. The DEM map shows safer areas were vital installations should be planned or developed. Further, most of the inundation in land use area due to sea water is reached by creeks only. Normally, areas adjacent to the creeks are having flat and much developed landforms with settlement and agricultural activities. Creeks are posing a higher risk and vulnerability to these valuable land resources and provides easy pathways for the ingression of the sea water. It would be better to block the possible creeks by strong and broad and structurally advance fence to make the land use safer and protected. This will be less costly than to protect the valuable land by individual efforts as it is being practiced in several parts of South Andaman Islands and further, the valley type topography of landform will be much helpful to accomplish this endeavour and that will be much helpful for risk reduction towards further land devastation due to water logging. Looking at the present scenarios the island is much vulnerable and under the edge of several risks if proper mitigation plan will not be implemented than the risks and vulnerability will be triggered and that can transform into a critical crisis related to developmental aspects.

7.8 CONCLUSION

Risk assessment needs both quantitative and qualitative terms, for implementation into mitigation strategies. However in PRA, the qualitative nature of these assessments can lead to inconsistency and imprecision in risk characterization that make risk prioritization difficult and risk aggregation impossible (Figure 7.1). An essential database and site information much essential for PRA studies further lack of information the assessment much deviates yet data gathered from remote sensing, GIS and field survey and thematic layers were generated to assess the impact of tsunami on land use. For the protection of land resources and minimization of the probability of critical consequences much study needed to achieve augmentation on different aspects of disaster management. In addition to development plans, projects should attend a proper understanding of the fragility of the South Andaman environment and socio-economic setting.

7.9 ACKNOWLEDGMENTS

The authors are thankful to Andaman and Nicobar Archives Secretariat, Directorate of Economics and Statistics, Andaman & Nicobar Administration, Port Blair for providing basic references and information during the preparation of this manuscript.

References

Atwood, C.L., La Chance, J.L. Martz, H.F., Anderson, D.L., Englehardte, M., Whitehead, D. and Wheeler, T. (2003).*Handbook of Parameter Estimation for Probabilistic Risk Assessment.*, Washington DC: U.S. Nuclear Regulatory Commission Office of Nuclear Regulatory Research.

Claudio, S., and Horst, S. (2007). A GIS-based vulnerability assessment of coastal natural hazards, State of Pará, Brazil. *Journal of Coast Conservation, 11,* pp.53-66.

Deirdre, S., and Jong, P. (2008). Disaster Risk Management for Coastal Tourism Destinations Responding to Climate Change: A Practical Guide for Decision Makers. Paris: UNEP.

Ghosh, J.K., Delampady, M., and Samanta, T. (2006).*An Introduction to Bayesian Analysis: Theory and Methods.* LLC, USA: Springer Science+Business Media.

Inham, D., and Nordstrom, C. (1971). On the tectonic and morphological classification of coastal. *Journal of Geology.79,* pp.1–27.

Klein, R., and Maciver, D. (1999). Adaptation to climate variability and change: Methodological issues. *Mitigation andAdaptation Strategies to Global Change, 4,* pp.189–198.

Laxminarayana, K., and Kar, A. (2010). *Post-Tsunami groundwater management studies in Neil, Havelock and Little Andaman Islands, S. Andaman District, A & N Islands.* Kolkata: Central Ground Water Board Eastern Region.

Mishra, O.P., Singh, O.P., Chakrabortty G.K., Kayal J.R., and Ghosh, D. (2007).*Aftershock Investigation in the Andaman-Nicobar Islands: An Antidote to Public Panic? Seismological Research Letters 78,* pp. 591-599.

Hamlina, T.L., Michael, A., Cangaa, R., Boyera, L. and Thigpen, E.B. (2009). 2009 Space Shuttle Probabilistic Risk Assessment Overview, NASA, Houston.

Thompso, W.A. (1981). On the Foundations of Reliability. *Technometrics 23*(1), pp.1-13.

Velmurugan, T.P. and Swarnam, R.N. (2006).Assessment of tsunami impact in South Andaman using remote sensing and GIS. *Journal of the Indian Society of Remote Sensing, 34,* pp.193-202.

8

The Role of Land use Planning in Chemical Disaster Risk Management

Christian Jochum

8.1 INTRODUCTION

Chemical risk management always starts in the plant itself, where primarily loss of containment has to be prevented by technical and organisational safeguards. In addition (and as the second layer in the "onion skin" model) any reasonable effort has to be undertaken to mitigate possible releases of hazardous chemicals. A number of sad experiences however, demonstrate that safety and mitigation measures may fail. Toxic clouds may travel far outside the facility as in the Bhopal tragedy, inflammable vapours may ignite beyond the facility's border as in Buncefield /GB and Mexico City or an explosion on site may cause a shockwave killing people in the neighbourhood of the plant as in Toulouse/France or Enschede /NL. All those events have one commonality: the bigger the distance to the source of the hazardous chemical the smaller are the effects. Safety distances therefore are essential safety measures. They are also a key issue for disaster risk management. The further away people are from the place of a disaster, the lesser the casualties will be. The fewer people live around a major hazard facility the easier will be an evacuation.

Initially in Europe, most chemical sites were built far away from other developments. Although pollution was the driving force, safety distances were also maintained. This common-sense approach was rapidly undermined in most cases; workers wanted to live close to the facility, other businesses appreciated the vicinity of chemical facilities for different reasons, traffic lines had to be built anyway, etc. Especially in densely populated areas of Germany and other countries of the European Union and other parts of the world space became so valuable that eventually most major chemical sites are closely surrounded by residential areas, schools, hospitals, railway stations, highways, etc., creating a serious potential for disasters.

8.2 EU REGULATION ON LAND USE PLANNING AROUND MAJOR HAZARD SITES

Process safety and emergency response experts always had been aware of this intrinsic risk. However, even in the European Union, which traditionally tends to be risk sensitive, it was not before 1996 that land use planning around major hazard sites became regulated till the Seveso II

Disaster Management and Risk Reduction: Role of Environmental Knowledge; Editors Anil K. Gupta, Sreeja S. Nair, Florian Bemmerlein-Lux and Sandhya Chatterji; Copyright © 2013, Narosa Publishing House, New Delhi

Directive (Council Directive 96/82/EC of 9 December 1996 on the control of major-accident hazards involving dangerous substances):

Article 12 (Land-use planning): Member States shall ensure that the objectives of preventing major accidents and limiting the consequences of such accidents are taken into account in their land-use policies and/or other relevant policies. They shall pursue those objectives through controls on (a) siting of new establishments, (b) modifications to existing establishments covered by Article 10, (c) new developments such as transport links, locations frequented by the public and residential areas in the vicinity of existing establishments, where the siting or developments are such as to increase the risk or consequences of a major accident.

Member states shall ensure that their land-use and/or other relevant policies and the procedures for implementing those policies take account of the need, in the long term, to maintain appropriate distances between establishments covered by this directive and residential areas, buildings and areas of public use, major transport routes as far as possible, recreational areas and areas of particular natural sensitivity or interest and, in the case of existing establishments, of the need for additional technical measures in accordance with article 5 so as not to increase the risks to people.

1. The Commission is invited by 31 December 2006, in close cooperation with the member states, to draw up guidelines defining a technical database including risk data and risk scenarios, to be used for assessing the compatibility between the establishments covered by this directive and the areas described in paragraph 1. The definition of this database shall as far as possible take account of the evaluations made by the competent authorities, the information obtained from operators and all other relevant information such as the socioeconomic benefits of development and the mitigating effects of emergency plans.

2. Member states shall ensure that all competent authorities and planning authorities responsible for decisions in this area set up appropriate consultation procedures to facilitate implementation of the policies established under paragraph 1. The procedures shall be designed to ensure that technical advice on the risks arising from the establishment is available, either on a case-by-case or on a generic basis, when decisions are taken.

8.3 GERMAN GUIDANCE ON LAND-USE PLANNING

In most EU countries different authorities responsible for spatial planning and implementation of the Seveso Directive are slow in the process. It needed the explosion of a fertilizer storage factory at Toulouse France in 2001 with more than 10 fatalities outside the site, and there after the implementation of regulation became effective. In Germany, the Commission on Process Safety which advises the German Federal Government issued guidance on land-use planning in 2005. It covers the planning of industrial zones, for which the concrete usage is not yet known (Greenfield Developments), and even more important in Germany, developments in the neighbourhood of existing major-hazard sites. The guidance has recently been updated, taking into account the experience of the first years of implementation. The new version (KAS-18) is available at www.kas-bmu.de. A short version of the first issue in English is available (SFK/TAA-GS-1 short version).

8.3.1 Separation Distance Recommendations for Land-use Planning without Detail Knowledge

The guidance first deals with the setting out of new industrial and trading areas, for which the concrete usage is not yet known, but are designated by the local administration under the German Federal Emission Control Act, for the permissible siting of establishments. This "land-use planning without detail knowledge" would be applicable for instance if a city which is responsible for spatial planning decides to develop a new industrial area. In this case the distance to existing or also planned sensitive areas, e.g. residential areas, is known. The guidance now defines precautionary "separation distances" for the possible use of certain hazardous substances in the planned industrial area. At that point of time typically no detailed information exists about the chemical plants which may be built there. This tool should help in the decision which substances could be used in major-hazard installation in that area and which not. Eventually this information can be included in the legally binding land-use plan, restricting the future use of this area to minimize the consequences of industrial disasters.

For explosives and ammonium nitrate specific German regulations, which already set safety distances, are to be applied. In recommending separation distances for all other hazardous substances it was decided to use a precautionary deterministic approach, in line with the major hazards legislation as practiced in Germany. Long term operating experience and the registered major accidents in the last 15 years in Germany have been reviewed. As a result of these studies,

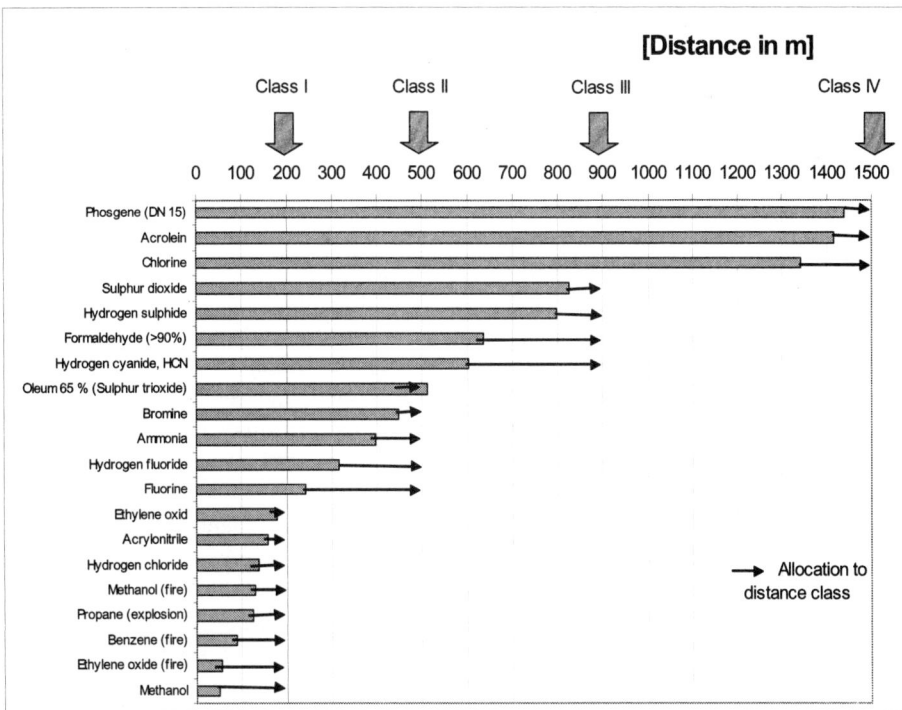

Figure 8.1: Recommended separation distances for land-use planning without detailed knowledge *(source: KAS-18 at www.kas-bmu.de)*

as basis for a consequence assessment as source term for a release a leakage of 490 mm² was adopted (equivalent to the cross-sectional area of a DN 25 pipe). For highly hazardous substances such as phosgene a DN 15 leakage was chosen, as the above mentioned reviews demonstrated that this comes closer to reality, where such substances are handled with extreme care. It has to be noted that a number of other European states use a probabilistic approach to land-use planning instead of this deterministic approach.

As scenarios, fire, vapor cloud explosion with immediate ignition and release of toxic substances were chosen. As permissible thresholds were used for thermal radiation 1.6 kW/m², for explosions 0.1 bar and for toxic substances the concentration guidance value ERPG-2. The dispersion model used was the VDI Guideline 3783. The dispersion conditions chosen for the hazardous substances were average meteorology (including a wind speed of 3 m/s) in a typical industrial topology (uniform buildings).

There is no simple relationship between toxicity, thermal radiation or explosion pressure wave and the recommended separation distance s. Substance characteristics (e.g. vapour pressure) and typical operating conditions lead to differing release rates, which had to be taken into account. However, consequence modelling of these scenarios cannot be considered as completely realistic, but as a model typification. Taking this into account, the substances have been clustered into distance classes to make the application easier. The guidance had to be limited to the most prominent substances (see figure 8.1). However, it sets out a procedure which would allow the development of such separation distances for other substances, too. It has to be noted that the separation distance recommendations are only related to people as the subject to be protected. For sensitive environment, e.g. natural reserves, no guidance exists up to now.

8.3.2 Separation Distance Recommendations for Land-use Planning with Detailed Knowledge

Ironically, Greenfield development of industrial sites occurs rather seldom in Germany nowadays and more and more development is coming closer to industrial and especially chemical sites due economic viability (figure 8.2). Consumer markets would like to make use of the good traffic connections near industrial sites, new roads or railway stations may be planned, etc. Quite frequently other industrial sites near major-hazard facilities may have had to close down and their land is a valuable asset which attracts investors. Here the guidance gives "separation distance recommendations for land-use planning with detailed knowledge".

For existing major-hazard facilities the substances, their licensed quantities and the technical installations in which they are handled are already known. In this case a specific individual case study with a systematic hazard analysis is possible. For the determination of separation distances the following recommendations are made in the guidance:

1. If the distance to sensitive areas is less than the separation distance recommendation for land use planning without detail knowledge (see Figure 8.1), then an individual case study is necessary.
2. If other legal requirements prescribe a minimum distance for the type of installation (e.g. explosives law, technical regulations) then these are to be respected.
3. For the individual case study the following recommendations are made with regard to the scenarios which are to be considered:

The loss of the complete inventory, the loss of the largest contiguous volume, bursting of a vessel and the shearing of a very large pipe are not to be considered for land-use planning, as they are too improbable when the state of the art technology is adhered·to.

(a) For storage in barrels and drums, and storage in gas bottles the release of the contents of a barrel/drum or bottle is to be considered.

(b) For process installations and in storage facilities it is to be assumed that leaks from pipe-work, vessels, safety equipment, etc. can occur.

    (i) In general the starting point is the consideration of a leak area of 490 mm² (equivalent to DN 25).

    (ii) In the individual case study the leak area is determined according to the technical systems actually in place.

    (iii) As a minimum assumption it is recommended that a leak of no less than 80 mm² (equivalent to DN 10) is chosen. (This corresponds to the flange leakage of larger pipe-work, the shearing of a small line, the leakage due to corrosion damage, etc.)

    (iv) Measures to limit the effects of a release are to be considered in so far as they are not damaged by the events in the scenario considered.

    (v) It is explicitly noted that these recommendations cannot replace the consideration of the particular situation in the individual case study.

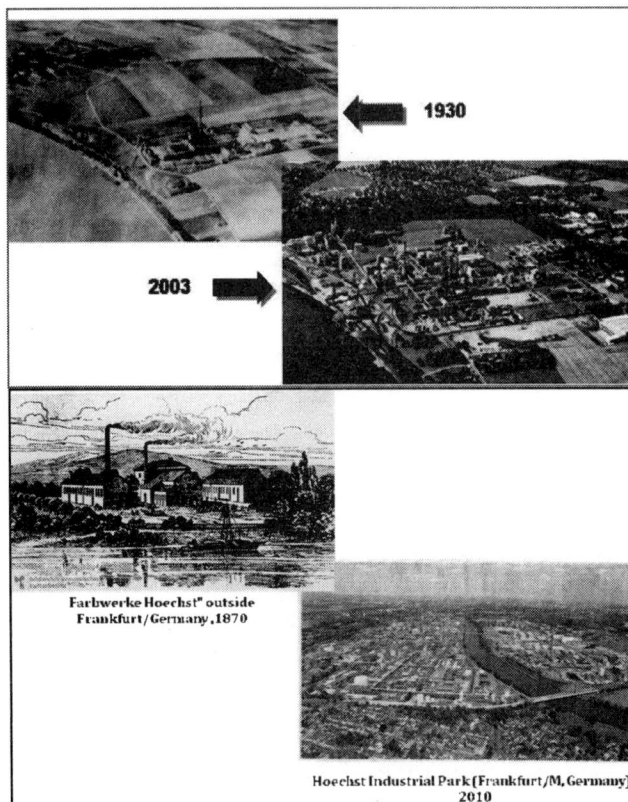

Figure 8.2: Industrial sites: Original design and today

In nutshell, whenever the (precautionary) "separation distances for land-use planning without detailed knowledge" cannot be met, the necessary separation distance between the existing major-hazard facility and new sensitive developments have to be assessed individually, following similar criteria as developed for Greenfields. As all safety measures within the facility have to be taken into account, the individual separation distance may be considerably smaller as in the Greenfield case.

Theoretically the separation distance could be zero, if the assessment of the installation's inner layers of protection (inherently safe processes, high integrity of equipment, very efficient mitigation measures) lead to the conclusion that there is no major hazard outside the fence. However, the precautionary approach will usually lead to a separation distance. Also other planning rules prevent the direct neighbourhood between industrial sites causing hazards, pollution or noise, and residential areas.

8.3.3 Other Planning Scenarios

A third quite common situation both in Europe as in India are "mixed situations" of major-hazard facilities and sensitive neighbourhood, grown more or less unplanned over many years. Here the directive of the European Union is only setting the long term goal to maintain appropriate separation distances and, on a case by case decision, the request for additional technical measures. Those cases are legally very complex, as operators have been granted licenses to operate as they do. The rather schematic and strictly precautionary approach of the above mentioned separation distances in Germany can here only serve as a very first approach. Eventually the interests and rights of all stakeholders have to be balanced in light of the history of these developments, and this will frequently end at the court. Relocation either of hazardous installations or sensitive neighbourhood is rarely possible, so that here the focus is on improved safety measures and emergency preparedness.

The fourth case is the planning of a new major-hazard facility close to existing developments. Here the authority has to make an individual assessment, if such a facility does not pose an unacceptable risk to its neighbourhood before they grant a license. Although the procedure for "separation distance recommendations for land-use planning with detailed knowledge" is not legally applicable here, it may serve as a reference.

8.4 SEPARATION DISTANCES AND EMERGENCY PLANNING

On reaching or exceeding the recommended separation distances, it may be generally assumed that the effects of a major accident within an establishment, based on the assumptions made, will not lead to a serious hazard for the population. The "intermediate zone resulting from the separation distance recommendation should not be understood as an area free of buildings. Within these distances less sensitive usage may be planned. This could be e.g. other industries or office buildings with a clearly defined population which may be included in the major-hazard facility's alarm system and trained how to act in emergency situations.

Outside the separation distances emergency planning still is a must. Although the consequences for people are minimised by proper land-use planning, the underlying assumptions are "credible worst case scenarios" only. As Bhopal and recently Fukushima have demonstrated it could come worse. Even if the probability of such catastrophes is very low, emergency responders have to take that into account. Also reaching e.g. the ERPG-2 value for toxic substances, which is one of the definitions of the separation distances, has to trigger emergency procedures.

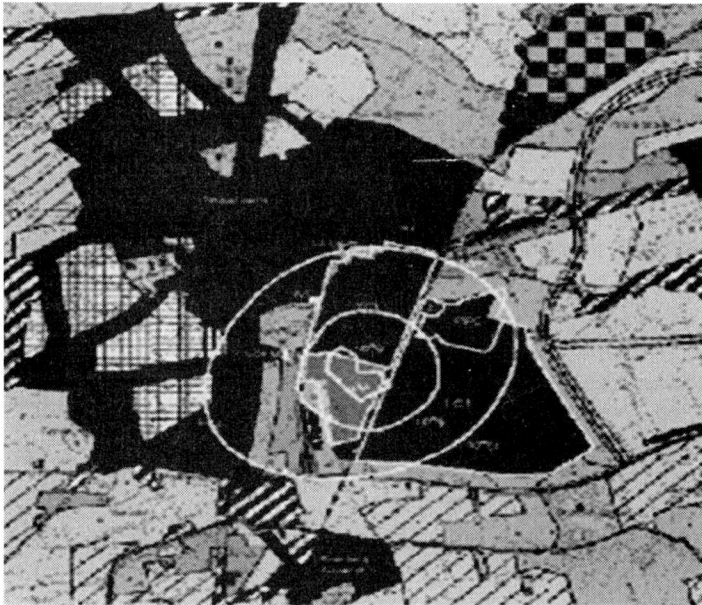

Figure 8.3: Zoning around major-hazard facilities

As a result, 3 different zones have to be assumed (see Figure 8.3). Looking from inside the facility, its fence forms the first zone and the area of the on-site emergency plan. Then comes a zone characterised by the separation distances with limited use, followed by the outer zone without limitations of use, but also subject to the off-site emergency planning.

8.5 IMPLEMENTATION OF THE GUIDANCE ON LAND-USE PLANNING

Land-use planning always is a trade-off between different justified interests: safety and health, ecology, economy, culture and people. For all these interests different authorities may be responsible. This makes implementation of land-use regulations difficult. In Germany for instance, land-use planning is the responsibility of municipalities. The EU Seveso II Directive forces them now to consult the authorities competent for major-hazard installations. However, a local mayor may be more interested in major investments as in major hazards and therefore may not take the obligatory consultation seriously enough. Although the cooperation between the different authorities improved in the recent years, a growing number of cases ended up at court. The guidance of the German Commission on Process Safety, although legally not binding, has been widely accepted by all parties as well as the courts as a reference. One typical case is pending at the European High Court and a fundamental judgement can be expected. Anyhow experience in Germany shows that a critical precondition for effective land use planning is a good cooperation between the different authorities responsible for spatial planning and environmental safety.

References

European Commission (1996). *Chemical Accidents (Seveso II) – Legislation.* Council Directive 96/82/EC of December 9, 1996 on the control of major-accident hazards involving dangerous substances. Retrieved from http://ec.europa.eu/environment/seveso/legislation.htm.

European Commission (2008). *Directive 2008/1/EC.* European Parliament and of the Council of January 15, 2008 concerning integrated pollution prevention and control. Retrieved from http://ec.europa.eu/environment/air/pollutants/stationary/ippc/index.htm.

Federal Law Gazette I (n.d). *Federal Immission Control Act (An Act on the Prevention of Harmful Effects on the Environment caused by Air Pollution, Noise, Vibration and Similar Phenomena) – Excerpts.* Retrieved from http://www.iuscomp.org/gla/statutes/BImSchG.htm.

Jochum, C. (2011). The Role of Land-use Planning in Chemical Disaster Risk Management. In Anil K. Gupta and Sreeja S. Nair, (eds), *Environmental Knowledge for Disaster Risk Management* (background papers of the workshop), pp.42-50.

Kommission für Anlagensicherheit (2005). *Guidance on Land-use Planning.* (KAS-18) (Commission on Process Safety). Retrieved from www.kas-bmu.de.

9

Bridging Industrial Risk Management Deficits in India using a Geo-ICT based Tool

Debanjan Bandyopadhyay, Nilanjan Paul, Anandita Sengupta, Cees van Westen and Anne van der Veen

9.1 INTRODUCTION

Dealing with the issue of industrial risk is a complex task, which requires collective action from a range of actors. A platform for sharing pertinent risk related information and building collaboration between relevant risk factors can be an important contribution in this respect. Such a platform supports decisions on key functional aspects, regulatory support, risk assessment, emergency management, risk informed land use planning and risk communication (Fedra 1998; Neuvel, Scholten et al. 2010).In a number of developed countries, the need for a framework to share risk information between actors like governments, industries and the public has been mandated through policy and regulations which form the basis for risk management initiatives (Walker, Simmons et al., 1999). In order to attain such objectives, regulatory authorities and institutions have established databases that provide information about facilities storing or emitting hazardous chemicals - systems like the Toxic Release Inventory (TRI) of Chemicals and the Major Accident Reporting System(MARS) and the Seveso Plant Information Retrieval System (SPIRS) have been made operational(JRC). Several other tools ranging from commercial software like SAFETI and EFECTS to applications like ARIPAR and ALOHA have been promoted by research and regulatory agencies to provide decision support capabilities to key risk actors (Technica 1984; Pe 2005; Boot, Veld et al. 2006). Understanding that risk has a strong spatial dimension, Geo-ICT modules for risk management have also been integrated with certain tools like XENVIS, IRIMS and HARIA (Contini, Bellezza et al., 2000). In addition, online maps delivered through GIS platforms and integrated with country wide risk data are emerging as a new frontier for sharing risk information at national level, as in the Risicokaart initiative of the Netherlands (Risicokaart ; Moen and Ale 1998; Basta, Neuvel et al. 2007). Nonetheless, very few integrated tools and information systems have been designed to be able to serve the needs of a majority of risk actors in this broad spectrum.

Disaster Management and Risk Reduction: Role of Environmental Knowledge; Editors Anil K. Gupta, Sreeja S. Nair, Florian Bemmerlein-Lux and Sandhya Chatterji; Copyright © 2013, Narosa Publishing House, New Delhi

In spite of the Bhopal accident in 1984 and several others which occurred subsequently in different parts of India, significant weaknesses currently prevail in the risk management framework of the country. This is in spite of a rapid rise in the number of major accident hazard (MAH) industries, which are distributed across more than 100 industrial clusters in the country. Though a comprehensive set of regulations have been formulated and responsibilities assigned to several competent authorities, yet the implementation of risk prevention and mitigation measures have been weak. Recently, a study conducted by the National Disaster Management Authority (NDMA), has identified the key gaps in risk management - the findings include lack of accessible information on potential chemical hazards to planners and emergency planning personnel and the administration, absence of harmonized criteria for undertaking risk assessment, no provisions for risk guided land use planning in regulations or the planning framework and the inadequate risk communication to the communities at risk (NDMA 2007). Although, India has witnessed considerable progress in ICT in the last decade, the amount of effective risk information that is available in the public domain and to different actors is negligible. Due to these deficiencies, industrial accidents from hazardous facilities continue to cause considerable loss of human lives and damage to property, as exhibited most recently by the Jaipur accident (MoPNG 2009).

Recently, efforts have been initiated by the government and other stakeholders to improve information availability on industrial risk sources and the hazardous substances they store. The Ministry of Environment and Forests (MoEF) commissioned a project to develop a GIS-based Emergency Planning and Response System (GEPR) for MAH industry clusters in select industrialized States in the Country (Gahlout, Guha et al., 2009). The GEPR was aimed at providing an inventory of hazardous substance storages and displaying consequences of a potential accident in the form of a spatial footprint on a digital map. However, a review of its capabilities point to certain weaknesses - the low resolution of spatial data is not adequate for local emergency response and lack of integration with an intrinsic hazard modelling tool that requires the system to function based on a number of predefined hazard scenarios that may not match with an actual accident scenario. This limits the application of GEPR in emergency situations where rapid hazard simulations matching the scale of the accident are desired. Another novel initiative was taken up by an industry organization through the Environmental Risk Reporting and Information System (ERRIS) project, which encouraged industries to voluntarily report hazard related information to a web-based Risk Reporting Information System (Bandyopadhyay and Paul 2008). The ERRIS was made accessible to selected stakeholders for two industrial clusters in the eastern part of the country. At this time though, there is no integrated ICT solution in India that can reinforce and consolidate risk management activities through better sharing of information and provide guidance for informed decision making to several actors.

This paper makes an effort to demonstrate the capabilities of the Risk Management Information System (RMIS) tool, which has been developed as a follow up of the ERRIS project. Taking into account the present management framework for industrial risks in India, we explore how a system like RMIS can bridge some of the existing gaps with regard to sharing of risk information and using it for informed decision making. Through this paper, we hope to stimulate discussion and debate on the application of information systems to better manage industrial risks in India.

9.2 THE GEO-ICT INTEGRATED SOLUTION: RMIS

The Risk Management Information System (RMIS) is an evolving Geo-ICT based tool which combines in one platform a risk information system, a modelling toolbox and risk oriented planning support capabilities. First developed as a web-based risk information system under the ERRIS project, further research and development based on feedbacks from potential users, has now resulted in a robust and versatile information system and planning support tool that can cater to the requirements of a host of risk management actors including regulators, planners, emergency responders and the community in general.

With the need for drawing and assembling data from diverse sources, the RMIS integrates both spatial and non-spatial data using different data capture mechanisms, assimilates them through a logical data model and makes relevant set of information available through an easy to use, intuitive map based user interface. The conceptual data model that links up data from these sources is presented in Figure 9.1 below.

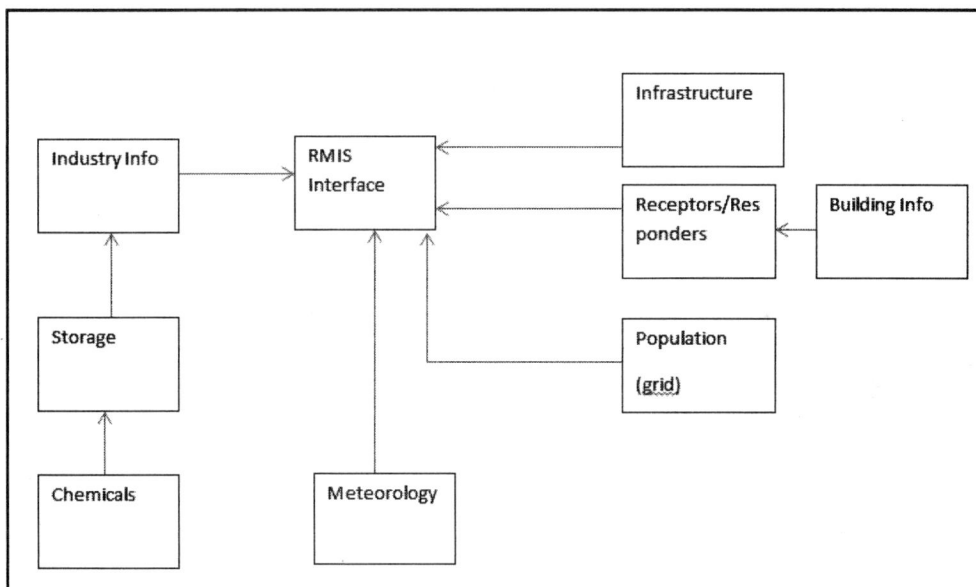

Figure 9.1: RMIS conceptual data model

The RMIS comprises of a web-map based graphical interface which provides the end user with a display of spatial and attributes data on a map along with the option to perform spatial queries. Consequence modelling and risk estimation algorithms for simulating probable damages from a potential accident scenario have been integrated with the RMIS along with live meteorological data feed. The RMIS can be accessed by end-users from remote locations once they connect to the application over the internet/intranet. The distributed web-GIS architecture ensures easy accessibility of hazard information in a spatial context and improves the capability of decision makers, emergency responders and local administration to effectively manage technological accidents originating from hazardous facilities. The RMIS has also been implemented on scalable hardware and software infrastructure so that it remains easily maintainable and can accommodate future innovations in hardware and communication technologies.

As several actors having different set of information needs are expected to use the RMIS for meeting their information requirements, the level of access to information or rights to alter data in several linked databases is provided based on the role a particular user has been assigned by the system administrator through an integrated user access control and role authorization module.

9.3 BRIDGING THE GAPS

9.3.1 Regulatory Support

Regulatory mechanisms play an important role in risk management. Several regulatory styles have been practiced by nations, which differ in the way how risk regulations are evolved and implemented. The prevailing approach for regulating risk from hazardous installations is based on a prescriptive 'Command and Control' approach (Mohan and Aggarwal 1990) which focused on licensing and monitoring based control of hazardous installations. However, with the occurrence of the Bhopal disaster in early 1980's, many countries initiated a review of their respective regulatory systems for dealing with industrial risk. The European system made a shift towards a more performance and goal oriented regulatory approach through the Seveso II Directive, which laid considerable stress on risk assessment and management as the guiding principles of risk regulations and ensure that population is not exposed to any undue risk or safety concerns (EC 1996; Wettig and Porter 1998). In the US, a progression has been noted to a more risk-informed approach – a regulatory philosophy, where risk insights are considered together with other factors to establish requirements that better focus licensee and regulatory attention on design and operational issues, prioritized based on their importance to health and safety (Genn 2003). The success of these evolving regulatory approaches depended to a considerable extent on the availability of up to date and valid information on hazards, dangerous substances, and vulnerabilities existing in the surroundings of hazardous installations (Walker, Simmons et al., 1999).

In India, after Union Carbide disaster in Bhopal and followed closely by the Sriram Gas leak incident in Delhi in the 1980s, the provisions relating to industrial safety in the Factories Act was strengthened and legislation specific to prevention and management of accidents arising out of hazardous installations was formulated in the form of the Manufacture, Storage, Import of Hazardous Chemicals Rules (MSIHC), 1989 and subsequently amended in 2000 (MoEF 2000). According to MSIHC Rules, industries storing and handling hazardous chemicals are required to submit a safety report and an Onsite Emergency Plan to the regulatory agencies containing vital industry specific information that would enable the regulators attain swift control of an industrial accident. The Factories Act (GoI 1987) also lays down specific requirements for hazardous industries in terms of laying down adequate safeguards within the installation, preparation of on-site emergency measures and information disclosure about potential accidents to the workers and the public. However, with the prescriptive and authoritative approach being still prevalent, industries appear to be secretive and not unwilling to divulge information on hazards and risks, fearing punitive actions and penalties. As a result, competent authorities often do not have up-to-date information about nature of hazardous chemicals stored, specific amounts and locations of such storages which also leads to inability of crisis management groups and emergency responders to initiate coordinated action and demonstrate a coherent response to managing risk from hazardous installations.

Understanding the requirement for updated information on hazard sources, chemical substance and other risk related data, a need has been felt to make a transition towards more risk-informed regulations enabling transparent sharing and exchange of information between the competent authority and the hazardous facilities, and wherever required with the stakeholders. Using the RMIS, competent authorities can readily access regulatory information and reports for all hazardous industries within a MAH industry cluster along with full understanding of spatial locations of such storages and activities. The presence of consolidated information in a single database can aid in regulatory planning activities - like understanding overall hazard potential in an area. This is illustrated through a search in the RMIS to identify MAH industries which store flammable hazardous chemical of quantity more than 10,000 metric tons, as in Figure 9.2 below.

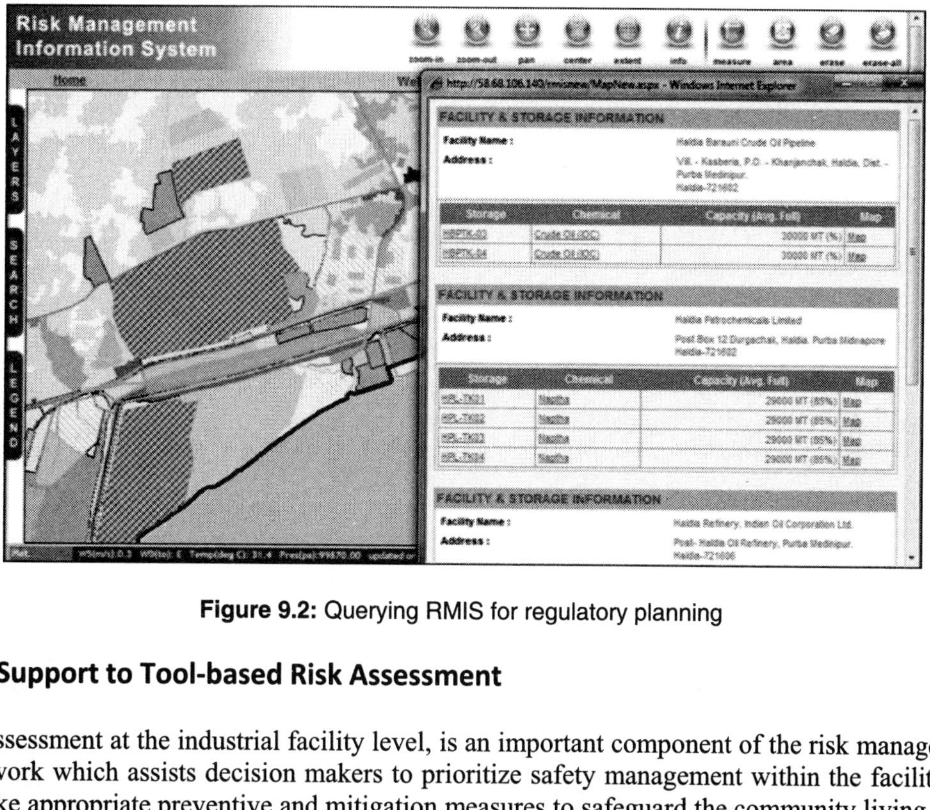

Figure 9.2: Querying RMIS for regulatory planning

9.3.2 Support to Tool-based Risk Assessment

Risk assessment at the industrial facility level, is an important component of the risk management framework which assists decision makers to prioritize safety management within the facility and also take appropriate preventive and mitigation measures to safeguard the community living in the vicinity of such a facility from a potential industrial accident. Risk assessments are data intensive, large number of variables and criteria are often required along with a number of assumptions that need to be made to undertake quantitative or semi-quantitative risk assessments. Taking these complexities into account, many countries have laid down standardized metrics for undertaking such risk assessment. Countries like the Netherlands, UK and USA have laid down detailed guidance for undertaking facility level risk assessments. For example, the coloured books in Netherlands along with the BEVI Guidance lay down methods, set limits on input parameters, specify models to be used and endpoints for hazard calculation (VROM 1997; VROM 1999; RIVM 2004). Applying such a standardized framework would ensure that risks from individual facilities are verifiable, reproducible and comparable and can be used for cumulative risk summation on a common platform.

In India, though basic templates for safety audits and emergency management plans are set in regulations, there is no framework document that delineates the risk philosophy to be followed and neither are there specific guidance's for undertaking a regulatory risk assessment. As a result, risk assessment is undertaken by hazardous industries based on methodology laid down by different risk consultants. A review of more than 80 cluster level hazard analysis studies undertaken by different consultants, as a part of a MoEF project in 2001, revealed substantial differences in consequence modelling results for similar accidents scenarios involving the same hazardous chemical (ERM 2002).Thus, it is difficult to use such results for drawing up area level risk minimization measures. In response to the need for standardization and harmonization, there is the need for the use of a common tool for risk assessment, which would reduce uncertainty in results and intrinsically standardize risk assessment to a significant extent when such risk assessments are undertaken following standard set of guidance and within specified boundary conditions.

The RMIS can be proposed as a tool similar to the RMP Comp software (USEPA 1999) in the US which is used by industries for standardized mapping and submission of RA results to regulatory authorities. Following the US model, the algorithms underlying such consequence calculations have been intentionally kept simple in the RMIS, so that the data requirements are not very intensive. It is possible to be transparently validated against spreadsheets which are available in public domain (AIChE 2000) ensuring that consequence modelling calculations performed by the tool are traceable. The RMIS has an in-built database of hazardous chemicals, and also a database on hazardous chemical storages in an area and can accept feed from an online meteorological station, providing data on aspects like average wind speeds and resultant vector for predominant wind direction thus automating the calculation process, considerably. Moreover, as the system is integrated with a geo-database, the results can be visualized on a referenced geographical frame and also tied up to vulnerability data.

9.3.3 Guidance to land-use planners

The accidents in hazardous facilities in Bhopal and Mexico highlighted the need for adopting controls through which separation can be maintained between such sites, residential areas and settlements using land use planning instruments. In some densely populated countries of the West like the Netherlands and UK, land use planning measures based on assessment of risk had been put into practice since 1960's and these aspects were further reinforced when the Article 12 of the Seveso II Directive in Europe made it obligatory for government of Member States to adopt provisions for land use planning controls when new installations are authorized or when further urbanization occurs in the vicinity of hazardous installations (EC 1996). The responsibility of drawing up such land use controls measures were accorded to the competent authorities in each of the EU Member States and several approaches have been adopted for this purpose (Christou and Mattarelli 2000). In a country like the Netherlands or UK, where risk consciousness is considered to be high, an external safety policy guides the risk informed land use planning framework taking into account national level criteria for acceptable risk in terms of two measures: Individual Risk (IR), the Societal Risk (SR) representing the expectation value of the number of people killed per year. These criteria are to be followed by the local planning authorities through restrictions in the development and land use plan (Ale 2001), and several existing tools like ARIPAR and SAFETI can assist in this process through calculation, analysis and visualization of results on a GIS based interface (Spadoni, Contini et al. 2003).

In India, land use planning is applicable to urban planning areas which are demarcated based on regulatory provisions laid down by the state level Town and Country Planning Acts. The consideration of industrial risk issues into land use planning has seldom occurred because of lack of appreciation of these aspects amongst the planning community. Site assessment and EIAs are required to be undertaken for new hazardous industries as per provisions of Factories Act and EIA Notification. In practice, however, seldom are alternative site options considered from the risk point of view for setting up an industry; instead the focus is to plan on a site where land is made available by the government or can be attained through purchase. As a result, the risk levels in areas housing a cluster of hazardous industries are quite alarming, and any accident can result in significant damage to life and property.

The RMIS now integrates a risk mapping module with the expectation that criteria for risk informed land use planning will soon be formulated to guide further development of new hazardous industries. Using this module in RMIS, the user can calculate two measures of risk – Individual Risk and Societal Risk and helps the planner to visualize it on a map of the concerned area. Individual Risk indicates the cumulative risk to a person from a combination of hazardous facilities in an area while Societal Risk is represented by Potential Loss of Life (PLL) to population over the planning area under consideration. The individual risk levels, with predefined acceptance criteria, can assist planners to understand the implications of setting up a new industry in the planning area by studying the escalation of risk that may be caused by the facility and residential areas which may be affected as result. The PLL can facilitate the consideration of alternatives for stabilization of risk levels through the adoption of preventive measures like restriction on further construction of residential houses in a particular area, which already shows high societal risk levels. Figure 9.3 below portrays RMIS in functioning in the risk mapping mode with PLL levels displayed for five different risk scenarios.

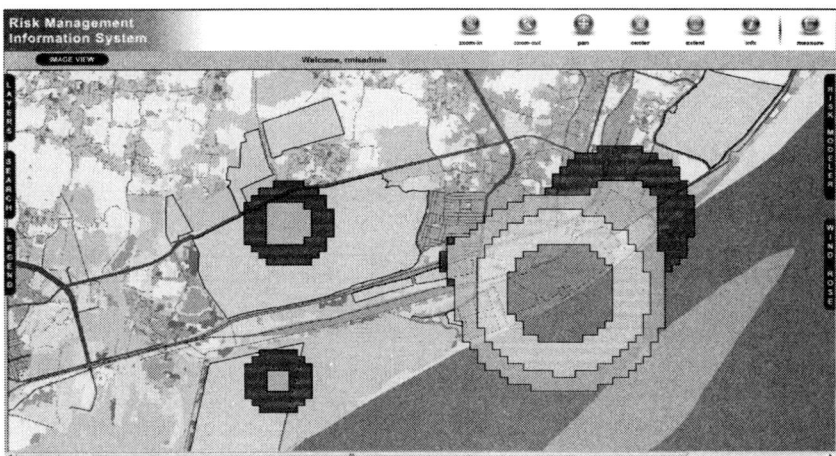

Figure 9.3: RMIS showing individual risk from number of reference scenarios

9.3.4 Enabling Emergency Response

With the rapid advances made in the field of ICT, there is recognition that having access to updated and time critical information is the key to effectively manage an emergency arising out of an industrial accident. Recent emergencies and emergency response exercises, however, have highlighted deficiencies in emergency management with information not reaching the right organizations and people at the right time, resulting in unnecessary loss of life and property (Kevany 2003)thus constraining effective mitigation, preparation, response and recovery.

At present, the practice of emergency management in India is still on the path to attaining maturity. The EPPRCA Rules envisage a four-tier crisis management system in the country – involving the Central Crisis Group, the State Crisis Group, the District Crisis Group and the Local Crisis Group to manage emergencies arising out of industrial operations. However, in actual terms, these groups are not functional in many industrial towns. In those where they function, they are often handicapped by the lack of sharing of information or its availability in the right form, and the coordination between different agencies and emergency responders is often weak. Discussions with administration and response personnel reveal that one of the key aspects that prevent efficient emergency management is the lack of updated and valid information available in the paper based on-site emergency management plans. Also, no computerized tool presently exists which can provide assistance to emergency responders to help tackle an emergency situation within short time, rather than reviewing a report to extract information which is often too time consuming.

Figure 9.4: RMIS functional view in emergency response mode

The RMIS is designed to assist emergency responders in devising response strategies and initiate appropriate preventive and mitigation actions. Once information about a certain accident, including identity of the storage involved and the amount of chemical likely to be involved, is conveyed to the local emergency centre, response personnel could initiate the hazard modeller module of the RMIS. The Consequence Modeller is designed to capture updated information from the data available in the databases and real time information on wind speed and wind direction from the plugged in automatic weather station. It is also guided by intrinsic rule-based expert system, through which the user can generate a spatial footprint to a certain endpoint in terms of toxic gas concentration, radiation or overpressure level, originating from the source point of the hazard, and aligned with the wind direction in the case atmospheric dispersion is involved, visualized on an automated GIS map on the screen of the system. The system also queries for relevant information on vulnerabilities underlying the footprint and brings up in a pop up window, information on the estimated number of people likely to be affected, susceptible receptors like schools hospitals in the area, other industrial hazards where cascading effects can occur and transportation infrastructure which may be severed by the impacts. To assist in better visualization, the linear infrastructures affected are highlighted in 'red' for the responders to take note. Figure 9.4 presents how RMIS displays the footprint of a toxic gas release scenario and the information it brings up and shows to the emergency responder to assist in emergency management.

9.3.5 Information to Communities

There has been growing recognition that communities should be involved, educated and made aware of industrial risk and how best to deal with it in the event of an accident. Acknowledging the societal concerns about risk arising from hazardous facilities and with the objective of helping communities to better cope with the consequences arising from a potential industrial accident in the neighbourhood, several countries have framed or included provisions in risk regulation whereby the role of the public and information sharing with regards to hazards have been clearly laid down. The Seveso Directive II (EC 1996) in Europe not only acknowledges the need for access to risk related information like the Safety reports, but also requires hazardous industries and competent authorities to involve in meaningful consultations during offsite emergency planning and land use planning decisions. In the US, the Emergency Planning and Community Right to Know Act (Congress 1986) establishes requirements for the government and the industries to report information on hazardous and toxic chemicals.

In India, learning lessons from the Bhopal disaster, regulation formulated to attain better control on industrial hazards recognized the right of the public in knowing information to improve their capacity to cope with impacts of possible accidents. Rule 15 of the MISHC Rules prescribes that facility managers should provide information to the public on the nature of the major accident hazard that can originate from a facility and the safety measures, including "do's and don'ts" which should be adopted by the public in such a situation (MoEF 2000). This is supplemented by Section 13 of the Rules on Emergency Planning, Preparedness and Response for Chemical Accidents (EPPRCA)titled "Information to the Public" which require the Local Crisis Group to assist the MAH installations in an industrial pocket to take appropriate steps to inform persons likely to be affected by a chemical accident (MoEF 1996). Section 41B of The Factories Act, 1949 as amended in 1987 also mandates compulsory disclosure of information by the occupier hazardous facility with regard to storage, transportation and handling of hazardous substances to the workers and general public living in the vicinity of hazardous facilities (GoI 1987). The Right to Information Act, 2005 also reinforces the right of citizens to seek information regarding hazardous substances and its effects. However, in practice, it has been seen that industries are seldom willing to share such information, primarily considering that it will give rise to public anxiety and will be detrimental to corporate image and reputation.

The RMIS, through a restricted access system, is in a position to make available minimum set of information to the public through the internet. This updated information can also be used in preparing information booklets and other paper guidance documents to educate the community on various precautionary and safeguard measures to be taken up by local communities who are exposed to industrial risk. Such information may include basic data on type of hazardous substances and a summary of the MSDS sheet in an easily understandable format, with visualization of the hazard source location being aided through a map based interface. It is expected that such active dissemination of information on hazards would lead to gradual improvement in safety culture and knowledge thereby improving the coping capacity of the community to respond to potential industrial accidents.

9.4 CONCLUSION

The RMIS can provide a host of benefits to different risk factors including regulators, administrators, emergency responders, planners and has the potential to be an ICT tool of choice for improving risk management in India. Not only can it provide valid and up-to-date information about risk sources, substances, vulnerabilities and local meteorology integrated from diverse sources, but also its decision and planning support functions enable informed decisions, making risk management more effective. The application is available through an interactive thin-client interface and provides visualization of risk on a spatial dimension, supported by wizard assisted analytical functionalities, helpful not only to experts in industrial risk assessment but also to administrators and decision makers trained on using the system through a focused capacity building exercise. Being developed on a distributed architecture, it is expected that the RMIS can play a vital role in facilitating dialogue and help in developing a shared understanding of chemical risk leading to the formulation of feasible risk reduction strategies and management plans. Trials of RMIS undertaken with various stakeholders in the Haldia region of West Bengal, demonstrate that the system can adapt to situational and operation requirements for risk information and decision support in India. Overall, the development and implementation of RMIS would be in line with the Disaster Management Policy defined by the Government of India which emphasizes on management of knowledge and information to build better coordination between different government agencies and other actors who play a role in disaster prevention and management. Presently, the RMIS is an evolving platform - further research and development is on-going to improve functionalities and capabilities of the system. They range from making the interface more user friendly to improving the system architecture for meeting the needs and requirements from multiple users located at several locations, building operational redundancy and improving communication bandwidth.

References

AIChE (2000). Guidelines for Chemical Process Quantitative Risk Analysis. (2nd Edition). *Center for Chemical Process Safety*.

Ale, B.J.M. (2001). Risk Assessment Practices in The Netherlands. *Safety Science 40*(1- 4)105-126.

Bandyopadhyay, D., and Paul, N. (2008). A GIS Based Framework for Managing Industrial Risks in India. Municipalika. *Making Cities Work*. Mumbai, India.

Basta, C., Neuvel, J.M.M., Zlatanova, S., and Ale, B. (2007). Risk-maps informing Land-Use Planning Processes: A survey on the Netherlands and the United Kingdom recent developments. *Journal of Hazardous Materials 145*(1-2)241-249.

Boot, F.H., Veld, H.V., and Kootstra, F. (2006). Riskcurves: A Comprehensive Program Package for Performing a Quantitative Risk Assessment. *AIChE Spring National Meeting* Orlando.

Christou, M. D.,and Mattarelli, M. (2000). Land-use planning in the vicinity of chemical sites: Risk-informed decision making at a local community level. *Journal of Hazardous Materials 78*(1-3) 191-222.

Congress, U. (1986).*Emergency Planning and Community Right-to-Know Act*. Title 42, Chapter 116.

Contini, S., Bellezza, F., Christou, M.D., and Kirchsteiger, C. (2000). The use of geographic information systems in major accident risk assessment and management. *Journal of Hazardous Materials 78*(1-3),223-245.

European Commission (1996). Seveso II - Council Directive 96/82/EC on the Control of Major-accident Hazards Involving Dangerous Substances.

ERM (2002). Review of Hazard Analysis Report and Preparation of Vulnerability Map of India.

Fedra, K. (1998). Integrated risk assessment and management: overview and state of the art. *Journal of Hazardous Materials 61*(1-3) 5-22.

Gahlout, S.S. (2009). GIS and Online Emergency Planning and Information Reporting System for Chemical Accidents in India. Paper presented at Second India Disaster Management Congress, New Delhi.

Genn, S. (2003). Safety goals in 'risk-informed, performance-based' regulation. *Reliability Engineering & System Safety 80*(2) p 163-172.

Government of India (1987). The Factories Act, 1948,as amended by the Factories (Amendment) Act, 1987.

JRC. MARS and SPIRS. Retrieved on 16.09.2011, fromhttp://mahb.jrc.it/index.php?id=39.

Kevany, M.J. (2003). GIS in the World Trade Center attack--trial by fire. *Computers, Environment and Urban Systems 27*(6) 571-583.

Kirchsteiger, C. (2002). Towards harmonising risk-informed decision making: the ARAMIS and compass projects. *Journal of Loss Prevention in the Process Industries 15*(3)199-203.

MoEF (1996). Rules on Emergency Planning, Preparedness and Response for Chemical Accidents.

MoEF (2000). S.O.57(E): Manufacture, Storage and Import of Hazardous Chemical (Amendment) Rules.

Moen, J.E.T.,and Ale,B.J.M. (1998). Risk maps and communication. *Journal of Hazardous Materials 61*(1-3) 271-278.

Mohan, R. and Aggarwal,V. (1990). Commands and controls: Planning for indian industrial development, 1951–1990. *Journal of Comparative Economics 14*(4) 681-712.

MoPNG (2009). IOC Fire Accident Investigation Report.

NDMA (2007). National Disaster Management Guidelines, National Disaster Management Authority, Govt. of India.

Jeroen M., Neuvel, M., Scholten, H.J., and Brink, A. van den (2012). From Spatial Data to Synchronised Actions: The Network-centric Organisation of Spatial Decision Support for Risk and Emergency Management. *Applied Spatial Analysis and Policy5* (1),51-72

NRC (2007). Successful Response Starts with a Map: Improving Geospatial Support for Disaster Management. Washington DC, National Research Council.

Pe, C.S P.S.M.D. (2005). Computer aids. Lees' Loss Prevention in the Process Industries (Third Edition). Burlington, Butterworth-Heinemann (1-5).

Risicokaart. Retrieved from http://risicokaart.nl/. on 26.08.2011

RIVM (2004). External Safety Establishments Regulations (BEVI).

SNDR (2002). A National Hazards Information Strategy : Reducing Disaster Losses Through Better Information. National Science and Technology Council.

Spadoni, G., and S. Contini, et al. (2003). The New Version of ARIPAR and the Benefits Given in Assessing and Managing Major Risks in Industrialised Areas. *Process Safety and Environmental Protection 81*(1) 19-30.

Technica (1984). *The SAFETI Package.* London, Technica Ltd.

USEPA (1999). Risk Management Program Guidance for Offsite Consequence Analysis.

Vrom, M.O. (1997). CPR14E - Methods for the calculation of physical effects due to the release of hazardous materials (liquids and gases), Yellow Book. C. J. H. van den Bosch and R. A. P. M. Weterings. Den Haag.

Vrom, M.O. (1999). CPR14E - *Guidelines for Quantitative Risk Assessment.* Purple Book. Den Haag.

Walkera, G., Simmonsb, P., Irwinc, A., and Wynneb, B. (1999). Risk communication, public participation and the Seveso II directive. *Journal of Hazardous Materials 65*(1-2) 179-190.

Wettig,J.,and Porter, S. (1998). Seveso Directive: Background, contents and requirements. *Industrial Safety Series.* M. D. C. Christian Kirchsteiger and A. P. Georgios, Elsevier (6) 27-68.

Vulnerability Assessment of IOCL Depot, Jaipur, Rajasthan, India

Shreya Roy, B.D. Bharath and B.S. Sokhi

10.1 INTRODUCTION

India is developing as a key global player in the industrial and technology sector. With these has increased the frequency and severity of technological hazards. Installations which are located in dense urban areas, working at capacities beyond their permissible threshold limits, lacking an adequate emergency and safety measure are more hazardous than those located in areas far from the locality. Sometimes situation may arise where a fire/explosion load generated by an accident in one unit triggers secondary and higher order accidents in other units. This type of domino effect is very much common in areas storing petroleum products (J. R. B. Alencar et al., 2005). Such chains of accidents have a greater propensity to cause damage than stand-alone accidents. They can cause havoc to human life, property, economy as well as to the environment. Heat and Smoke from burning of petroleum products disrupts the fire fighting system, serial blasting of storage tanks due to fire damages the buildings and properties besides spreading of metal flakes in surroundings and nearby areas (Environmental Impact of the fire in Indian Oil Corporation Depot Sitapura, Jaipur, February 2010). Prevalent atmospheric conditions like atmospheric stability, mixing height, wind speed, wind direction, rain wash etc also plays an important role in dispersion of burnouts.

A major accident involving Motor Spirit (MS) Vapor Cloud Explosion (VCE) and Fire occurred at Indian Oil Corporation's Petroleum Oil Lubricants (POL) Terminal, Sitapura, Jaipur on October 29, 2009 at 19:30 hrs. There were 11 storage tanks in the terminal that includes 5 for MS and 3 each for High Speed Diesel (HSD) and Kerosene containing 60,000 kl of product. (Environmental Impact of the fire in Indian Oil Corporation Depot Sitapura, Jaipur, February 2010). In the process of lining up the MS tank, to nearby BPCL terminal a huge leak of liquid MS took place from the Hammer Blind Valve assembly. As it is highly volatile at ambient temperature and sufficient amount of vapours accumulated and finally exploded after 75 minutes. A huge fireball covered the entire installation spreading to all other tanks and continued to rage for about 11 days. This is the first instance in the country and perhaps the third in the world in which storage and handling of these products under atmospheric pressure has resulted into such a catastrophic accident.

Disaster Management and Risk Reduction: Role of Environmental Knowledge; Editors Anil K. Gupta, Sreeja S. Nair, Florian Bemmerlein-Lux and Sandhya Chatterji; Copyright © 2013, Narosa Publishing House, New Delhi

This study is an attempt to integrate remote sensing and GIS techniques to analyse the nature of vulnerability due to this type of accidents which can help the planners, industry personnel to identify suitable areas for their installations.

10.2 STUDY AREA

The terminal is within the Sitapura Industrial Area which is about 16 kilometres south of the city of Jaipur and is within the administrative boundary of Jaipur Development Authority (JDA) (Figure 10.1).

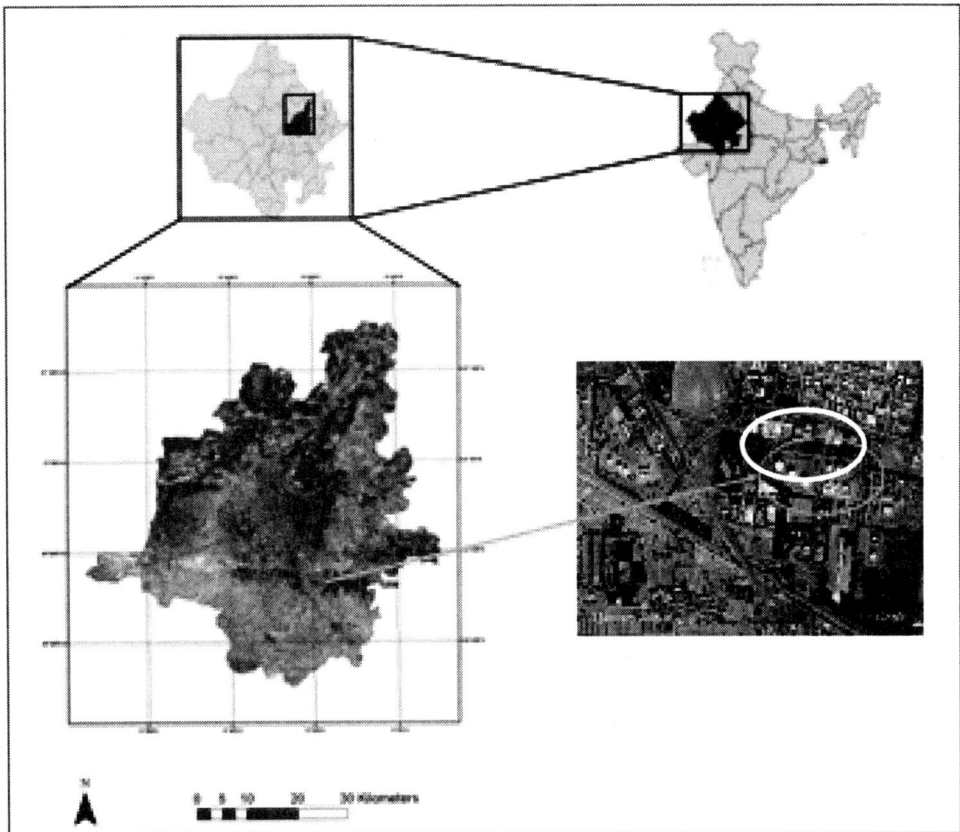

Figure 10.1: Study Area Map

Jaipur being located in the semi-arid zone witnesses hot summers and mild winters. As per the data from Indian Meteorological Department (IMD), the mean minimum temperature recorded is in the month of January (7.8 °C) and maximum in the month of May (40.3 °C). Annual mean rainfall recorded is 673.9 mm. July and August months record the maximum rainfall.

10.3 MATERIALS & METHODOLOGY

Flowchart depicting the methodology is given below (Figure 10.2).

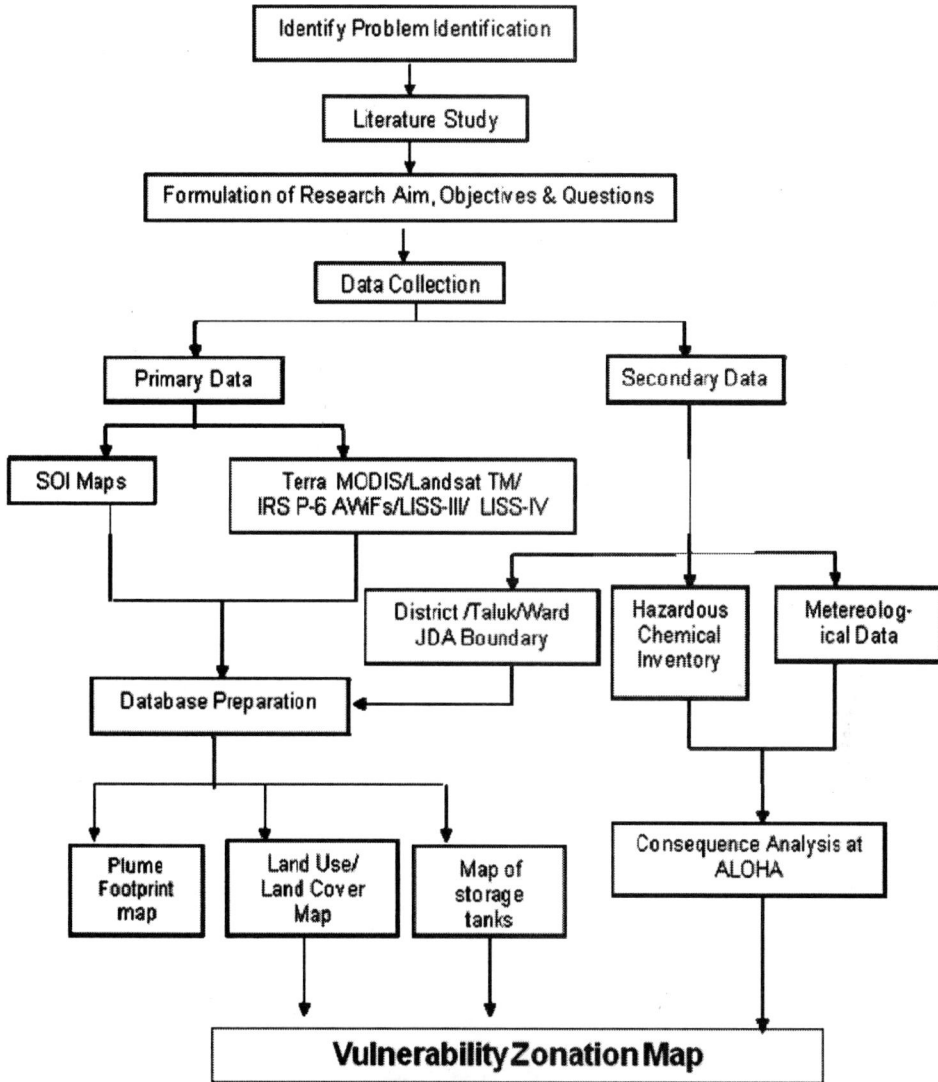

Figure 10.2: Flowchart depicting methodology

10.3.1 Data used

The following satellite data and ancillary datasets used in this study were given in Table 10.1 and Table 10.2 respectively and satellite images in Figure 10.3.

Table 10.1: Satellite data used

Sensor	Date of Acquisition	Path-Row	Resolution	No. of Bands	Source
Terra / MODIS (MOD09GQ)	October 30, 2009 October 31, 2009 November 02, 2009	-	250 m	1	MODIS Land Team via LP-DAAC
AWiFs	November 01,2009	094-050 (shift with LISS Row 52)	56 m	4	NRSC, Hyderabad
Landsat TM	November 01,2009	147/41	30m	7	USGS, GLOVIS
LISSIII	March 11, 2009 November 01,2009	095-052 095-052	23.5 m	4	NRSC, Hyderabad
LISSIV	October 13,2008	202-066	5.8 m	3	NRSC, Hyderabad

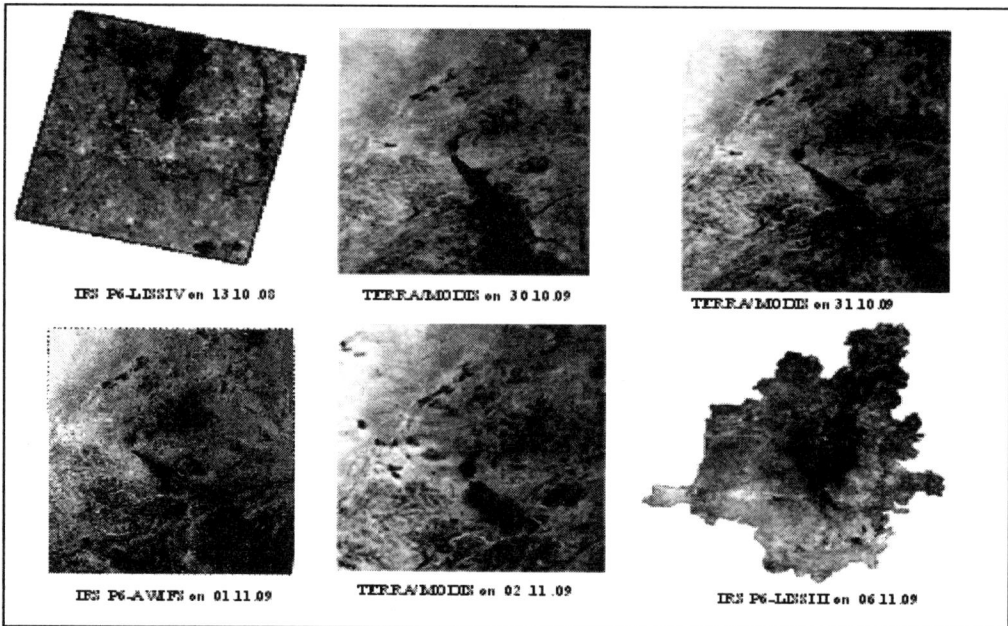

Figure 10.3: Satellite data used

Table 10.2: Ancillary data used

Ancillary Data	Source
Toposheet (SOI-45N/13), Jaipur Guide Map	Survey of India
District & Taluk Boundary	Survey of India
Boundary of Jaipur Development Authority (JDA),Ward Boundary	JDA , Jaipur
Meteorological Data during Peak Fire Period as well as during Worst Case Scenario	IOC Environment Report, February 2010
Hazardous Chemical Inventory	IOC Environment Report, February 2010 IOC Accident Investigation Report.

10.3.2 Preparation of Database

(i) MODIS data were obtained in EOS HDF data format in sinusoidal projection system. After importing to ERDAS image format, layer stack was done to generate FCC image. They were then re-projected to UTM/WGS84 Zone 43 and subset operation was performed with AWiFs data. AWiFS data was geo-referenced with Landsat-TM of the same date. These were used for generation of Plume Footprint Map in the Arc GIS Environment.

(ii) Plume footprint map of November 06, 2009 (shift in wind direction after November 05, 2009) was generated from IRS P6-LISS III after geo-referencing it with Landsat-TM.

(iii) LU/LC map was generated from IRS-P6 LISS-III dated March 11, 2009 by supervised classification after geo-referencing it with Landsat-TM. Subset was carried out using JDA boundary.

(iv) Roads & railways layer were generated using IRS-P6 LISS-III and attributes were added with the help of Toposheet, Guide Map and Google Maps.

(v) A polygon map of the oil storage tanks was prepared from IRS-P6 LISS-IV image dated October 13, 2008 after geo-referencing and building footprint map was prepared for the vulnerable areas.

(vi) In order to assess the vulnerability of fire and vapour cloud explosion on the areas surrounding the oil storage terminal after the accident four possible scenarios of Vapour Cloud Explosion & Pool Fire (2 for MS and 2 for Diesel) were generated in ALOHA taking into account the stock present at the time of the accident, nature of failure, prevalent atmospheric conditions, dispersion distances (Lower Flammability Limit or LFL) of the substance released, the probable radiation from any fire that results and the possible shock wave upon equipment and structures that could arise from an explosion.

(vii) To generate hazard scenarios hexane was considered as input chemical (for motor spirit) having flammable concentration limits to be 1.2 per cent and 7.4 per cent by volume and N-Dodecane was considered as input for running Diesel.

(viii) The first scenario was run for motor spirit vapour cloud explosion considering a single direct source and leakage rate of 237.46 cubic meters per minute for 75 min. The wind speed was taken as 2.1 meters/second from 340° true at 6.6 feet; Air Temperature as 31.4° C; Stability Class: E; and Relative Humidity as 19 per cent.

(ix) The VCE then triggered a pool fire (IOC Accident Investigation Report). The maximum puddle diameter of 200 m and the maximum amount present in the puddle as 10000 cubic metres were considered in this case. The initial puddle temperature was taken as ambient temperature since no data was available. The atmospheric conditions prevailing at that time were: Wind Speed: 1.5 meters/second from 320° true at 6.6 feet; Cloud Cover: 0 tenths; Stability Class: E; No Inversion Height, Relative Humidity: 25 per cent.

(x) The third scenario was run for Diesel Flash Fire considering a single direct source with a leakage rate of 668 cubic meters/min. The date and time of accident was taken as October 30, 2009 at 11.30 a.m. The prevalent atmospheric conditions at that time were as follows: Wind Speed: 3.1 meters/second from 320° true at 6.6 feet; Air Temperature: 31.2° C; Stability Class: D and Relative Humidity as 19 %.

(xi) The fourth scenario was for Diesel Pool Fire. The parameters were same like scenario two. At normal conditions diesel doesn't catch fire as its flash point is much higher than ambient temperature. But there were high temperatures at the vicinity which probably led to a pool fire.

(xii) In case of worst case scenario, tanks were considered fully loaded at their highest daily temperature. For that purpose the month of May recording a temperature of 46° C was chosen. The wind speed during that time period was 3.3 m/sec. Though the predominant wind direction during that time period was from NNW sometimes the wind blows from W-SW/S also. Since at the time of the accident the wind direction was from N-NW it was not considered for worst case scenario. The stability class was taken as B.

(xiii) Hexane was taken as input for MS with a single direct source with the entire amount leaking at a rate of 560 cubic meters/min (total: 33,600 cubic meters). VCE and Pool Fire Hazard Scenarios were generated. Dodecane was taken as input for Diesel with a single direct source with the entire amount leaking at 900 cubic meters/min (total: 54,000 cubic metres). VCE and Pool Fire Hazard Scenarios were generated.

(xiv) The results were finally imported in Arc GIS using ALOHA import tool to prepare a Vulnerability Zonation Map.

10.3.3 Ground Truth

A field survey was conducted to understand the nature as well as the extent of damage due to fire and vapour cloud explosion. Google Earth Imagery was used to understand the areas which were vulnerable and verified from the local people present at the time of the accident. GPS readings were also taken to identify them later in LISS-IV data.

BPCL TERMINAL   ENTRY OF IOC TERMINAL   HPCL TERMINAL

DEVASTATION AT THE IOC TERMINAL (PHOTO TAKEN FROM THE N.E. SIDE OF THE PLANT)

IMPACT ON THE GREEN BELT OF THE IOC SITE AFTER THE BLAST

Figure 10.4: Field photographs.

10.3.4 Enabling Emergency Response

Coarse resolution multi temporal satellite data (Terra MODIS) was used to find the area affected by smoke plume and medium to high resolution satellite data (LISS-III / IV) was used to identify the most vulnerable zones. ALOHA tool developed by USEPA was used to find out threat / vulnerable zones and finally integrating them in GIS environment to find out the buildings in different vulnerability zones. Results derived from the study are presented as following:

Table 10.3: Area Statistics of Plume Coverage on October 30, 2009

District	Taluk	Taluk Area (km²)	Plume Coverage Area (km²)	% Covered
Jaipur	Sanganeer	471.84	23.36	4.9
Jaipur	Chaksu	845.77	225.78	26.7
Tonk	Niwai	861.77	286.09	33.2
Tonk	Uniara	1,268.88	107.15	8.4
Sawai Madhopur	Sawai Madhopur	1,505.73	702	46.6
Sawai Madhopur	Bonali	1,074.86	139.4	13.0
Bundi	Keshorai Patan	917.05	78.25	8.5
Kota	Pipalda Kalan	1,105.65	552.38	50
Kota	Digod	1,018.31	166.07	6.31
Kota	Mangrol	1,109.11	75.54	6.8

Table 10.4: Area Statistics of Plume Coverage on October 31, 2009

District	Taluk	Taluk Area (km²)	Plume Coverage Area(km2)	% Covered
Jaipur	Sanganeer	471.84	10.12	2.14
Jaipur	Chaksu	845.77	41.2	4.87

Table 10.5: Area Statistics of Plume Coverage on November 01, 2009

District	Taluk	Taluk Area (km²)	Plume Coverage Area (km²)	% Covered
Jaipur	Sanganeer	471.84	19.57	4.14
Jaipur	Chaksu	845.77	14	1.66
Jaipur	Bassi	469.18	10.96	2.33

Table 10.6: Area Statistics of Plume Coverage on November 02, 2009

District	Taluk	Taluk Area (km²)	Plume Coverage Area (km²)	% Covered
Jaipur	Sanganeer	471.84	7.7	1.63
Jaipur	Chaksu	845.77	40	4.73

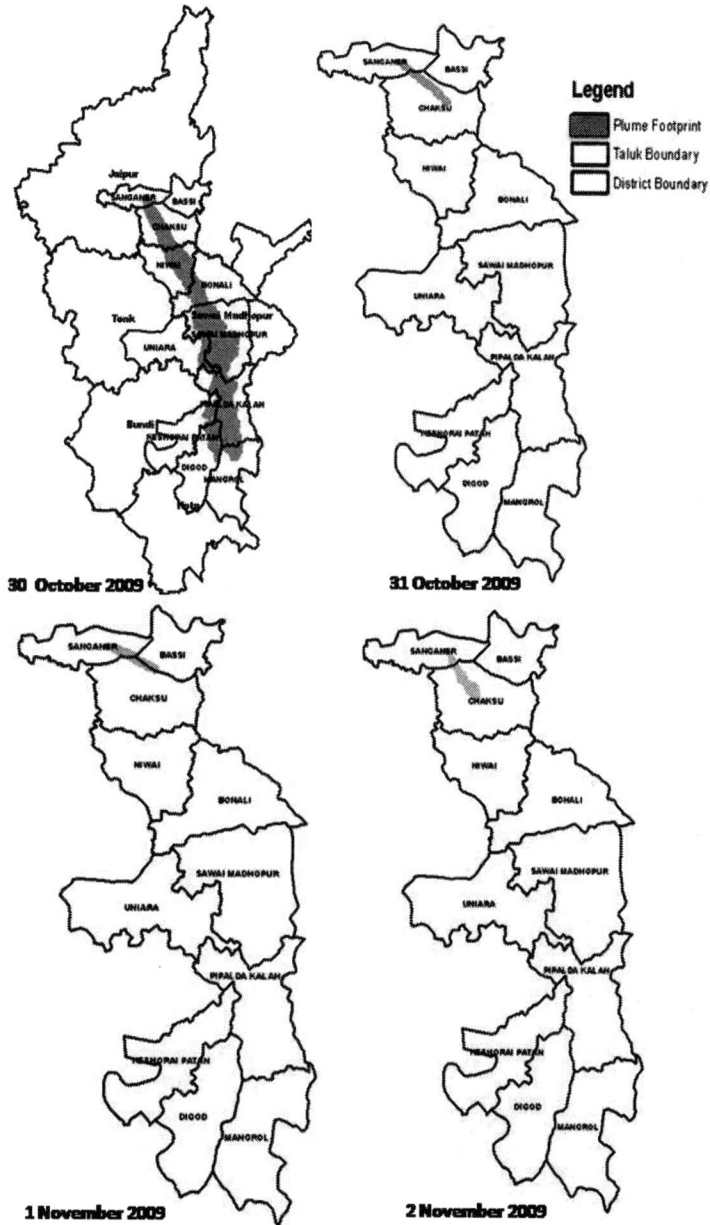

Figure 10.5: Plume footprint map on October 30; October 31, November 01, November 02, 2009

The movement of the plume is a function of wind speed, wind direction, degree of dispersion which in turn is a function of atmospheric stability conditions. The more turbulent the atmosphere the more rapid the spread of the plume in the direction transverse to the direction of propagation of the plume. Given below are the area statistics of plume coverage on different dates after the accident.

On October 30, 2009 the day immediately after the accident the predominant wind direction was towards S-SE (Environmental Impact of the fire in Indian Oil Corporation Depot Sitapura, Jaipur, February 2010). Thus the major impact of the plume was in the areas lying to the S-SE portions of the IOC terminal. Also the atmosphere was highly unstable so the plume spread to a larger distance covering the following areas:

On October 31, 2009 it shifted more towards SE having a coverage area of 51.32km^2 and was restricted to the Jaipur District only. On November 01, 2009 it drifted more towards ESE covering the Taluk of Bassi and the coverage area reduced to 44.53 km^2. On November 02, 2009 it again shifted towards SSE and the coverage area was around 47.7 km^2. On 6th November the wind direction changed towards NW and the plume drifted towards the Sanganeer side covering an area of 1.5 km^2. However it has limited impact on the main city of Jaipur which is about 16 km N from the IOC terminal. Chaksu and Sanganeer Taluks were the most vulnerable areas after the accident. IOC terminal was in the Sanganeer Taluk itself and Chaksu was on the immediate SSE part of the terminal where the plume moved during the peak fire days (Figure 10.5).

The most predominant land use traversed by the plume was agricultural land followed by vacant land and built up areas. Due to the presence of agricultural fields and vacant lands on its way the plume travelled unobstructed to large distances which were further enhanced by the unstable conditions of the atmosphere (Figure 10.6).

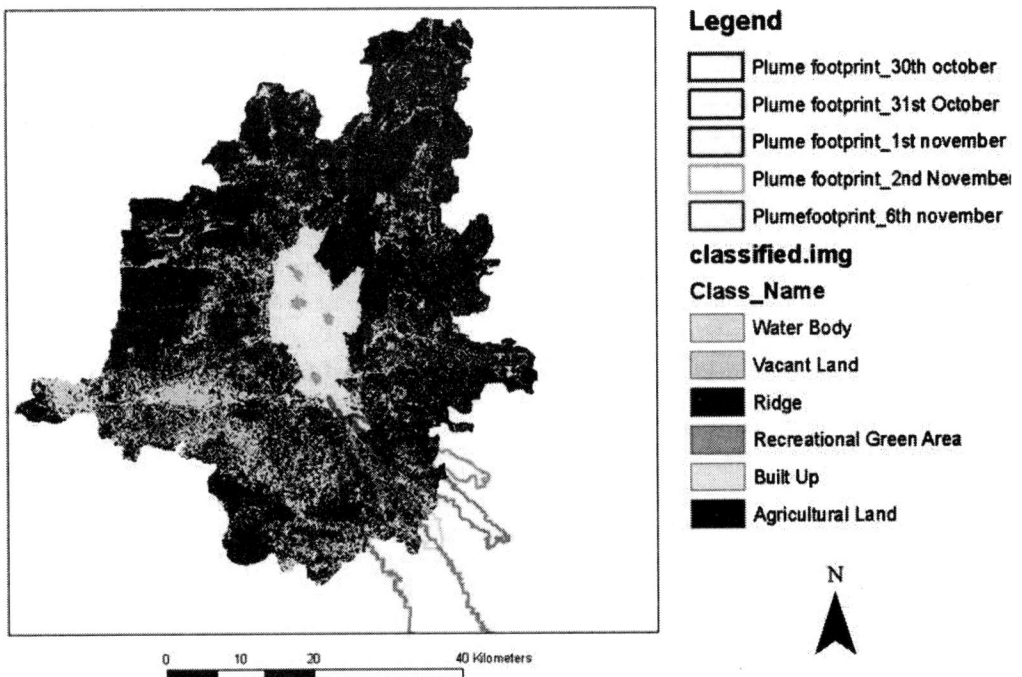

Figure 10.6: Plume footprint maps overlaid on mand use/mand cover map

Table 10.7: Consequence analysis at the time of the accident

Scenario	Name of the chemical	Threat modelled	Amount released / amount burnt	Threat zones
One	Motor Spirit	Flammable Area of Vapour Cloud	9,263,990 kg	Red :329 m (76800 ppm)=UEL
				Orange:1.6 km(10500 ppm)=LEL
				Yellow:2.1 km(6,300 ppm)= Flame Pockets
		Overpressure (Blast Force) from VCE)		Red : 1.6 km (5 psi=Structural Damage)
				Orange : 1.6 km (3.5 psi=Serious Injury)
				Yellow: 2.0 km (1.5 psi=Shatters Glass)
Two	Motor Spirit	Thermal Radiation from Pool Fire	6,502,140 kg Flame length :193m	Red: 503 m (10.0 Kw/sqm = Potentially Lethal)
				Orange: 697 m (5.0 Kw/sqm= 2nd Degree burns within 60 sec)
				Yellow:1.1 km (2.0 Kw/sqm = pain within 60 sec)
Three	Diesel	Flammable Area of Vapor Cloud	29,736,972 kg	Red :573 m (49000 ppm)=UEL
				Orange:2.2 km(6000 ppm)=LEL
				Yellow:2.9 km(3600 ppm)= Flame Pockets
Four	Diesel	Thermal Radiation from Pool Fire	7,419,404 kg Flame length:141 m	Red: 396 m (10.0 Kw/sqm = Potentially Lethal)
				Orange: 543m (5.0 Kw/sqm= 2nd Degree burns within 60 sec)
				Yellow:826 m (2.0 Kw/sqm = pain within 60 sec)

Due to the volatile nature of Motor Spirit substantial amount of vapour was formed which exploded after 75 minutes due to the formation of considerable amount of an "Explosive Mass" within "Lower Flammability Limit" (LFL) and "Upper Flammability Limit" (UFL) of motor spirit. The quantity of explosives mass basically determines the effect of the blast overpressure and the time required for forming the explosive mass depends on the rate of vaporization which, in turn depends on "Vapour Pressure of the liquid which "vaporizes" and the "liquid pool area" resulting from "loss of containment" and also to an extent on the "nature of loss of containment (IOC Accident Investigation Report). It was assumed that due to considerable time gap the vapour cloud formed was within its flammable limits which after coming into contact with a source of ignition downstream gave rise to a flash fire which flashed back to the source of the leak in the MS Tank outlet giving rise to a secondary pool fire as well as Vapour Cloud Explosion. Also the flame-front of the VCE triggered the pool.

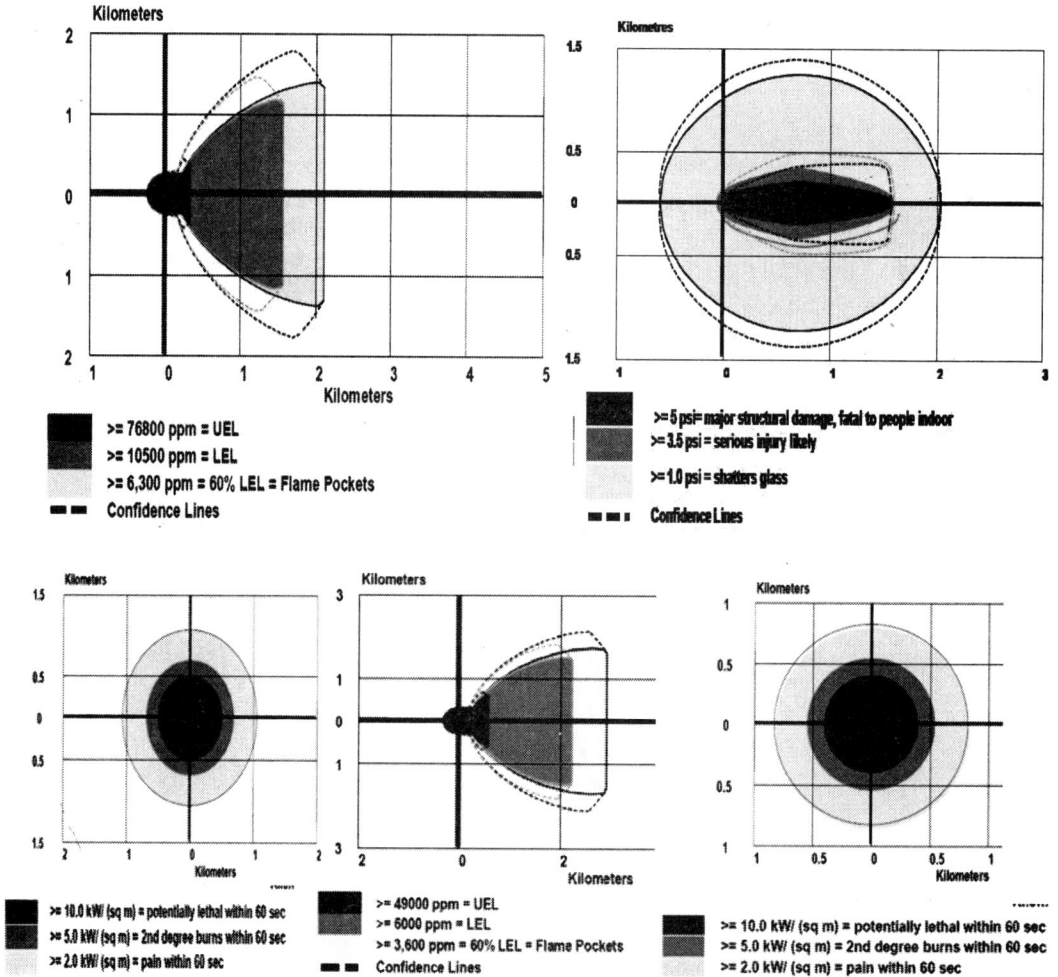

Figure 10.7: Consequence Analyses in ALOHA

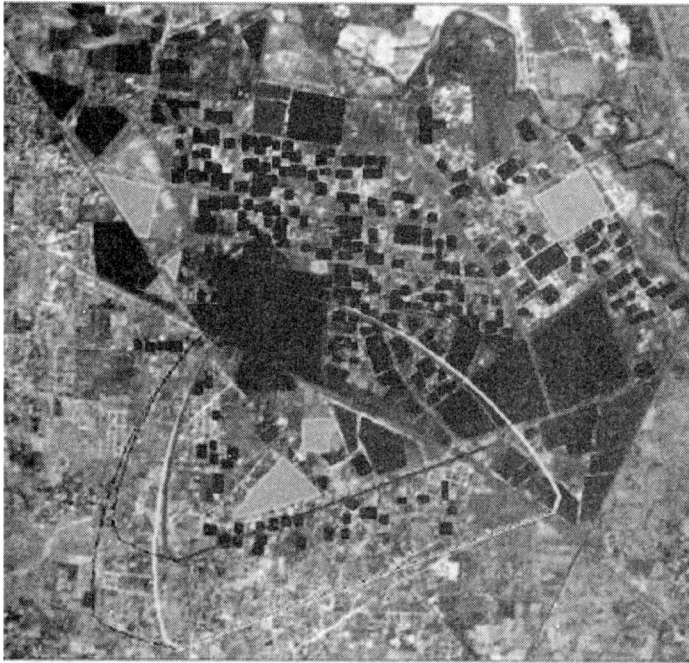

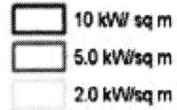

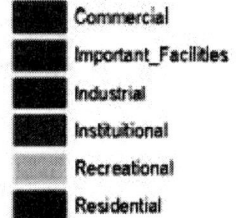

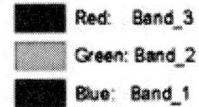

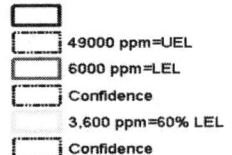

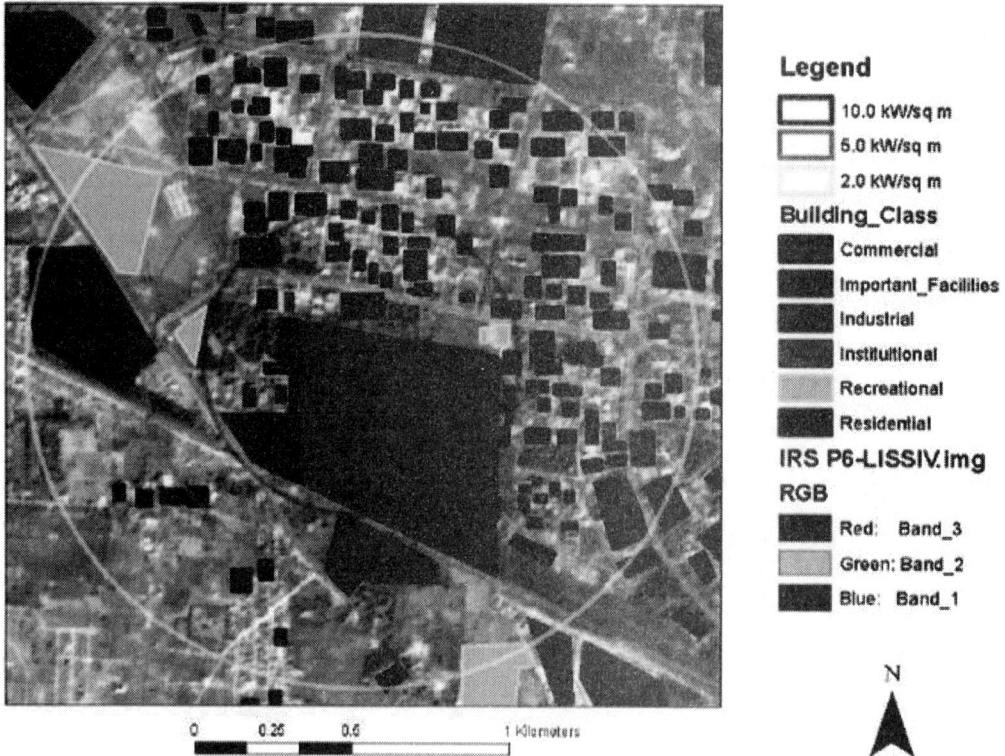

Figure 10.8: Threat zones of VCE & Pool Fires overlaid on building footprint maps

Since fuel from the spill had been accumulating for more than an hour, the radiation from the Pool Fire also caused some amount of damage to the tanks in the vicinity as well as the explosion gave rise to a consequential dyke fire and tank roof failure, on account of pressurization of the tank by the dyke fire .In this way all the 11 tanks caught fire in series leading to formation of thick black smoke which traversed for long distances due to the unstable atmospheric condition prevalent at that time.

Since the terminal was situated within the Sitapura Industrial Area the vulnerability was limited to the neighbouring industrial and commercial facilities including some recreational areas like Chokhi Dhani, Ayur International, Amber Villa etc. NH 12 (Tonk Road) and a section of Western Railway (Sawai Madhopur Laharu Section) lie on the southern part of the terminal. Industrial Facilities like Genus, J V S Flora, Flora-o- Foods, Laxmi Bhog Aata Factory lying within 500 to 700 meters were severely affected. Since the explosion took place in the evening around 7.30 p.m. there was very little casuality. But buildings, roofs were heavily damaged, glass window panes were broken. These were kept in highly vulnerable areas. Facilities beyond 700 m up to 2km suffered glass window pane breakages. So they were placed in moderately vulnerable areas. Out of 291 buildings 48 buildings were highly vulnerable, 205 were moderately vulnerable and the remaining 38 were under low vulnerable category.

Legend

Vulnerability Zonation Map

Nature of Vulnerability

- High
- Moderate
- Low

Road

- Byepass Road
- National Highway
- Ring Road
- Secondary Road
- State Highway
- Railway

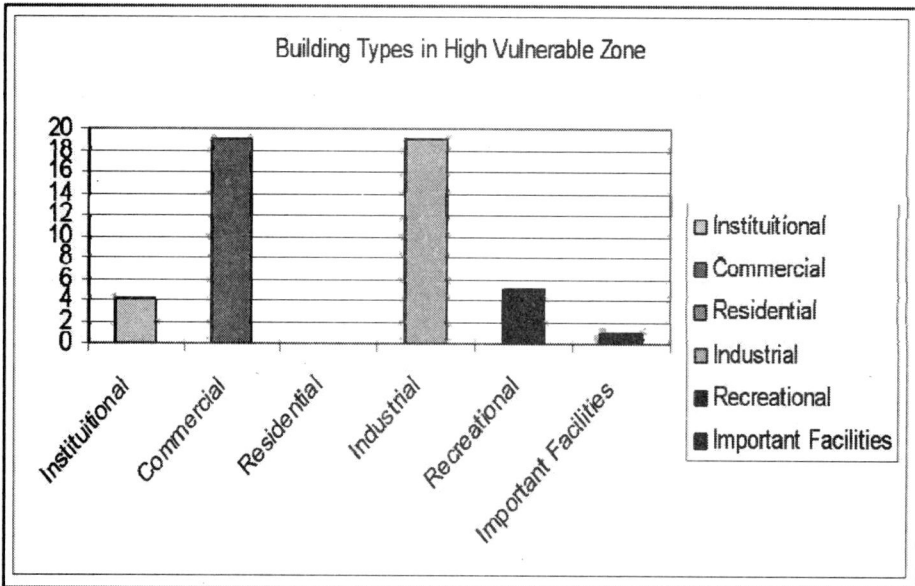

N

0   0.5   1   2 Kilometers

Building Types in High Vulnerable Zone

- Instituitional
- Commercial
- Residential
- Industrial
- Recreational
- Important Facilities

Building Types in Moderately Vulnerable Zones

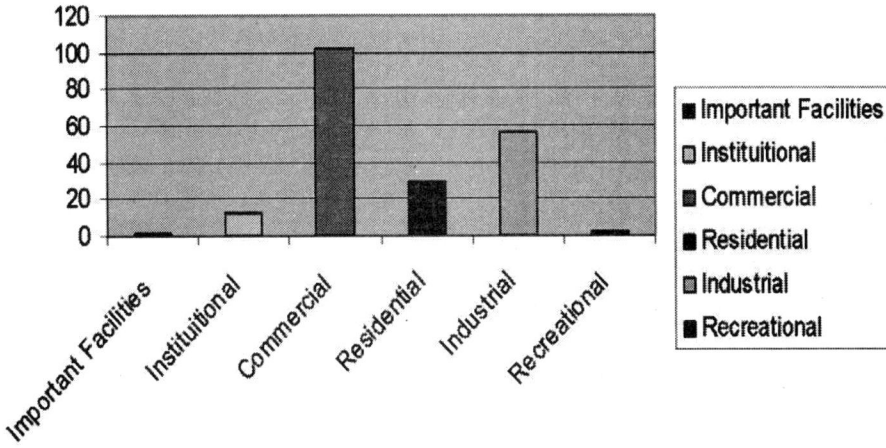

Building Types in Low Vulnerable Zones

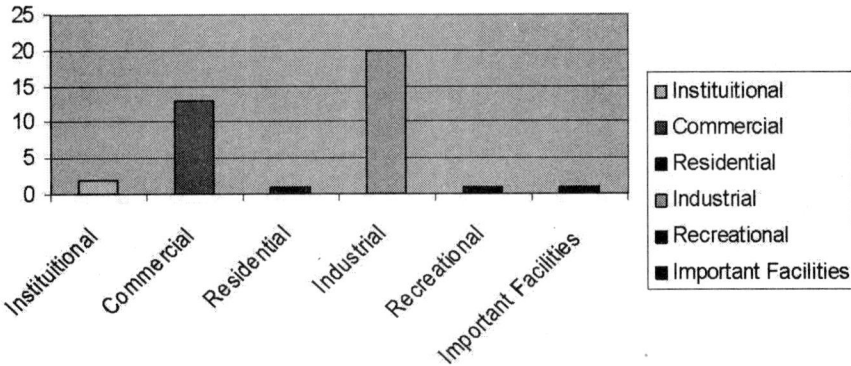

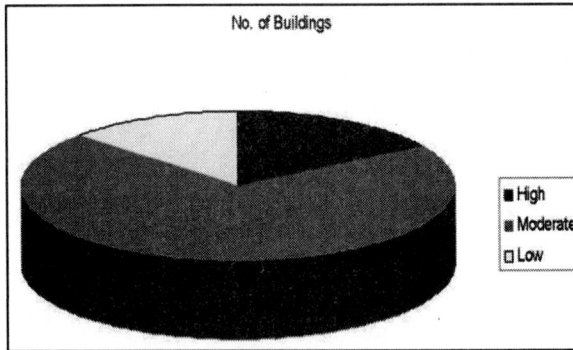

Figure 10.9: Vulnerability Zonation Map and graphs depicting the building types in different vulnerable zones at the time of the accident

Table 10.8: Consequence Analysis for worst case scenario

Scenario	Name of the chemical	Threat modelled	Amount released/ amount burnt	Threat zones
One	Motor Spirit	Flammable Area of Vapour Cloud	21,392,330 kg	Red :419 m (76800 ppm)=UEL Orange:1.8 km(10500 ppm)=LEL Yellow:2.4 km(6,300 ppm)= Flame Pockets
		Overpressure (Blast Force) from VCE)		Red : 1.8 km (5 psi=Structural Damage) Orange : 1.8 km (3.5 psi=Serious Injury) Yellow: 2.3 km (1.5 psi=Shatters Glass)
Two		Thermal Radiation from Pool Fire		Red: 536 m (10.0 Kw/sqm = Potentially Lethal) Orange: 746 m (5.0 Kw/sqm= 2nd Degree burns within 60 sec) Yellow:1.1 km (2.0 Kw/sqm = pain within 60 sec)
Three	Diesel	Flammable Area of Vapour Cloud	39,498,645 kg	Red : 640 m (49000 ppm)=UEL Orange :2.5 km(6000 ppm)=LEL Yellow: 3.3 km(3600 ppm)= Flame Pockets
Four		Thermal Radiation from Pool Fire		Red: 412 m (10.0 Kw/sqm = Potentially Lethal) Orange: 567 m (5.0 Kw/sqm= 2nd Degree burns within 60 sec) Yellow: 864 m (2.0 Kw/sqm = pain within 60 sec)

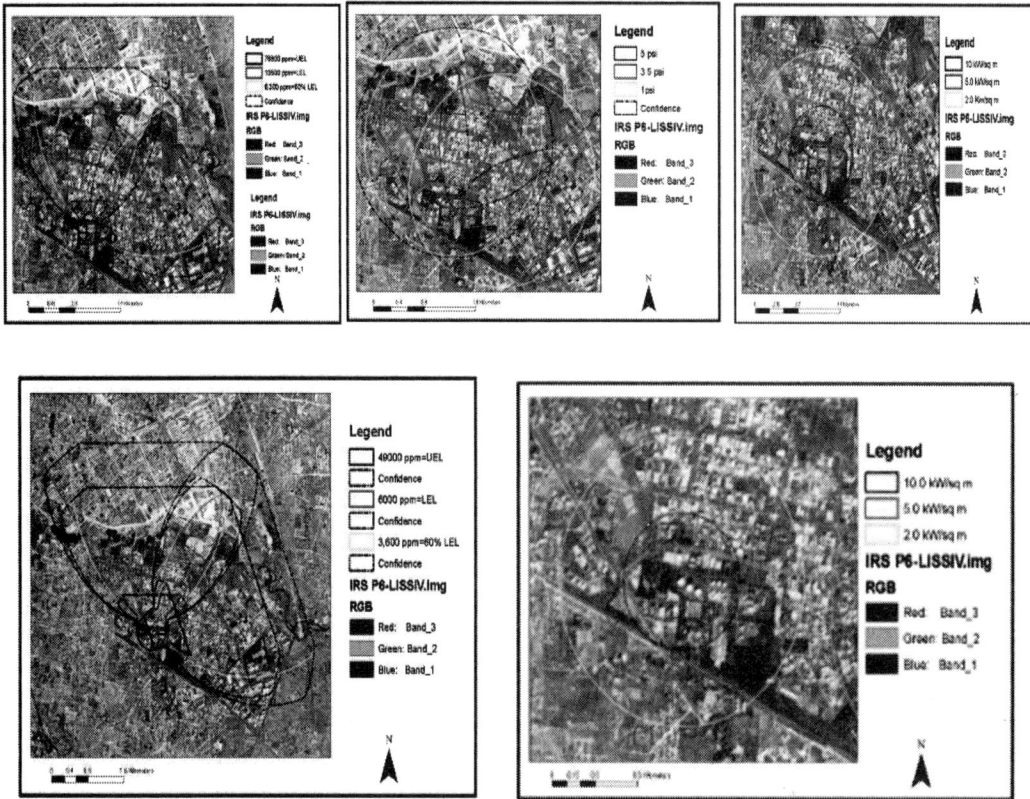

Figure 10.10: Threat zones of VCE & Pool Fires overlaid on building footprint maps in case of a worst case scenario

It was observed that if the accident took place with the conditions mentioned above then the vulnerability of the urban areas will be more since the predominant wind direction in that case be dense residential areas of Pratap Nagar. However the maximum vulnerability in this case is also limited within the Sitapura Industrial Area itself. Since the main city of Jaipur was around 16 km N , it suffered little impact apart from few puffs of smoke after 6th November, a week after the accident. Thus the sitting of hazardous installations like this far from the main urban areas reduces vulnerability as well as risks to a great extent.

10.4 CONCLUSION

The study was an attempt to utilize coarse resolution multi temporal satellite data like Terra MODIS to visualize post hazard smoke plume for larger geographical areas in order to prepare plume footprint maps having larger coverage area, to identify the most vulnerable zones underlying the plume in LISS III/IV and to generate different scenarios of the accident occurred at the IOC storage terminal that day as well as to find the threat in case of a worst condition in ALOHA, a software developed by USEPA. The smoke plume traversed maximum distance on the day immediately after the accident due to atmospheric conditions favourable for dispersion and ultimately got diluted in successive days. Accordingly the plume was visible in coarse resolution data initially and gradually reduced to two-three pixels. However it can be visualized in medium or high resolution satellite imagery if available for that time period. In this study satellite imagery dated of IRS LISSIII was available for November 6, a week after the fire. It was found that the plume also changed its direction from SSE to NW and spread towards the city of Jaipur. Since it was diluted by that time it had little impact there.

ALOHA predicted areas within 500 to 700 meters from the IOC storage terminals as the most vulnerable zones immediately after the incident followed by facilities beyond 700 m to 2km where glass window pane breakages were observed. This was confirmed by the local peoples during field visit. The worst case scenario was run for full capacity and highest daily maximum temperature during peak summer month of May, and was found that if the accident occurred in that condition then the dense areas of Pratap Nagar would have been highly affected. These types of studies can provide a good insight for the decision makers to identify the possible threat zones from hazards like fire, explosions before siting any storage installations handling flammable products like gasoline, diesel, and kerosene. Remote sensing plays an important role in providing real time as well as past information of any place on the earth surface and GIS provides an interface to analyse multiple data in different formats simultaneously and finally integrating them to give a single output map which enables the user community to visualize them spatially.

References

Alencar, J.R.B., Barbosa, R.A.P., and de Souza, M.B. (Jr.) (2005). Evaluation of accidents with domino effect in LPG storage areas. Engenharia Térmica (Thermal Engineering), Vol. 4(1).

Dandrieux, A., Dusserre, G., and Thomas, O. (2003). The DVS model: a new concept for heavy gas dispersion by water curtain. Environmental Modeling & Software 18: 253-259.

Desai, D. (2008). Industrial Risk Assessment for Planning and Emergency Response: A case of Ahmedabad. M.Sc thesis. (ITC), Enschede, The Netherlands.

Gupta, A.K., Nair, S.S., and Shard, S. (2009) (Eds.).Chemical Disaster Management: Proceeding Volume of the National Workshop, 30 September – 1 October 2008, New Delhi. Ministry of Environment and Forest and National Institute of Disaster Management, Ministry of Home Affairs, Govt. of India.

Harbawi, M.El. et al. (2008).Rapid analysis of risk assessment using developed simulation of chemical industrial accidents software package. International Journal of Environmenal Science Technology, 5(1), 53-64.

Havens, J.A., and Spicer, T.O. (1985). Development of an Atmospheric Dispersion Model for Heavier than air gas mixtures. Report CG-D-22-85 to U.S.1

Oil Indutry Safety Directorate (IOCL Fire Accident Investigation Report. Accessed from http://www.scribd.com/doc/47021609/Report-of-Committee-on-Jaipur-Incident.

Joint Committee Report (February, 2010). Environmental Impacts of the Fire in Indian Oil Corporation Depot Sitapura, Jaipur, Rajasthan.

Kaur, K. (2008). Vulnerability and Risk Zonation for Hazardous Installation (Petroleum and LPG Stations) in Dehradun City. Project Report, Joint Programme of IIRS, Dehradun & ITC, Enschede The Netherlands.

Khan, F., and Abbasi, S. (1999). HAZDIG: a new software package for assessing the risks of accidental release of toxic chemicals. Journal of Loss Prevention Proc., 12. 167-181.

Khan, F.I., and Abbasi, S. A. (2001). Estimation of probabilities and likely consequences of a chain of accidents (domino effect) in Manali Industrial Complex. Journal of Cleaner Production 9(6):493-508.

Nordin, J.S. (2003). Technical Discussion Flashpoints, Flammables, Combustibles, LEL, UEL, and Fires.

Sengupta, A. (2007). (Unpublished). Industrial Hazard, Vulnerability and Risk Assessment for Land use Planning: A Case Study of Haldia Town, West Bengal, India. M.Sc thesis, Joint Programme of IIRS, Dehradun & ITC, Enschede The Netherlands.

U.S. Environmental Protection Agency (1987). Technical Guidance for Hazard Analysis Emergency Planning for Extremely Hazardous (EHS) Substances, U.S. Environmental Protection Agency (USEPA), Federal Emergency Management Agency (FEMA), US Department of Transportation.

Disasters due to Unplanned Urbanization:
A Case Study of Greater Bangalore

T. V. Ramachandra

11.1 INTRODUCTION

Urbanization is a form of metropolitan growth that is a response to often bewildering sets of economic, social, and political forces and to the physical geography of an area. The 20th century is witnessing "the rapid urbanisation of the world's population", as the global proportion of urban population rose dramatically from 13 per cent (220 million) in 1900, to 29 per cent (732 million) in 1950, to 49 per cent (3.2 billion) in 2005 and is projected to rise to 60 per cent (4.9 billion) by 2030 (World Urbanization Prospects, 2005). Urban ecosystems are the consequence of the intrinsic nature of humans as social beings to live together (Sudhira et al., 2003; Ramachandra and Uttam Kumar, 2008). Urbanization and urban sprawl have posed serious challenges to the decision makers in the city planning and management process involving plethora of issues like infrastructure development, traffic congestion, and basic amenities (electricity, water, and sanitation), etc. (Kulkarni and Ramachandra, 2006). Apart from this, major implications of urbanisation are:

(i)   Loss of wetlands: Urbanization has tolling influences on the natural resources such as decline in number of wetlands and / or depleting groundwater table.

(ii) Floods: Common consequences of urban development are increased peak discharge and frequency of floods, as land is converted from fields or woodlands to roads and parking lots, and loses its ability to absorb rainfall. Conversion of water bodies to residential layouts has compounded the problem by removing the interconnectivities in an undulating terrain. Encroachment of natural drains, alteration of topography involving the construction of high rise buildings, removal of vegetative cover, and post 2000, reclamation of wetlands are the prime reasons for frequent flooding even during normal rainfall.

Disaster Management and Risk Reduction: Role of Environmental Knowledge; Editors Anil K. Gupta, Sreeja S. Nair, Florian Bemmerlein-Lux and Sandhya Chatterji; Copyright © 2013, Narosa Publishing House, New Delhi

(iii)   Decline in groundwater table: Studies reveal the removal of water bodies has led to a decline in the water table. After the reclamation of lake and its catchment area for commercial activities, the water table has declined to 300 m from 28 m over a period of 20 years. The groundwater table in intensely urbanized area such as Whitefield has now dropped from 400 to 500m.

(iv)   Loss of tree cover: Drastic reduction in tree cover is observed due to the removal of lane tress and conversion of plantations to residential layouts, etc.

(v)   Heat island: Surface and atmospheric temperatures increase by anthropogenic heat discharge due to energy consumption, increased land surface coverage by artificial construction material having high heat capacities and conductivities, and the associated decreases in vegetation and water surfaces, which reduce surface temperature through evapo-transpiration.

(vi)   Increased carbon footprint: Due to the adoption of inappropriate building architecture, the consumption of electricity has increased in certain corporation wards drastically. The building design conducive to tropical climate would have reduced the dependence on electricity. Higher energy consumption, enhanced pollution levels due to the increase of private vehicles, and traffic bottlenecks have contributed to increasing carbon emissions significantly. Apart from these, mismanagement of solid and liquid wastes has aggravated the situation.

Unplanned urbanisation has drastically altered the drainage characteristics of natural catchments or drainage areas, by increasing the volume and rate of surface runoff. Drainage systems are unable to cope with the increased volume of water and are often blocked due to indiscriminate disposal of solid wastes. Encroachment of wetlands, floodplains, etc. obstructs flood-ways causing loss of natural flood storage. Damages from urban flooding could be categorized as: direct damage – typically material damage caused by water or flowing water, and indirect damage – e.g. traffic disruptions, administrative and labour costs, production losses, spreading of diseases, etc.

Studies on the phenomenon of Urban Heat Island (UHI) using satellite derived land surface temperature (LST) measurements have been conducted using various satellite data products acquired in thermal region of the electromagnetic spectrum. Currently available satellite thermal infrared sensors provide different spatial resolution and temporal coverage data that can be used to estimate LST. The Geostationary Operational Environmental Satellite (GOES) has a 4 km resolution in the thermal infrared, while the NOAA-Advanced Very High Resolution Radiometer (AVHRR) and the Terra and Aqua-MODIS have 1-km spatial resolutions. Significantly high resolution data come from the Terra-Advanced Space borne Thermal Emission and Reflection Radiometer (ASTER) which has a 90-m pixel resolution, the Landsat-5 Thematic Mapper (TM) which has a 120-m resolution, and Landsat-7 Enhanced Thematic Mapper (ETM) which has a 60-m resolution. However, these instruments have a repeat cycle of 16 days (Li et. al., 2004; Ramachandra and Uttam Kumar, 2009).

11.2 STUDY AREA

Greater Bangalore (77°37'19.54'' E and 12°59'09.76'' N) is the principal administrative, cultural, commercial, industrial, and knowledge capital of the state of Karnataka with an area of 741 sq.

km. Bangalore city administrative jurisdiction was widened in 2006 by merging the existing area of Bangalore city spatial limits with 8 neighbouring Urban Local Bodies (ULBs) and 111 Villages of Bangalore Urban District (Ramachandra and Uttam Kumar, 2008; Sudhira et al., 2007). Thus, Bangalore has grown spatially more than ten times since 1949 (69 square kilometres) and is a part of both the Bangalore urban and rural districts (Figure 11.1). Now, Bangalore is the fifth largest metropolis in India currently with a population of about 7 million (Figure 11.2). The mean annual total rainfall is about 880 mm with about 60 rainy days a year over the last ten years. The summer temperature ranges from 18° C – 38° C, while the winter temperature ranges from 12° C – 25° C. Thus, Bangalore enjoys a salubrious climate all round the year. Bangalore is located at an altitude of 920 meters above mean sea level, delineating four watersheds, viz. Hebbal, Koramangala, Challaghatta and Vrishabhavathi watersheds. The undulating terrain in the region has facilitated creation of a large number of tanks providing for the traditional uses of irrigation, drinking, fishing and washing. This led to Bangalore having hundreds of such water bodies through the centuries. Even in early second half of 20th century, in 1961, the number of lakes and tanks in the city stood at 262 (and spatial extent of Bangalore was 112 sq km). However, number of lakes and tanks in 1985 was 81 (and spatial extent of Bangalore was 161 sq km).

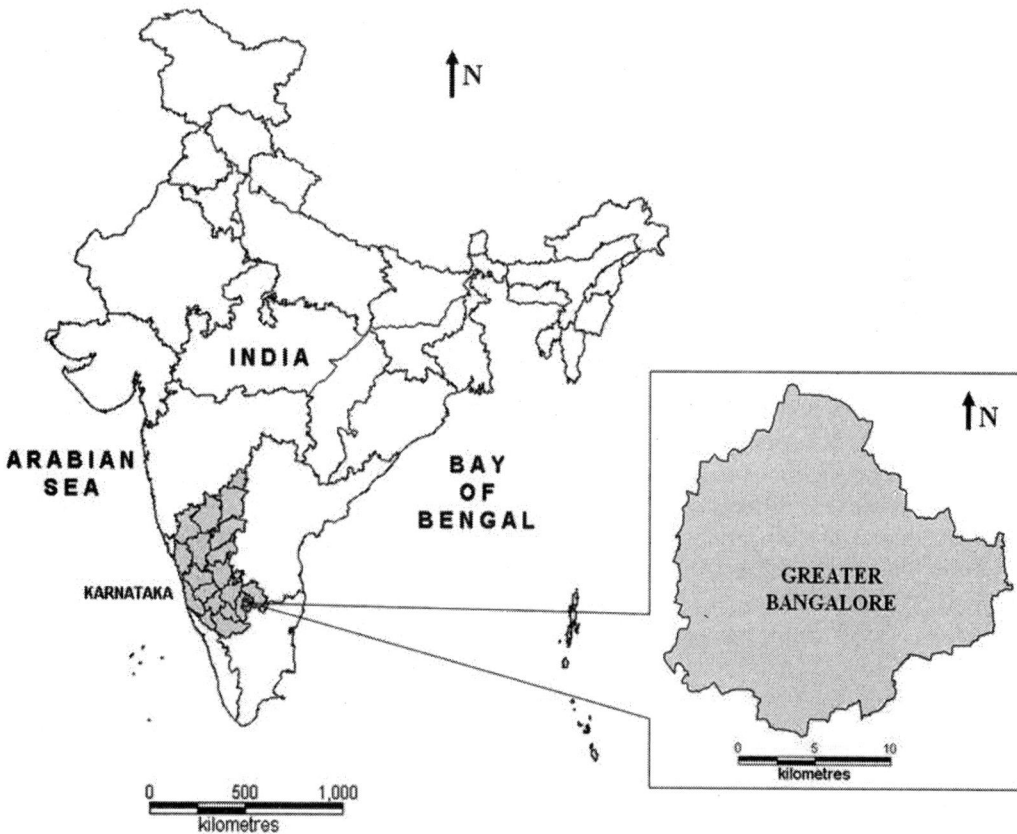

Figure 11.1: Study area – Greater Bangalore

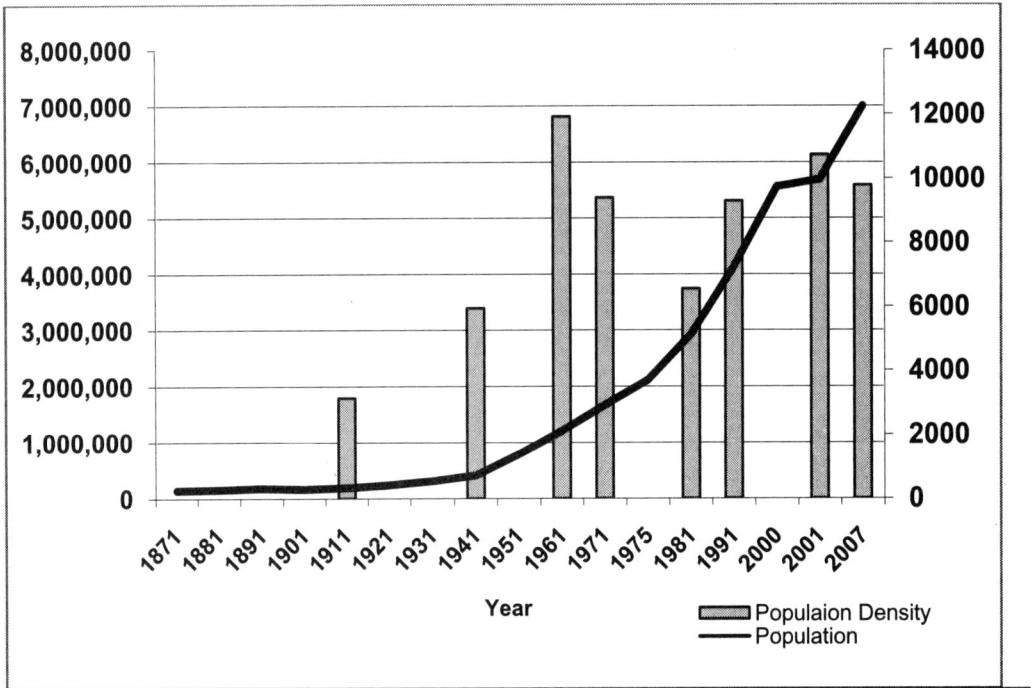

Figure 11.2: Population growth and density over the period 1871-2007

11.3 MATERIALS AND METHODS

Survey of India (SOI) toposheets of 1:50000 and 1:250000 scales were used to generate base map. Field data was collected using handheld GPS. Remote sensing data used for the study are: Landsat MSS (1973), Landsat TM (1992), Landsat ETM+ (2000 and 2009) [Landsat data downloaded from http://glcf.umiacs.umd.edu/data/], IRS (Indian Remote Sensing) LISS (Linear Imaging Self Scanner)-III of (1999 and 2006), MODIS (Moderate Resolution Imaging Spectro-radiometer) Surface Reflectance 7 bands product (http://edcdaac.usgs.gov/main.asp) of 2002, MODIS Land Surface Temperature/Emissivity 8-Day L3 Global and Daily L3 Global (V004 product) (http://lpdaac.usgs.gov/modis/dataproducts.asp#mod11).

Google Earth data (http://earth.google.com) served in pre and post classification process and validation of the results. The methods adopted in the analysis involved:

(i)   Georeferencing of acquired remote sensing data to latitude-longitude coordinate system with Everest 56 datum: Landsat bands, IRS LISS-III MSS bands, MODIS bands 1 and 2 (spatial resolution 250 m) and bands 3 to 7 (spatial resolution 500 m) were geo-corrected with the known ground control points (GCP's) and projected to Polyconic with Everest 1956 as the datum, followed by masking and cropping of the study area.

    (a) Band 1, 2, 3 and 4 of Landsat 1973 data to 79 m.

    (b) Band 1, 2, 3 and 4 of Landsat TM of 1992 to 30 m.

    (c) Band 1, 2, 3, 4, 5 and 7 of Landsat ETM+ to 30 m.

    (d) MODIS bands 1 to 7 to 250 m.

(e) IRS LISS-III band 1, 2 and 3 to 23.5 m.

(f) Thermal band of TM (resampled to 120m), ETM+ (to 60m) and MODIS (to 1 km) and Panchromatic bands of ETM+ (resampled to 15 m).

(ii) Supervised Classification using Bayesian Classifier: In supervised classification, the pixel categorisation process is done by specifying the numerical descriptors of the various LC types present in a scene. It involves (i) training, (ii) classification and (iii) output.

(iii) Accuracy assessment: Accuracy assessments were done with field knowledge, visual interpretation and also referring Google Earth (http://earth.google.com).

(iv) Computation of Normalised Difference Vegetation Index (NDVI):It separates green vegetation from its background soil brightness and retains the ability to minimize topographic effects while producing a measurement scale ranging from −1 to +1 with NDVI-values < 0 representing no vegetation.

Derivation of Land Surface Temperature (LST)

LST from Landsat TM: The TIR band 6 of Landsat-5 TM was used to calculate the surface temperature of the area. The digital number (DN) was first converted into radiance L_{TM} using

$$L_{TM} = 0.124 + 0.00563 * DN \qquad ....(Equation\ 11.1)$$

The radiance was converted to equivalent blackbody temperature $T_{TMSurface}$ at the satellite using

$$T_{TMSurface} = K_2/(K_1 - lnL_{TM}) - 273 \qquad .....(Equation\ 11.2)$$

The coefficients K_1 and K_2 depend on the range of blackbody temperatures. In the blackbody temperature range 260-300K the default values (Singh, S. M., 1988) for Landsat TM are K_1 = 4.127 and K_2 = 1274.7. Brightness temperature is the temperature that a blackbody would obtain in order to produce the same radiance at the same wavelength (λ = 11.5 μm). Therefore, additional correction for spectral emissivity (ε) is required to account for the non-uniform emissivity of the land surface. Spectral emissivity for all objects are very close to 1, yet for more accurate temperature derivation emissivity of each LC class is considered separately. Emissivity correction is carried out using surface emissivity for the specified LC (Table 11.1) derived from the methodology described in Snyder et al., (1998) and Stathopoulou et al., (2006).

Table 11.1: Surface emissivity values by LC type

LC type	Emissivity
Densely urban	0.946
Mixed urban (Medium Built)	0.964
Vegetation	0.985
Water body	0.990
Others	0.950

The procedure involves combining surface emissivity maps obtained from the Normalized Difference Vegetation Index Thresholds Method (NDVI[THM]) (Sobrino and Raissouni, 2000) with LC information. The emissivity corrected land surface temperature (Ts) were finally computed as follows (Artis and Carnhan, 1982)

$$T_s = \frac{T_B}{1+(\lambda \times T_B / \rho)\ln\varepsilon} \qquad\qquad .....\text{(Equation 11.3)}$$

where, λ is the wavelength of emitted radiance for which the peak response and the average of the limiting wavelengths (λ = 11.5 μm) were used, ρ = h x c/σ (1.438 × 10^{-2}mK), σ = Stefan Bolzmann's constant (5.67 × 10^{-8} $Wm^{-2}K^{-4}$ = 1.38 × 10^{-23} J/K), h = Planck's constant (6.626 × 10^{-34}Jsec), c = velocity of light (2.998 × 10^{8} m/sec), and ε is spectral emissivity.

LST from Landsat ETM+: The TIR image (band 6) was converted to a surface temperature map according to the following procedure (Weng et al., 2004). The DN of Landsat ETM+ was first converted into spectral radiance L_{ETM} using equation 11.4, and then converted to at-satellite brightness temperature (i.e., black body temperature, $T_{ETMSurface}$), under the assumption of uniform emissivity (ε ≈ 1) using equation 11.5 (Landsat Project Science Office, 2002):

L_{ETM} = 0.0370588 × DN   .....   (Equation 11.4)

$T_{ETMSurface}$ = K2/ln (K_1/ L_{ETM} + 1) .....   (Equation 11.5)

where, $T_{ETMSurface}$is the effective at-satellite temperature in Kelvin, L_{ETM} is spectral radiance in watts/(meters squared ×ster×μm); and K_2 and K_2 are pre-launch calibration constants. For Landsat-7 ETM+, K_2 = 1282.71 K and K_1 = 666.09 $mWcm^{-2}sr-1μm^{-1}$ were used (http://ltpwww.gsfc.nasa.gov/IAS/handbook/handbook_htmls/chapter11/chapter11. html). The emissivity corrected land surface temperatures Ts were finally computed by equation 11.3.

11.4 RESULTS AND DISCUSSION

The supervised classified images of 1973, 1992, 1999, 2000, 2002, 2006 and 2009 with an overall accuracy of 72 per cent, 75 per cent, 71 per cent, 77 per cent, 60 per cent, 73 per cent and 86 per cent were obtained using the open source programs (i.gensig, i.class and i.maxlik) of Geographic Resources Analysis Support System (http://wgbis.ces.iisc.ernet.in/ grass) as displayed in Figure 11.3. The class statistics is given in Table 11.2. The implementation of the classifier on Landsat, IRS and MODIS image helped in the digital data exploratory analysis as were also verified from field visits in July, 2007 and Google Earth image. From the classified raster maps, urban class was extracted and converted to vector representation for computation of precise area in hectares.

There has been a 632 per cent increase in built up area from 1973 to 2009 leading to a sharp decline of 79 per cent area in water bodies in Greater Bangalore mostly attributing to intense urbanisation process. Figure 11.4 shows Greater Bangalore with 265 water bodies (in 1972). The rapid development of urban sprawl has many potentially detrimental effects including the loss of valuable agricultural and eco-sensitive (e.g. wetlands, forests) lands, enhanced energy consumption and greenhouse gas emissions from increasing private vehicle use (Ramachandra and Shwetmala, 2009). Vegetation has decreased by 32 per cent from 1973 to 1992, by 38 per cent from 1992 to 2002 and by 63 per cent from 2002 to 2009. Disappearance or sharp decline in the number of water bodies in Bangalore is mainly due to intense urbanisation and urban sprawl. Many lakes (54%) were encroached for unauthorised illegal buildings. Field survey (during July-August 2007) shows that nearly 66 per cent of lakes are sewage fed, 14 per cent surrounded by slums and 72 per cent showed loss of catchment area. Also, lake catchments were used as dumping grounds for either municipal solid waste or building debris. The surrounding of these lakes have illegal constructions of buildings and most of the time, slum dwellers occupy the adjoining areas. At many sites, water is used for washing and household activities and even fishing was observed at one site. Multi-storied buildings have come up on some lake beds and totally changed the natural catchment flow, leading to a sharp decline and deteriorating in the quality of water bodies. Some of the lakes have been restored by the city corporation and the concerned authorities in recent times.

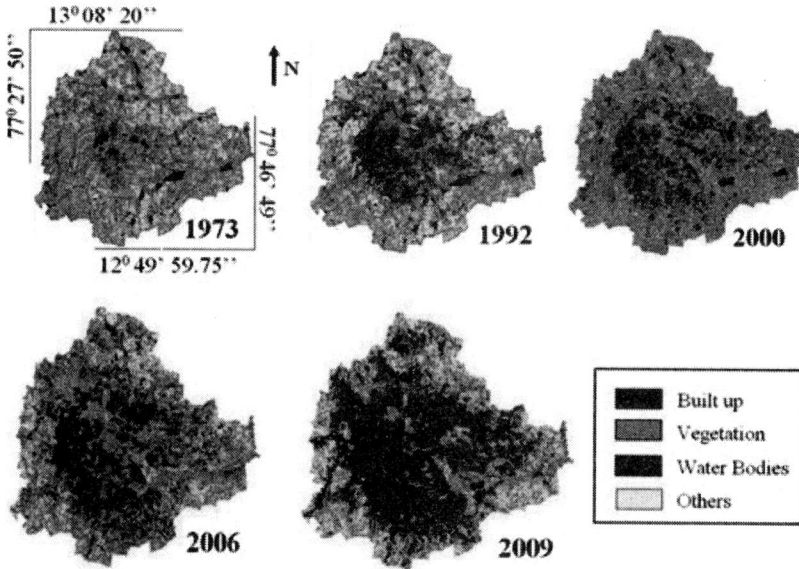

Figure 11.3: Greater Bangalore in 1973, 1992, 1999, 2000 and 2009

Table 11.2: Greater Bangalore LC statistics

Class → Year ↓		Built-up	Vegetation	Water Bodies	Others
1973	Ha	5448	46639	2324	13903
	%	7.97	68.27	3.40	20.35
1992	Ha	18650	31579	1790	16303
	%	27.30	46.22	2.60	23.86
1999	Ha	23532	31421	1574	11794
	%	34.44	45.99	2.30	17.26
2000	Ha	24163	31272	1542	11346
	%	35.37	45.77	2.26	16.61
2002	Ha	26992	28959	1218	11153
	%	39.51	42.39	1.80	16.32
2006	Ha	29535	19696	1073	18017
	%	43.23	28.83	1.57	26.37
2009	Ha	39910	11153	489	16785
	%	58.40	16.32	0.72	24.56

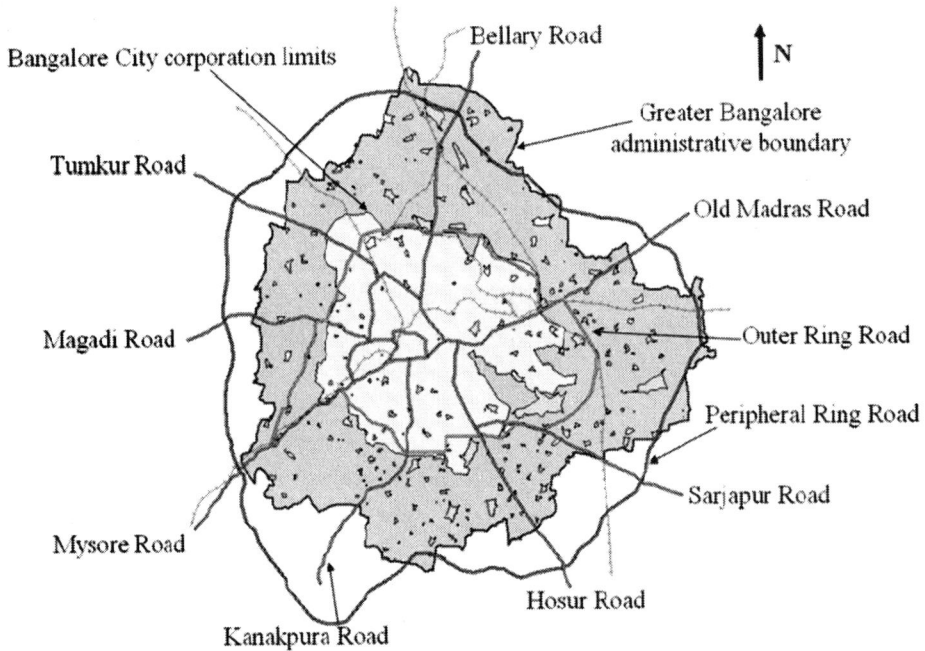

Figure 11.4: Greater Bangalore with 265 water bodies

LST were computed from Landsat TM and ETM thermal bands. The minimum and maximum temperature from Landsat TM data of 1992 was 12 and 21 with a mean of 16.5±2.5 while for ETM+ data was 13.49 and 26.32 with a mean of 21.75±2.3. MODIS Land Surface Temperature/Emissivity (LST/E) data with 1 km spatial resolution with a data type of 16-bit unsigned integer were multiplied by a scale factor of 0.02 (http://lpdaac.usgs.gov/modis/dataproducts.asp#mod11). The corresponding temperatures for all data were converted to degree Celsius. Figure 11.5 shows the LST map and NDVI of Greater Bangalore in 1992, 2000 and 2007. The minimum (min) and maximum (max) temperatures were computed as 20.23, 28.29 and 23.79, 34.29 with a mean of 23.71±1.26, 28.86± 1.60 for 2000 and 2007 respectively. Data was calibrated with *in-situ* measurements. NDVI was computed to study its relationship with LST. The Landsat TM NDVI had a mean of 0.04±0.4543, ETM+ data had a mean of 0.0252±0.5369 and MODIS had a mean of -0.0917±0.5131.

The correlation between NDVI and temperature of 1992 TM data was 0.88, 0.72 for MODIS 2000 and 0.65 for MODIS 2007 data respectively, suggesting that the extent of LC with vegetation plays a significant role in the regional LST. Respective NDVI and LST for different land uses is given in Table 11.3 and further analysis was carried out to understand the role of respective land uses in the regional LST's.

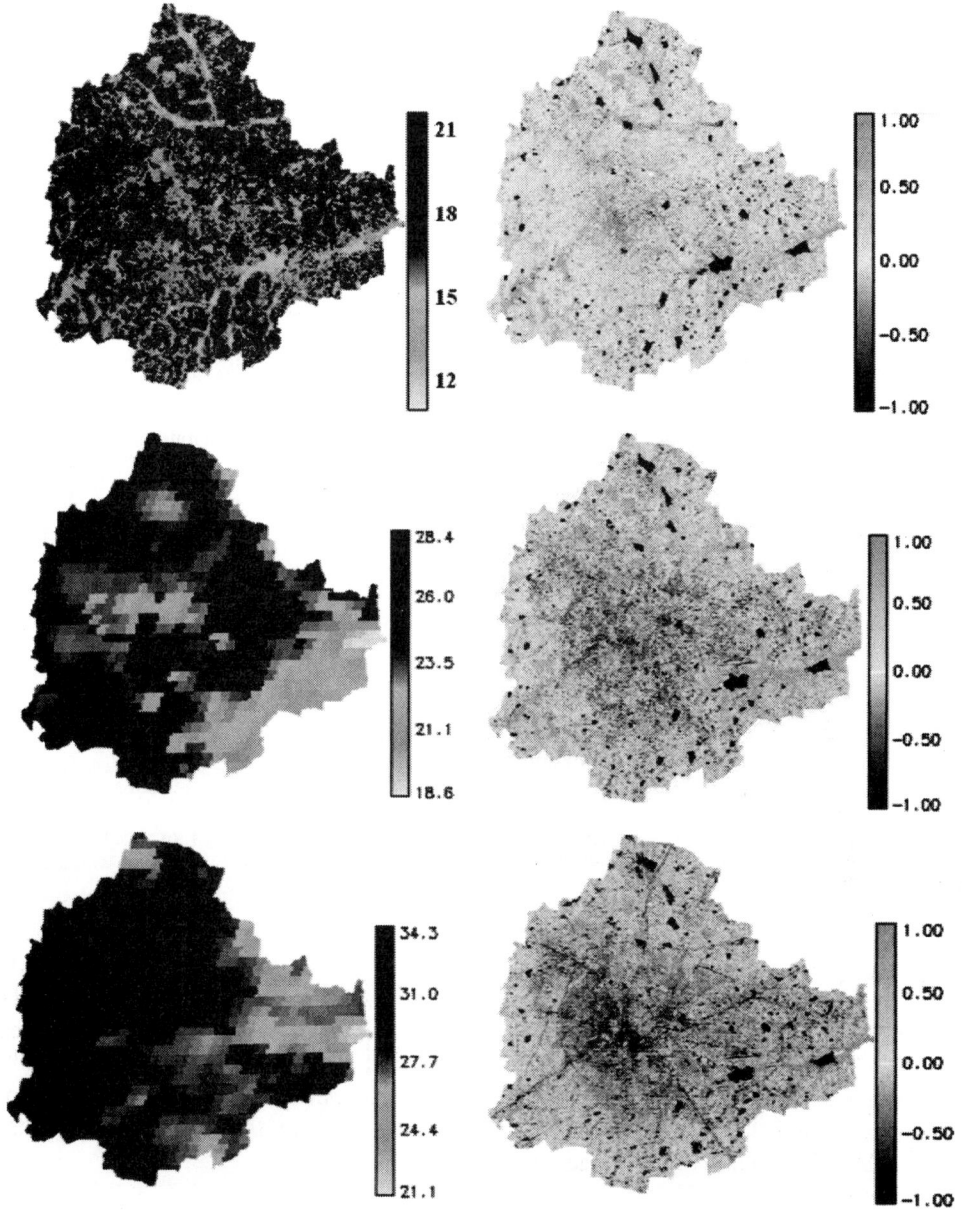

Figure 11.5: LST and NDVI from Landsat TM (1992), MODIS (2002 and 2007)

(Note: pixelisation of MODIS 2002 and 2007 is mainly due to coarse spatial resolution ~ 1 Km)

Table 11.3: LST (°C) and NDVI for various land uses

Land use	1992 (TM)		2000 (MODIS)		2007 (MODIS)	
	LST ± SD	NDVI ±SD	LST ±SD	NDVI ±SD	LST ± SD	NDVI ±SD
Built-up	19.03 ±1.47	-0.162 ±0.096	26.57 ±1.25	-0.614 ±0.359	31.24 ±2.21	-0.607 ±0.261
Vegetation	15.51 ±1.05	0.467 ±0.201	22.21 ±1.49	0.626 ±0.27	25.79 ±0.44	0.348 ±0.42
Water bodies	12.82 ±0.62	-0.954 ±0.055	21.27 ±1.03	-0.881 ±0.045	24.20 ±0.27	-0. 81 ±0.27
Open ground	17.66 ±2.46	-0.106 ±0.281	24.73 ±1.56	-0.016 ±0.283	28.85 ±1.54	-0.097 ±0.18

It is clear that urban areas that include commercial, industrial and residential land exhibited the highest temperature followed by open ground. The lowest temperature was observed in water bodies across all years and vegetation. Spatial variation of NDVI is not only subject to the influence of vegetation amount, but also to topography, slope, solar radiation availability, and other factors (Walsh et al., 1997). The relationship between LST and NDVI was investigated for each LC type through the Pearson's correlation coefficient at a pixel level and are listed in Table 11.4. The significance of each correlation coefficient was determined using a one-tail Student's t-test. It is apparent that values tend to negatively correlate with NDVI for all LC types. NDVI values for built up ranges from -0.05 to -0.6. Temporal increase in temperature with the increase in the number of urban pixels during 1992 to 2009 (113%) is confirmed with the increase in 'r' values for the respective years. The NDVI for vegetation ranges from 0.15 to 0.6. Temporal analyses of the vegetation show a decline of 65 per cent, with a consequent increase in the temperature.

Table 11.4: Correlation coefficients between LST and NDVI by LC type (p=0.05)

Land use	1992	2000	2007
Built up	-0.7188	-0.7745	-0.7900
Vegetation	-0.8720	-0.6211	-0.6071
Open ground	-0.6817	-0.5837	-0.6004
Water bodies	-0.4152	-0.4182	-0.4999

A closer look at the values of NDVI by LULC category (Table 11.3) indicates that the relationship between LST and NDVI may not be linear. Clearly, it is necessary to further examine the existing LST and vegetation abundance relationship using fraction as an indicator. The abundance images using linear unmixing from ETM+ bands were further analysed to see their contribution to the UHI by separating the pixels that contains 0-20 per cent, 20-40 per cent, 40-60 per cent, 60-80 per cent and 80-100 per cent of urban pixels. Table 11.5 gives the average LST for various land use classes.

Table 11.5: Mean LST for various land use classes for different abundances

Class → Abundance ↓	Mean Temperature± SD of dense urban	Mean Temperature± SD of mixed urban	Mean Temperature± SD of vegetation
0-20%	21.99±2.37	21.57±2.36	17.91±2.19
20-40%	22.06±2.15	21.58±2.36	17.39±1.37
40-60%	22.27±2.00	21.67±2.41	17.22±0.89
60-80%	22.33 ±2.22	22.28±2.02	17.13±0.85
80-100%	22.47±1.96	22.37±2.17	17.12±0.91

Eight transacts were laid across the city in different directions (north [N], north-east [NE], east [E], south-east [SE], south [S], south-west [SW], west [W] and north-west [NW]) and LST was analysed as shown in Figure 11.6, to understand the temperature dynamics.

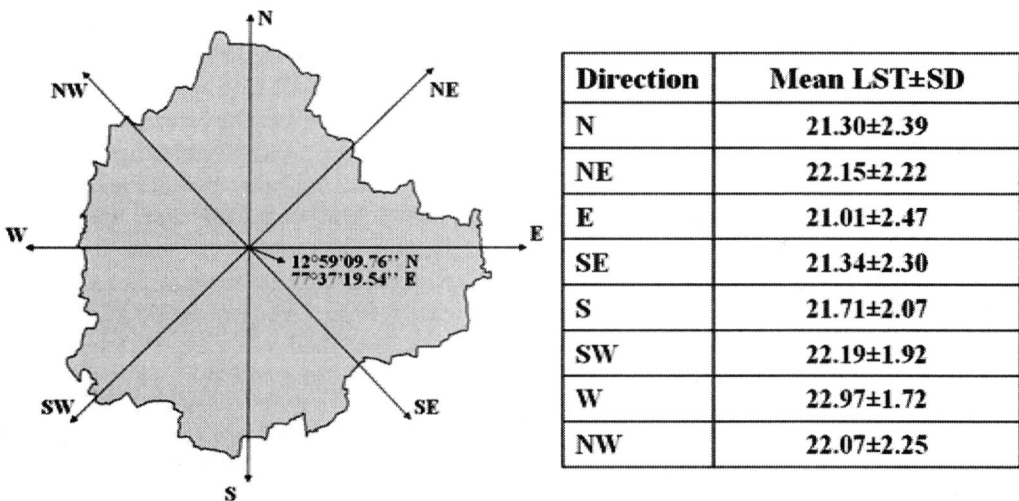

Direction	Mean LST±SD
N	21.30±2.39
NE	22.15±2.22
E	21.01±2.47
SE	21.34±2.30
S	21.71±2.07
SW	22.19±1.92
W	22.97±1.72
NW	22.07±2.25

Figure 11.6: Transect lines superimposed on Greater Bangalore boundary along with LST in various directions

The temperature profile was analysed by overlaying the LST map on the Baye's classified map to visualise the effect of vegetation, built-up, water bodies and open ground. The temperature profile plot fell below the mean when a vegetation patch or water body was encountered on the transact beginning from the centre of the city and moving outwards along the transact. The corresponding graphs are shown in Figure 11.7. The major natural green area and water bodies responsible for temperature decline are marked with circle. The spatial location of these green areas and water bodies are shown in Figure 11.8.

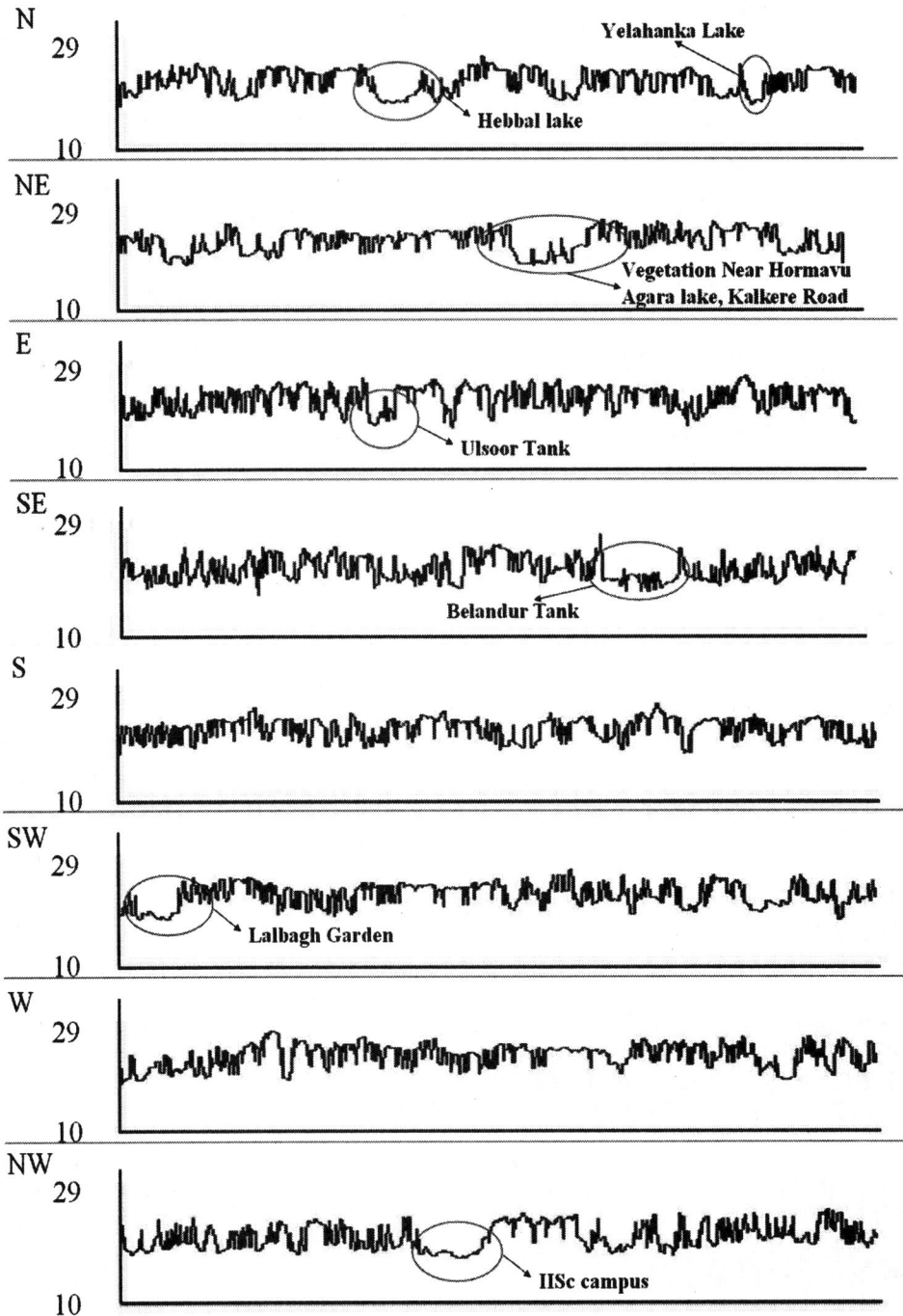

Figure 11.7: Temperature profile in various directions. X-axis – Movement along the transacts from the city centre, Y-axis - Temperature (°C)

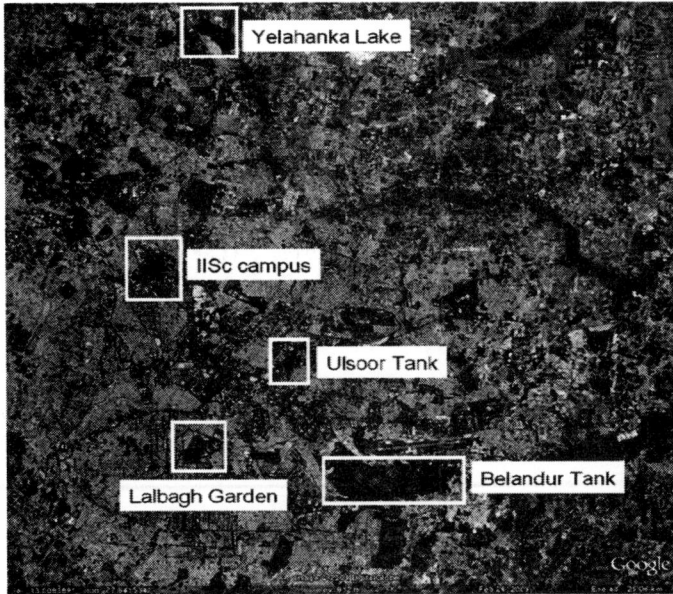

Figure 11.8: Google Earth image showing the low temperature areas (refer figure 11.7)
[Source: http://earth.google.com/]

11.5 CONCLUSION

Urbanisation and the consequent loss of lakes has led to decrease in catchment yield, water storage capacity, wetland area, number of migratory birds, flora and fauna diversity and ground water table. As land under vegetation cover is converted to built-up areas, it loses its ability to absorb rainfall. The relationship between LST and NDVI investigated through the Pearson's correlation coefficient at a pixel level and the significance tested through one-tail Student's t-test, confirms the relationship for all LC types. Also, increased urbanisation has resulted in higher population densities in certain wards which incidentally have higher LST due to high level of anthropogenic activities. The growth poles are towards N, NE, S and SE of the city indicating the intense urbanization process due to growth agents like setting up of IT corridors, industrial units, etc. Newly built-up areas in these regions consisted of maximum number of small-scale industries, IT companies, multistoried building and private houses that came up in the last decade. The growth in northern direction can be attributed to the new International Airport, encouraging other commercial and residential hubs. The southern part of the city is experiencing new residential and commercial layouts and the north-western part of the city outgrowth corresponds to the Peenya industrial belt along with the Bangalore-Pune National Highway 4.

11.6 ACKNOWLEDGEMENT

We thank the Ministry of Environment and Forests, Government of India, Indian Institute of Science and the Ministry of Science and Technology, DST, Government of India for the sustained financial and infrastructure support to energy and wetlands research.

References

Artis, D. A., and Carnahan, W.H. (1982). Survey of emissivity variability in thermography of urban areas. *Remote Sensing of Environment, 12*(4), 313-329.

Kulkarni, V.,and Ramachandra, T.V. (2006). *Environmental Management,* Commonwealth of Learning, Canada and Indian Institute of Science, Bangalore.

Landsat Project Science Office.(2002). Landsat 7 Science Data user's handbook.Goddard Space Flight Center. Retrived from http://ltwww.gsfc.nasa.gov/IAS/handbook/ handbook_toc.html.

Li, F., Jackson,T.J.,Kustas,W.,Schmugge,T.J., French, A.N., Cosh, M.L.,& Bindlish, R.(2004). Deriving land surface temperature from Landsat 5 and 7 during SMEX02/SMACEX.*Remote Sensing of Environment, 92*(4),521-534.

Nikolakopoulos, K.G., Vaiopoulos, D.A., and Skianis, G.A. (2003). Use of multitemporal remote sensing thermal data for the creation of temperature profile of Alfios river basin.*Geoscience and Remote Sensing Symposium, 21-25July, 2003, IGARSS '03. Proceedings, IEEE International, 4,* 2389-2391.

Ramachandra, T.V., and Kumar, U. (2008). Wetlands of Greater Bangalore, India: Automatic Delineation through Pattern Classifiers. *The Greendisk Environmental Journal,* 26. (International Electronic ournal.(http://egj.lib.uidaho.edu/index.php/egj/article/view/3171).

Ramachandra, T.V., and Kumar, U. (2009). Land surface temperature with land cover dynamics: multi-resolution, spatio-temporal data analysis of Greater Bangalore, *International Journal of Geoinformatics, 5* (3),43-53.

Ramachandra, T.V., and Shwetmala (2009). Emissions from India's Transport sector: State wise Synthesis. *Atmospheric Environment, 43* (34), 5510–5517.

Singh, S.M. (1998). Brightness Temperatures Algorithms of Landsat Thematic Mapper Data. *Remote Sensing of Environment, 24,* 509-512.

Snyder, W. C., Wan, Z., Zhang, Y., and Feng, Y.Z. (1998). Classification based emissivity for land surface temperature measurement from space. *International Journal of Remote Sensing* 19, 2753-2774.

Sobrino, J.A.,and Raissouni, N. (2000). Toward remote sensing methods for land cover dynamic monitoring : Application to Morocco. *International Journal of Remote Sensing, 21*(2), 353-366.

Stathopoplou, M., Cartalis, C., and Petrakis, M. (2006). Integrating CORINE land cover data and Landsat TM for surface emissivity definitions: An application for the urban area of Athens, Greece, *International Journal of Remote Sensing, 20,* 2367–2393

Stathopoulou, M., and Cartalis, C. (2007). Daytime urban heat island from Landsat ETM+ and Corine land cover data: An application to major cities in Greece. *Solar Energy,81*(3), 358-368.

Streutker, D.R. (2002). A Remote Sensing study of the Urban Heat Island of Houston, Texas. *International Journal of Remote Sensing,23*(13), 2595-2608.

Sudhira, H.S., Ramachandra, T.V., and Bala Subramanya, M.H. (2007). City Profile: Bangalore. *Cities, 24*(4), 379-390.

Sudhira, H.S., Ramachandra,T.V., and Jagadish, K.S. (2004). Urban sprawl: metrics, dynamics and modelling using GIS. *International Journal of Applied Earth Observation and Geoinformation, 5*(1), 29-39. doi: http://dx.doi.org/10.1016/j.jag.2003.08.002.

Walsh, S.J., Moody, A., Allen, T.R., and Brown, D.G. (1997). Scale dependence of NDVI and its relationship to mountainous terrain. In D. A. Quattrochi, and M. F. Goodchild (Eds.) *Scale in Remote Sensing and GIS,* pp. 27-55, Boca Raton, FL: Lewis Publishers.

Weng, Q. (2001). A remote sensing-GIS evaluation of urban expansion and its impact on surface temperature in the Zhujiang Delta, China. *International Journal of Remote Sensing,22*(10), 1999-2014.doi:10.1080/713860788.

Weng, Q. (2003). Fractal analysis of satellite-detected urban heat island effect. *Photogrammetric Engineering and Remote Sensing, 69* (5), 555-566.

Weng, Q., Lu, D., and Schubring, J. (2004). Estimation of land surface temperature -vegetation abundances relationship for urban heat island studies. *Remote Sensing of Environment, 89* (4), 467-483.

World Urbanization Prospects (2005). Revision, Population Division, Department of Economic and Social Affairs, UN.

Sustainable Urban Development:
Integrating Land Use Planning and Disaster Risk Reduction

Priti Attri, Anil K. Gupta, Smita Chaudhry and Subrat Sharma

12.1 INTRODUCTION

Rapid urbanization, coupled with global environmental change, is turning an increasing number of human settlements into potential hotspots for disaster risk. The 2005 South Asian earthquake, in which 18,000 children died when their schools collapsed, and the Indian Ocean Tsunami in 2004 that wiped out many coastal settlements in Sri Lanka, India and Indonesia, illustrate the risk that has accumulated in towns and cities and that is released when disaster strikes (UN-Habitat, 2007). Recent earthquakes in Haiti, Chile, China and Japan have been stark reminders of the increasing disaster risk faced by urban settlements around the world. Climate change will magnify this challenge, putting many cities at risk to multiple hazards. Urban risk reduction is a crucial component of wider development plans and will also help achieve the Millennium Development Goals. Disasters are defined as those events where human capacity to withstand and cope with a natural or human-made hazard is overwhelmed. Hazard is a potentially damaging physical event, phenomenon or human activity that may cause the loss of life or injury, property damage, social and economic disruption or environmental degradation. Vulnerability can be defined as the conditions determined by physical, social, economic, and environmental factors or processes, which increase the susceptibility of a community to the impact of hazards. Risk is the probability of harmful consequences, or expected losses resulting from interactions between natural or human-induced hazards and vulnerable conditions. A community is said to be 'at risk' when it is exposed to hazards and is likely to be adversely affected by the impact of those hazards when they occur.

Sustainable development means attaining a balance between environmental protection and human economic development and between the present and future needs. It means equity in development and sectoral actions across space and time (Cruz et al., 2007). The United Nations (UN) Habitat and the UN Human Settlements Programme defines a sustainable city as "a city where achievements in social, economic, and physical development are made to last. It has a

Disaster Management and Risk Reduction: Role of Environmental Knowledge; Editors Anil K. Gupta, Sreeja S. Nair, Florian Bemmerlein-Lux and Sandhya Chatterji; Copyright © 2013, Narosa Publishing House, New Delhi

lasting supply of the natural resources on which its development depends (using them only at a level of sustainable yield)".

Cities are particularly vulnerable to the effects of natural and human-made disasters due to a complex set of interrelated processes, including a concentration of assets, wealth and people; the location and rapid growth of major urban centres in coastal areas; the often unwise modification of the urban built and natural environment through human actions; the expansion of residential areas for the poor into hazard-prone locations; and the failure of urban authorities to regulate building standards and implement effective land-use planning strategies. In developed and developing countries urban planning should play a key role in enhancing urban safety by taking on issues of disaster preparedness, post-disaster and post-conflict reconstruction and rehabilitation, as well as urban crime and violence.

12.2 EXPLOSION OF URBAN POPULATION

Population shift is one of the main factors affecting urbanization. Urbanization is much more advanced in the developed parts of the world. Here, about 74 per cent of the population lives in cities. This trend is expected to continue as 86 per cent of the population is expected to be urban by 2050 (UN-Habitat, 2009).

Table 12.1: Global trends in urbanization (1950-2050)

Region	Urban population (million)					Percentage urban				
	1950	1975	2007	2025	2050	1950	1975	2007	2025	2050
World	737	1518	3294	4584	6398	29.1	37.3	49.4	57.2	69.6
Developed regions	427	702	916	995	1071	52.5	67.0	74.4	79.0	86.0
Less developed regions	310	817	2382	3590	5327	18.0	27.0	43.8	53.2	67.0
Africa	32	107	373	658	1233	14.5	25.7	38.7	47.2	61.8
Asia	237	574	1645	2440	3486	16.8	24.0	40.8	51.1	66.2
Europe	281	444	528	545	557	51.2	65.7	72.2	76.2	83.8
Latin America and the Caribbean	69	198	448	575	683	41.4	61.1	78.3	83.5	88.7
North America	110	180	275	365	402	63.9	73.8	81.3	85.7	90.2
Oceania	8	13	24	27	31	62.0	71.5	70.5	71.9	76.4

(Source: UN Habitat, 2008)

About 44 per cent of the population of developing countries lives in urban areas; this is expected to reach 67 per cent by 2050. Asia is home to some 3.7 million people, which is more than 60 per cent of the world's population. The region constitutes one of the world's most rapidly urbanizing regions. The urban population increased from 237 million (17 %) in 1950 to 1.65 billion (41 %) in 2007. By 2050, it is expected that more than two-thirds of the population will be urban (Table 12.1). The number of cities and towns in India increased from 4,651 in 1991 to 5,161 in 2001, with a significant increase in the number of cities with population above 1 million (12 in 1981 to 35 in 2001) (NIPFP, 2007).

The Registrar General of the census has projected the urban population will reach 534 million in 2026 (38% of India's total), up from 286 million in 2001 (28%) during the last census (Table 12.2). It is interesting to know that 67 per cent of the total population growth in India during the next 25 years is expected to take place in urban areas. In addition, three mega urban regions: Mumbai-Pune, the National Capital Region of Delhi and Kolkata will be among the largest urban concentrations in the world (Revi, 2008).

Table 12.2: Projected Urban and Total Population in India

	Year			
	2001	**2011**	**2021**	**2026**
Total Population (million)	1,028.611	1,192.501	1,339.741	1,399.83
Urban Population (million)	286.12	357.94	432.61	534.80
% Urban	27.82	30.02	32.29	38.21
Total AEGR (%)*	1.48	1.32	1.23	1.16
Urban AEGR*	2.24	2.07	2.50	1.89

* Annual Exponential Growth Rate
(Source: Population Projections for India, 2001–26, Registrar General of India, 2006)

12.3 URBAN RISKS

Factors contribute to the urban risk include:

(i)   The spatial arrangement of population and assets in an urban environment creates a different hazard profile for urban areas as compared to rural, in terms of the population at risk, the nature of hazards and the chances for interaction between different hazards (example floods and epidemic outbreaks).

(ii)   Cities also often expand in ways that may degrade natural buffer systems, for example mangroves (provide protection from the sea) and creeks (for drainage) and construct impermeable land surfaces that prevents percolation of water into the soil (Satterthwaite et al., 2007).

(iii) Constraints on the availability of land as a resource in urban areas often results in proliferation of slums and informal settlements on public and private land (Shaw, 2009).

12.3.1 Factors Contributing to the Increasing Vulnerability of Urban Areas

Cities are particularly vulnerable to natural and human-made disasters due to a complex set of interrelated processes. Root causes include poverty and other social vulnerabilities and weak urban governance. The rapid growth of urban areas, unplanned Urbanization and the modification of the built and natural environment through human actions are drivers of vulnerability (Figure 12.1). In paper 2, the authors (Kesavan and Swaminathan, 2008) have elaborated how environmental degradation leads to social disintegration through mass exodus of rural families (i.e. 'environmental refugees') to eke out a living in urban areas and the burden of poverty and responsibility of feeding the household members fall on the shoulders of women, often quite young. That is the 'feminization of poverty' (Figure 12.1).

The Progression of Vulnerability

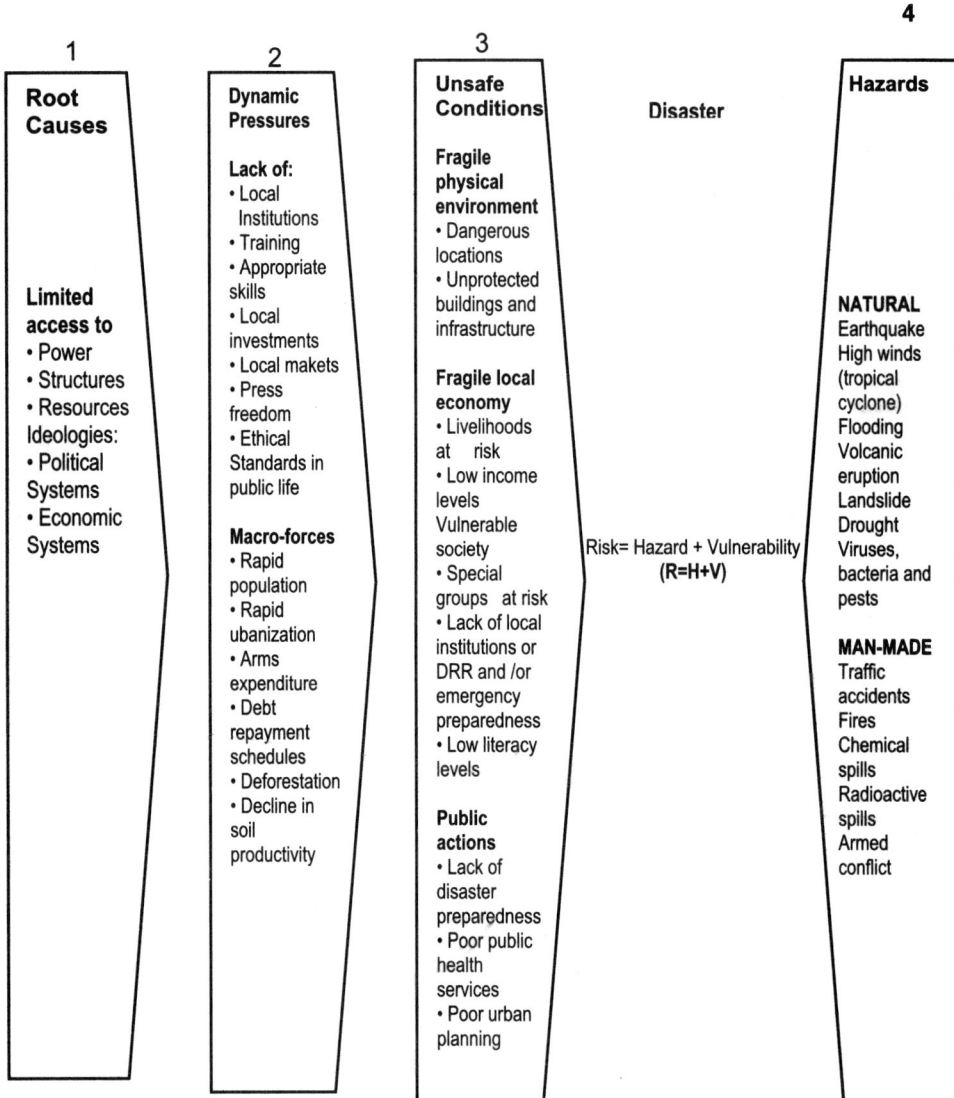

1	2	3		4
Root Causes	**Dynamic Pressures**	**Unsafe Conditions**	**Disaster**	**Hazards**

1 Root Causes

Limited access to
• Power
• Structures
• Resources
Ideologies:
• Political Systems
• Economic Systems

2 Dynamic Pressures

Lack of:
• Local Institutions
• Training
• Appropriate skills
• Local investments
• Local makets
• Press freedom
• Ethical Standards in public life

Macro-forces
• Rapid population
• Rapid ubanization
• Arms expenditure
• Debt repayment schedules
• Deforestation
• Decline in soil productivity

3 Unsafe Conditions

Fragile physical environment
• Dangerous locations
• Unprotected buildings and infrastructure

Fragile local economy
• Livelihoods at risk
• Low income levels Vulnerable society
• Special groups at risk
• Lack of local institutions or DRR and /or emergency preparedness
• Low literacy levels

Public actions
• Lack of disaster preparedness
• Poor public health services
• Poor urban planning

Disaster

Risk= Hazard + Vulnerability (R=H+V)

4 Hazards

NATURAL
Earthquake
High winds (tropical cyclone)
Flooding
Volcanic eruption
Landslide
Drought
Viruses, bacteria and pests

MAN-MADE
Traffic accidents
Fires
Chemical spills
Radioactive spills
Armed conflict

Figure 12.1: Blaikie's framework of the progression of vulnerability, with examples from an urban area

12.4 URBANIZATION AND DISASTER RISK

Disasters are frequently caused by human actions, such as uncontrolled or inadequately planned human settlements, lack of basic infrastructure and the occupation of disaster-prone areas. Disasters contribute to social, economic, cultural and political disruption in urban and rural contexts, each in its specific way. Large-scale urban concentrations are particularly fragile because of their complexity and the accumulation of population and infrastructures in limited areas. Cities are highly vulnerable to the effects of human-made and natural disasters. This is due to interrelated factors such as location and rapid growth of major urban centres in coastal locales; the human modification of the built and natural environments; expansion of settlements into hazard-prone zones; and the failure of authorities to regulate building standards and land-use planning strategies. A look at the major disasters over the last ten years highlights that large-scale disasters, which affect a whole region, usually have an urban component to them (Table 12.3). Some disasters could also be called 'urban' disasters, since the great majority of damages are concentrated in city centres. This is made evident by the recent earthquake in Port-au-Prince, Haiti and Hurricane Katrina in New Orleans, where urban areas sustained most of the losses and damages.

Location is a major determinant of the type and frequency of natural hazards in cities, 8 of the world's 10 most populous cities are on earthquake fault-lines, while 90 per cent of these cities are in region vulnerable to destructive storms.

Table 12.3: Large disaster events over the period 2000–2010 impacting on cities

Popular name	Main countries affected	Date of event	Type of hazard	Main cities affected	Total number of deaths	Total number of affected	Total damages US$ billion
Haiti Earthquake	Haiti	January 12, 2010	Earthquake	Port-au-Prince	222,570	3,400,000	n / a
Sichuan Earthquake	China	May 12, 2008	Earthquake	Beichuan, Dujiangyan, Shifang, Mianzhu, Juyuan, Jiangyou, Mianyang, Chengdu, Qionglai, Deyang	87,476	45,976,596	85
Cyclone Nargis	Myanmar	May 2, 2008	Tropical Cyclone	Yangon	138,366	2,420,000	4
Java Earthquake	Indonesia	May 27, 2006	Earthquake	Yogyakarta	5,778	3,177,923	3.1
Kashmir	Pakistan	October	Earthquake	Muzaffara-	73,338	5,128,000	5.2

Popular name	Main countries affected	Date of event	Type of hazard	Main cities affected	Total number of deaths	Total number of affected	Total damages US$ billion
Earthquake		8,2005		bad			
Hurricane Katrina	United States	August 29,2005	Tropical Cyclone	New Orleans	1,833	500,000	125
Mumbai Floods	India	July 26, 2005	Flood	Mumbai	1,200	20,000,055	3.3
South Asian Tsunami	Indonesia, Sri Lanka, India, Thailand, Malaysia, Maldives, Myanmar	December 26,2004	Earthquake and tsunami	Banda Aceh, Chennai (some damages)	226,408	2,321,700	9.2
Bam Earthquake	Iran	December 26,2003	Earthquake	Bam	26,796	267,628	500
European Heatwave	Italy, France, Spain, Germany, Portugal, Switzerland	Summer 2003	Extreme heat	Various	72,210	Not reported	NA
Dresden Floods	Germany	August 11,2002	Flood	Dresden	27	330,108	11.6
Gujurat Earthquake	India	January 26,2001	Earthquake	Bhuj, Ahmedabad	20,005	6,321,812	2.6

(Source: EM-DAT: The OFDA / CRED International Disaster Database (www.emdat.be))

Japan's location in one of the world's most active crustal zones puts its cities at risk of many natural hazards, including earthquakes, storms and floods (Table 12.4). India is also one of the most multi-hazard risk prone countries in the world (IFRC, 2005). Of the 35 cities in India having population over a million-18 are in the coastal states. These include Rajkot, Ahmedabad, Vadodara, Surat, Greater Mumbai, Pune, Nagpur, Nashik, Bangalore, Kochi, Hyderabad, Vishakhapatnam, Vijayawada, Chennai, Coimbatore, Madurai, Asansol, and Kolkata (GoI, 2004).

Cities, with their concentration of people, buildings, infrastructure and economic activities, are the locus of both large and small-scale disasters.

Table 12.4: Ten most populous cities and associated disaster risk, 2005

City	Population (million)	Disaster risk					
		Earthquake	Volcano	Storms	Tornado	Flood	Storm surge
Tokyo	35.2	X		X	X	X	X
Mexico City	19.4	X	X	X			
New York	18.7	X		X			X
Sao Paulo	18.3			X		X	
Mumbai	18.2	X		X		X	X
Delhi	15.0	X		X		X	
Shanghai	14.5	X		X		X	X
Kolkata	14.3	X		X	X	X	X
Jakarta	13.2	X				X	
Buenos Aries	12.6			X		X	X

(Source: Chafe, 2007)

12.5 LAND USE PLANNING AND DISASTER RISK REDUCTION

Urban planning is a self-conscious collective effort to imagine or re-imagine a town, city, urban region or wider territory and to translate the result into priorities for area investment, conservation measures, new and upgraded areas of settlement, strategic infrastructure investments and principles of land-use regulation (Healey, 2004). Land-use planning is an important contributor to sustainable development. It involves studies and mapping; analysis of economic, environmental and hazard data; formulation of alternative land-use decisions; and design of long-range plans for different geographical and administrative scales. Land-use planning can help to mitigate disasters and reduce risks by discouraging settlements and construction of key installations in hazard-prone areas, including consideration of service routes for transport, power, water, sewage and other critical facilities (UNISDR, 2009). Convergence of DRR and land use planning is essential not only for managing current risks but also potentially higher risks in future. Land use planning provides a set of useful planning tools for mainstreaming DRR into urban development processes, such as mapping, zoning and participatory planning. Land-use planning is important because the location of settlements and infrastructure is a key vulnerability factor. Land-use plans lay down regulations and guidelines for future urban developments, and can set controls on the expansion of existing settlements and infrastructure in disaster prone areas. Risk-sensitive land-use planning is informed by an assessment of risks (including hazards, vulnerability and capacity). Risks can be mapped throughout a city to show the zones with different levels of risk. If risk maps are overlaid on land-use maps, patterns of land use can be correlated with susceptibility to disasters (ADPC, 2010). Good land use and planning are essential for the prevention of disasters. In addition, land is fundamental to the recovery from disasters. It provides a site for shelter, a resource for livelihoods and a place to access services and infrastructure. Therefore, land issues – such as security of tenure, land use, land access and land administration - are important to key humanitarian sectors after a disaster (UN-Habitat, 2010). To ensure disaster resilient development in cities, there is need for better inter-agency coordination across Ministries and various departments at national, state and local level.

Risk-Sensitive Land Use Planning and Building a Competent Disaster Management Department in Kathmandu is one of the pilot applications of the mainstreaming disaster risk reduction in megacities project. The overall goal of the project is to ensure that the detailed land use plan of the Kathmandu Metropolitan City fully integrates disaster risk reduction within its spatial and physical development strategies including regulatory and non-regulatory planning tools, bylaws, regulations and procedures. The programme is an initiative of Earthquakes and Megacities Initiative (EMI), German Committee for Disaster Reduction (DKKV), Kathmandu Metropolitan City, Kathmandu Valley Town Planning Committee, and National Society for Earthquake Technology.

"Land-use Planning Appraisal for Risk Areas in Malaysia National Urbanization Policy" taken up by the Town and Country Planning Department, Ministry of Housing and Local Government, with the major goal of providing a systematic approach toward sustainable development by enhancing safe city planning initiatives; comprehensive safe city programmes; and enhanced public awareness programmes through a multi-sectorial approach (Shaw et al., 2009).

12.6 PRINCIPLES OF LAND-USE MANAGEMENT AND URBAN PLANNING FOR DRR

Land-use management plans form a shared basis for sustainable development and risk reduction strategies. As the physical and spatial projection of the social, economic, environmental and cultural policies of a country, land-use management includes various planning tools and management mechanisms. They are necessary for a productive but sustainable use of the national territory and provide for the successful regulation of the economic life of a country.

Land-use management operates at different geographical scales which require different ranges of management tools and operational mechanisms:

(i)   At the national level, sectoral economic policies are tied into the administrative framework of provincial or territorial jurisdictions.
(ii) At the metropolitan level, strategic plans are formulated for sustainable urban development.
(iii) At the municipal level, municipal ordinances and regulatory plans define local land-use management practices.
(iv) At the local or community level, plans encourage participatory management for community works and urban projects.

Land-use management involves legal, technical, and social dimensions. The legal and regulatory dimension includes laws, decrees, ordinances and other regulations adopted by national and local governments. The technical and instrumental dimension includes planning tools and instruments that regulate uses of land and strive for the best balance between private interests andthe public good. The social and institutional dimension includes those mechanisms which include citizen participation in land-use management practices, such as consultations, public hearings, open municipal sessions and plebiscites.

12.7 TOOLS FOR RISK–SENSITIVE URBAN LAND USE PLANNING

I.   Zoning

(i)   Identification of vulnerable areas within a region (areas prone to flood, cyclone, settlement of soil etc.).

(ii)   Check on density – strict vigilance is required on buildable area allowed and actually built (implement limits on building heights if necessary).

(iii)   Control or prohibit development in over developed areas - declare as Development Control Zone.

(iv)   Encourage a mix of land use to have a fair of distribution of population at all places at any point of time.

II.   Building regulation

(i)   Local authorities to enforce that all new and existing buildings meant for public use and high rise residential buildings must consider seismic safety in the design.

(ii)   Relocation of informal settlers and people residing in unsafe structures to properly engineered housing which are affordable.

(iii)   Incorporate retrofitting measures for disaster fighting in buildings of historical and cultural relevance.

III.   Land use Planning

(i)   Provide adequate space which can be used as refuge areas like an open space during earthquake, or raised shelters during a flood.

(ii)   Ensure strategic placement of essential physical infrastructure like hospitals to cater to the needs at various localities post disaster.

(iii)   Strengthen connectivity with neighbouring regions from where aids will be received post disaster.

(iv)   Provide roads to facilitate movement of emergency vehicles after a disaster.

IV.   Infrastructure Planning

(i)   Construction of a good drainage network system.

(ii)   Construction of a good water supply and sewerage network.

(iii)   Construction of a secure system for laying electrical or LV cables. (Sengupta et al., 2009).

12.8 HFA AND URBAN RISK REDUCTION

The "Hyogo Framework for Action 2005-2015: Building the Resilience of Nations and Communities to Disasters" (HFA) was adopted at the World Conference on Disaster Reduction (January, 2005, Kobe, Japan). The Hyogo Framework for Action (HFA) specifies that disaster risk is compounded by increasing vulnerabilities related to various elements including unplanned urbanization. Five priorities of the HFA are; (1) Making disaster risk reduction a priority, (2) Improving risk information and early warning, (3) Using knowledge and education to build a culture of safety and resilience, (4) Reducing the underlying risk factors, and (5) Strengthening preparedness for effective response. The HFA, explicitly specify land use planning and other technical measures to reduce the risk. The following are few provisions related to land-use and physical planning as specified in HFA.

(i) Incorporate disaster risk assessments into the urban planning and management of disaster-prone human settlements, in particular highly populated areas and quickly urbanizing settlements. The issues of informal or non-permanent housing and the location of housing in high-risk areas should be addressed as priorities, including in the framework of urban poverty reduction and slum-upgrading programmes.

(ii) Mainstream disaster risk considerations into planning procedures for major infrastructure projects, including the criteria for design, approval and implementation of such projects and considerations based on social, economic and environmental impact assessments.

(iii) Develop, upgrade and encourage the use of guidelines and monitoring tools for the reduction of disaster risk in the context of land-use policy and planning.

(iv) Incorporate disaster risk assessment into rural development planning and management, in particular with regard to mountain and coastal flood plain areas, including through the identification of land zones that are available and safe for human settlement.

(v) Encourage the revision of existing or the development of new building codes, standards, rehabilitation and reconstruction practices at the national or local levels, as appropriate, with the aim of making them more applicable in the local context, particularly in informal and marginal human settlements, and reinforce the capacity to implement, monitor and enforce such codes, through a consensus-based approach, with a view to fostering disaster-resistant structures.

12.9 INDIAN CONTEXT

In keeping with the objectives of the Yokohama Strategy and Plan of Action for a Safer World, a Vulnerability Atlas of India was developed in 1997. It has proved to be an innovative tool for assessing district-wide vulnerability and risk levels of existing building stock. The atlas has helped state governments and local authorities to strengthen regulatory frameworks. This was achieved by amending construction bylaws, regulations, master plans and land-use planning regulations for promoting disaster resistant design and planning processes. After the Gujarat earthquake in 2001 the relevance of the atlas has been highlighted and additional assessments in a more detailed scale are now being developed. India has been successful in modifying land use by seeking to address community requirements so as to gain wider commitment in executing land use changes. A national policy backed by local efforts is crucial to the success of these programmes.

Indian state governments are responsible for development plans, in particular those that contribute to natural hazards management, agriculture and land management. The first major initiative for preventing flood hazards in the Ganges plains was in 1960-1961 in the form of a soil conservation scheme in the catchment areas of the river valley projects as recommended by the National Flood Commission. The National Watershed Development Project for Rainfed Areas also aimed at promoting appropriate land use and the development of farming systems on a watershed basis. A national land-use policy outline adopted by the government presents a cohesive and coordinated strategy by government agencies and others to ensure the optimal use of land. In this connection, a national land-use and conservation board and state land-use boards have been established. The Indian experience has shown that measures to prevent disasters succeed to the extent that they focus on resource regeneration of the community living on the lands concerned. The approach needs to address both spatial and temporal dimensions of land use. Sustainability and effectiveness of interventions depend on appropriate land usage, for which peoples' participation in the planning and decision-making is a requirement.

India Urban Disaster Mitigation Project - Technological Hazard Mitigation in Baroda and Metropolitan Calcutta is another example. The objective of the project is to reduce the vulnerability of the population and infrastructure to technological/industrial hazards in selected municipalities or districts within the metropolitan Kolkata and Baroda areas. The first phase of the project consists of hazard mapping and vulnerability assessment for the target municipalities, as well as the development of guidelines for incorporating technological hazards into urban development planning. During the second phase, a full-scale mitigation strategy and off-site emergency preparedness plan will be prepared and implemented for one of the cities.

In partnership with the Government of India, the UNDP 'Urban Earthquake Vulnerability Reduction Programme' was aimed at strengthening he disaster mitigation, preparedness and response capacities of vulnerable communities, local bodies and municipal administrations in 38 Indian cities. The project provided a practical model for mainstreaming earthquake risk management initiatives at all levels, and helps to reduce seismic risk in India's most earthquake-prone urban areas. Further UNDP- DRR programmes (2008-2012) too have specific focus on Urban Disaster Reduction and Climate Resilience.

Government of India with support from UNDP is now implementing the GOI-UNDP Disaster Risk Reduction Programme (DRR) from 2009 to 2012.The GOI-UNDP DRR programme is envisaged to support Central and State Government programmes and initiatives by providing critical inputs that would enhance the efficiency and effectiveness of the efforts in Disaster Risk Reduction. The programme strives to strengthen the institutional structure to undertake Disaster Risk Reduction activities at various levels including risks being enhanced due to climate change and develop preparedness for recovery.

The first-ever global forum on adapting to the dramatic implications of climate, Resilient Cities 2010, ran 28-30 May in Bonn, Germany. The Resilient Cities Congress series is part of ICLEI's Action Pledge to meet the objectives and expected outcomes of the Nairobi Work Programme. Resilient Cities 2010 enabled exchange, learning, networking, debate and policy development on approaches and solutions to climate change adaptation for cities and local governments.

The NDMA has prepared Guidelines for Flood Management (FM), to assist the ministries and departments of the GOI, the state governments and other agencies in preparing Flood Management plans (FMPs). The guidelines rest on the following objectives cover all aspects of FM and aim at increasing the efficacy of the FMPs.

(i) Shifting the focus to preparedness by implementing, in a time-bound manner, an optimal combination of techno-economically viable, socially acceptable and eco-friendly structural and non-structural measures of FM.

(ii) Ensuring regular monitoring of the effectiveness and sustainability of various structures and taking appropriate measures for their restoration and strengthening.

(iii) Continuous modernisation of flood forecasting, early warning and decision support systems.

(iv) Ensuring the incorporation of flood resistant features in the design and construction of new structures in the flood prone areas.

(v) Drawing up time-bound plans for the flood proofing of strategic and public utility structures in flood prone areas.

(vi) Improving the awareness and preparedness of all stakeholders in the flood prone areas.

(vii)   Introducing appropriate capacity development interventions for effective FM.

(viii)   Improving the compliance regime through appropriate mechanisms.

(ix)   Strengthening the emergency response capabilities.

12.10 FUTURE CHALLENGES

A lack of current information about hazards and potential risks within specific areas is a common limitation that must be addressed within individual localities. This is tied to the resulting undesirable consequences of a community's inability to anticipate hazard events or to undertake necessary measures to minimize their potential effects. The often high costs and protracted nature of multidisciplinary involvement associated with the technical aspects of hazard mapping or vulnerability and risk assessment activities can be considered an impediment to establishing a systematic land-use programme. This can however, be overcome if a strategic approach is adopted which reviews plans and schedules various stages of activity over a period of time. Most fundamentally though, efforts need to be exerted to minimize local political interests or community tendencies which resist a wider acceptance of the beneficial rationale for land controls.

This may be associated with various related concerns such as competing economic valuations of properties or locations, weak or marginal interest in the enforcement of land-use policies, problematic licensing practices, and lax administrative procedures which invite noticeably corrupt practices in too many countries. Ultimately a crucial priority needs to be accorded in weighing private, individual or singular uses of land against a wider concern for public values and the more broadly applicable considerations of public safety and socially determined access. The determination of how that balance is struck and where it is actually displayed in physical terms remains an obligation for public expressions of interest and concern. Tensions or vested interests between government and private interests, national and local interests or instruments of the state and the population can occur. Dynamic factors such as population growth, migration, conflicts over the use, supply or demand of services will occur. There will be factors specific to risk management including the changing nature of vulnerability, major fluctuations in land values, urban services and environmental services.

References

ADPC (2010). *Urban Governance and Community Resilience Guides. Mainstreaming Disaster Risk Reduction book 4.*Bangkok: Asian Disaster Preparedness Center (ADPC).

ADRC (2002).*Data Book on Asian Natural Disasters, 2.* Japan: Asian Disaster Reduction Center.

ADRC (2009). Thematic overview of Urban Risk Reduction in Asia. Submitted for the Asia Regional Task Force on Urban Risk Reduction as input to the Global Assessment Report on Disaster Risk Reduction. Retrieved from http://www.adrc.asia/events/RTFmeeting/20080130/.

Blaikie, P., Cannon, T., Davis, I., and Wisner, B. (1994). Disaster Pressure and Release Model. In *At Risk: Natural Hazards, People's Vulnerability, and Disasters.* Routledge: London and New York, pp. 21- 45.

Chafe, Z. (2007). Reducing natural risk disasters in cities. In *2007 State of the World: Our Urban Future.* World watch Institute, Washington, DC.

CRED(n.d). Emergency Events Database (EM-DAT). Louvain, Belgium: CRED, Catholic University of Louvain. Retrieved from www.emdat.be.

Cruz, R.V., Harasawa, H., Lal, M., Wu, S., Anokhin, Y., Punsalmaa, B., Honda, Y., Jafari, M., Li, C., and Hu Ninh, N. (2007). Asia. Climate Change, 2007: Impacts, Adaptation and Vulner-ability. Contribution of Working Group II to *The Fourth Assessment Report the Inter-governmental Panel on Climate Change,* M.L. Parry, O.F. Canziani, J.P. Palutikof, P.J. van der Linden and C.E. Hanson (Eds).Cambridge, U.K.: Cambridge University Press, 469-506.

DESA, UN (2008). World Urbanization Prospects The 2007 Revision, New York: Department of Economic and Social Affairs (DESA), United Nations. Retrieved from http://www.un.org/esa/

Government of India (2004). Initial National Communications to the United Framework Convention on Climate Change. New Delhi: Ministry of Environment and Forests, Government of India.

Healey, P. (2004). The treatment of space and place in the new strategic spatial planning in Europe. *International Journal of Urban and Regional Research 28*(1)*,* 45–67.

IFRC (2005).World Disasters Report. Geneva: International Federation of Red Cross

NIPFP (2007). *India Urban Report- A Summary Assessment.* New Delhi: National Institute of Public Finance and Policy. Retrieved from http://www.nipfp.org.in/newweb/sites/ default/files/Annual Report07-08.pdf

Revi, A., (2008). Climate Change Risk: An Adaptation and Mitigation Agenda for Indian Cities. *Environment and Urbanization, Vol. 20 (1)*, 207-229.

Satterthwaite, D., Huq, S., Pelling, M., Reid, H., and Lankao, P.R. (2007). *Adapting to Climate Change in Urban Areas: The possibilities and constraints in low- and middle-income nations.* Human Settlements Discussion Paper Series. Theme: Climate Change and Cities – 1. working paper produced by the Human Settlements Group and the Climate Change Group at the International Institute for Environment and Development (IIED), London. Retrieved from www. iied.org/pubs/display.php?o=10549IIED, 2 March, 2010.

Sengupta, B.K., and Banerji, H. (2009). Disaster Mitigation Strategies through Land Use Planning and Zoning in an Urban Context, Retrieved from www.nidm.gov.in/idmc2/PDF/Presentations/.

Shaw, R., Matsuoka, Y., Tsunozaki, E., Sharma, A., and Imai, A. (2009). *Reducing Urban Risk in Asia: Status Report and Inventory of Initiatives.* Retrieved fromhttp://www.unisdr.org/publications.

UN-Habitat (2007). *Mitigating the Impacts of Disasters: Policy Directions Enhancing Urban Safety and Security Global report on Human Settlements 2007.* Abridged Edition Volume 3. London: Earth Scan.

UN-Habitat (2009). *Global Report on Human Settlement 2009,* United Nations Human Settlements Programme. London: Earth Scan. Retrieved from http://www.unhabitat.org/ downloads/docs/GRHS2009/GRHS.2009.pdf

UN-Habitat (2010). *Land and Natural Disasters: Guidance for Practitioners.* Nairobi, Kenya: United Nations Human Settlements Programme (UN-HABITAT).

UN-ISDR (2009). *Terminology for Disaster Risk Reduction.* Retrieved on 10 July 2010 from www.unisdr.org/we/inform/terminology.

Disaster Risk Management and Legal Framework:
Analysis of Indian Environmental Legislation

Sreeja S. Nair, Anil K. Gupta and Swati Singh

13.1 INTRODUCTION

Disasters are the events of environmental extremes and inevitable entities of this living world. Disaster management highlight the interdependence of the economy, environment and inclusive development. Need of linking disaster risk reduction with environmental management is globally recognised. The Hyogo Framework for Action (HFA) calls for efforts to "encourage the sustainable use and management of ecosystems, through better land-use planning and development activities to reduce risk and vulnerabilities." The framework promotes the implementation of "integrated environmental and natural resource management approaches that incorporate disaster risk reduction, including structural and non-structural measures, like the integrated flood management and appropriate management of fragile ecosystems." In view of the Hyogo Framework of Action (HFA), the UN-ISDR Global Joint Work programme for 2008-2009 sought to ensure that "national and local authorities are better equipped to protect environmental services in coastal areas, flood and fire-sensitive basins and mountain ecosystems"(UNEP & UNISDR, 2010).

Hazards and disasters are two sides of the same coin; neither can be fully understood or explained from the standpoint of either physical science or social science alone; and are inextricably linked to the ongoing environmental changes at global, regional and local levels. Environmental hazards exist at the interface between the natural events and human systems. Human responses to hazards can modify both the natural events, and the human use of, the environment (Figure 13.1, Burton et al., 1993).Environmental degradation is a process that reduces the capacity of the environment for meeting the social and ecological needs. The potential impacts of degradation vary and may contribute to increase in vulnerable conditions along and intensity in occurrence of natural hazards. Some examples include: land degradation, deforestation, desertification, wild land fires, loss of biodiversity, land, water and air pollution, climate change, sea level rise and ozone depletion etc.

Disaster Management and Risk Reduction: Role of Environmental Knowledge; Editors Anil K. Gupta, Sreeja S. Nair, Florian Bemmerlein-Lux and Sandhya Chatterji; Copyright © 2013, Narosa Publishing House, New Delhi

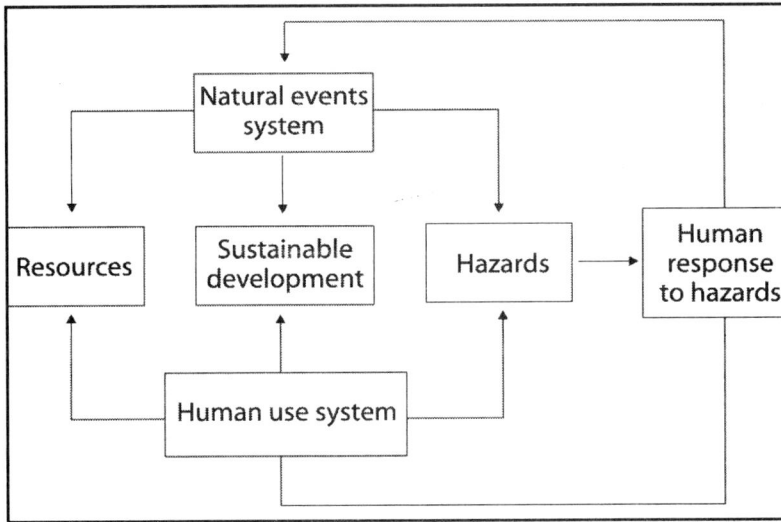

Figure 13.1: Environmental hazards and interface of natural events system with human use system (Burton et al., 1993)

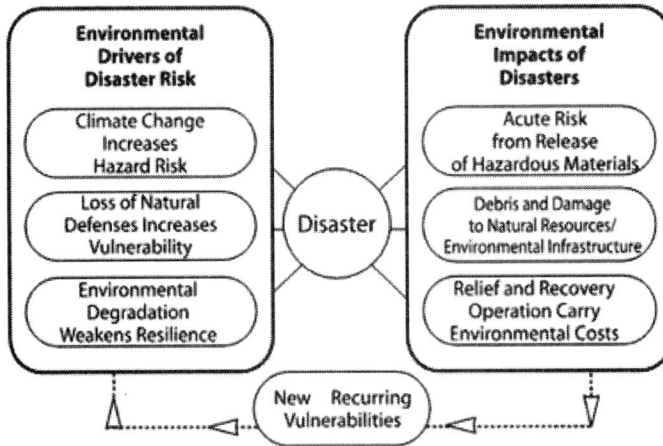

Figure13.2: Environmental causes and consequences of disasters (UNEP, 2010)

The major environmental changes driving hazards and vulnerabilities to disasters are climate change, land-use changes and degradation of natural resource (Gupta and Nair, 2011). Environmentally causes and consequences of disasters are illustrated in Figure 13.2.The inter-relationship between environment and disasters is now widely recognized in terms of the following interfaces:

(a) Environmental degradation leading to disasters: Environmental changes are known to generate or aggravate disasters especially of hydro-meteorological origin.

(b) Environmental degradation causes vulnerability: Environmental degradation causes vulnerability: People are going to be adversely affected due to decline in ecosystem services, i.e. the provisional, recreational, regulatory and supporting services.

(c) Disasters impact environment and ecology: Disasters cause primary and secondary impacts on the environment, affecting natural processes, resources and ecosystems, thereby creating new hazards and conditions for future disasters.

(d) Relief & Recovery compromise environmental sustainability: Aspects related to environment is compromised during the event of disaster relief operations and recovery process.

Due to improper disposal of disaster and relief waste, there is a mismanagement of natural resources such as water, or land, inappropriate use or management of land- mostly ecological sensitive zones or natural hazard prone zones–flood plains, and landscape modifications in the case of sanctuaries, national parks, bio-reserves with introduction of alien species or substances including organisms.

13.2 ENVIRONMENTAL APPROACH TO DRM – 2nd PARADIGM SHIFT

Globally, disaster management has voiced a paradigm shift from being 'response & relief' centric' in approach to becoming 'mitigation and preparedness' oriented, a lesson drawn from UN-IDNDR since 1990s. A 2nd paradigm shift is underway, driven by climate-change awareness and sustainability concerns in disaster management (figure 13.3) (Gupta, 2010). This has resulted in a wider acceptance of the 'Disaster Risk Reduction' (DRR) over 'Disaster Management', and calls for 'environmental approach to disaster management and risk reduction'. Integrating with environment and sustainable development is now of prime concern in disaster management strategies worldwide. Monitoring and observing environmental factors that signal the onset of a hazard are fundamental to early warning systems.

Environmental monitoring and assessment play an important role in generating relevant information that assists in identifying risks, vulnerabilities and opportunities to promote community resilience (UNEP-UNISDR-PEDRR, 2010). Environmental governance includes policies, legal and regulatory frameworks and institutional structures, and offers important opportunities for mainstreaming disaster risk reduction into environmental management, and for strengthening the environmental components of disaster risk reduction. Knowledge of environment is crucial in all stages of disaster risk management. Environmental services like shelter, water, food security, sanitation, waste management and disease control form crucial components of emergency relief.

13.3 NEED OF POLICIES AND LAW FOR DISASTER MANAGEMENT

Legislation for DRR is fundamental to the enhancement of human security. It is the first step towards mainstreaming disaster risk reduction into development. Legislation provides the framework around which strategies to build risk reduction into development and reconstruction activities can be empowered. The law can be used to provide penalties and incentives by enforcing standards in construction, land use, tenant's rights and by defining people's rights during relief and reconstruction (Pelling and Holloway, 2006).

Law and regulations are needed for implementation of provisioning, planning, operation and monitoring of the functions suggested as a set of guiding principles, whereas a policy is set of guiding principles and roadmap of actions towards a defined aim or goal. Hence, the goal of sustainable development through disaster reduction needs to be operationalized through a set of policies and an effective legal framework. Legal measures and provisions can play significant roles in following specific ways.

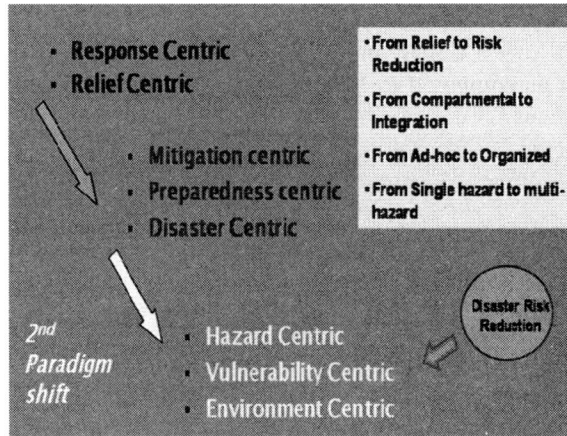

Figure 13.3: Paradigm shifts in disaster management (Gupta and Nair, 2009)

(a)   Legal provision are needed for envisaging disaster prevention through (i) hazard identification and vulnerability reduction, (ii) proactive planning, assessment and audit, (iii) standards methods and codes/standards, (iv) public information, mock-exercise and (v) capacity development and institutional framework. (Risk Management).

(b)   Legal framework on disaster incident management through (i) incident command/ response management, (ii) resource mobilization and relief coordination, (iii) damage & loss assessment and financing, (iv) services in emergencies – environmental-health (shelter, water & sanitation, food, waste/debris, disaster prevention), power, fuel, roads, timber, disposal, etc. (v) immediate compensation, penalties, responder's safety. (Response and Relief Management).

(c)   Legal provisions on post-disaster risk reduction: (i) liability for compensation and rehabilitation (ii) enquiries and litigations, (iii) reconstruction (sustainable and safer),(iv) recovery (sustainable livelihood, environment, development) (v) addressing future risk (Reinforcing Risk Reduction).

(d)   Legal provisions (policies, acts, rules and guidelines) are required to guide the actions and behaviours of the States, Institutions, Local bodies, communities and other actors for planning, provisioning, performance and monitoring of aforementioned aspects of the disaster management.

Looking to the need and roles of legal provisions in the area of disaster management and risk reduction, it would neither be possible nor feasible to enact new laws to address various aspects of multi-functional, multi-dimensional, multi-disciplinary approach to disaster management.

Therefore, constitutional and legal provisions and guidelines related with the following aspects, and their interpretations – functional and judicial, shall be of immense use:

(i)   Legal provisions on disaster, safety and emergency response.

(ii) Legal provisions on environment & natural resources.

(iii) Legal provisions on developmental planning.

(iv) Legal provisions on human rights.

13.4 ROLE OF ENVIRONMENTAL LAW IN DISASTER MANAGEMENT

Since the International Decade of Disaster Risk Reduction in 1990s (UN-IDNDR) and following the Hyogo Framework for Action (HFA) in 2005, countries started paying greater attention to Disaster Risk Management as compared to the emergency response and relief centric approach. The HFA 2005-2015 advocates for disaster risk reduction and mainstreaming in sectoral planning process, and provides impetus for environment based practices. Millennium Assessment Report (2005) also identified environmental degradation as a major factor leading to the increasing vulnerability.

Several laws related to protection and improvement of environment and natural resources existed in India before 1950's as well. However, the real momentum for bringing a framework dedicated to environmental protection came after the UN Conference on the Human Environment (Stockholm, June 1972). The National Council for Environmental Policy and Planning was set up in 1972 after the conference, within the Department of Science and Technology in India. The Council later evolved into a full-fledged Ministry of Environment and Forests (MoEF) in 1985.

Provision directly related with natural disaster management, viz. environmental clearance, EIA and audit, risk analysis, land-use and zoning, emergency preparedness; and environmental services - water, sanitation, waste disposal, preventive-health, including climate change mitigation and adaptation etc. relate to the environmental laws and policies in many countries including India. These regulations and policies primarily aiming at environmental quality, resource management and related procedures, have provisions and offer potential interpretations for disaster management and risk reduction (Gupta and Nair, 2012).

Environmental law includes provisions and regulations related to environment and its constituents, protection and management of natural resources, i.e., water, land, agriculture, forests, wildlife; habitats – protected areas, zoo, parks, reserves; procedures and planning to safeguard environment; resources and ecosystems. Environmental law can comprise of the following:

(i)   Constitutional provisions

(ii)   Common law

(iii)   Statutory law

(iv)   Customary law

On the basis of their primary objectives environmental laws can also be broadly grouped as below:

(i)   Laws on environment protection and conservation (natural resources).

(ii)   Laws on environmental quality (pollution and waste) management.

(iii)   Laws on chemical safety and emergency preparedness.

The laws on environmental protection and quality including natural resource conservation, pollution and waste management, are becoming more relevant in Natural Disaster Risk Reduction (DRR) in the wake of paradigm shift in disaster management. The current emphasis of environment based DRR aims at pre-disaster risk reduction and post-disaster sustainable recovery processes, along efficient relief management. Whereas the safety and emergency preparedness related laws provide for proper risk assessment, emergency planning and response organization aimed at minimizing the impacts of a disaster event in general and in particular of an accident involving hazardous substance or hazardous process. The growing emphasis on the 'greening disaster response' calls for a greater role of environmental law. Related standards and codes ensure preventive environmental-health (food safety and shelter provisions, water and sanitation, waste management and controls of disease outbreak) so as to avoid secondary disasters and/or complex emergencies.

13.4.1 Constitutional Provisions

Indian constitution comprises provisions related to environment and human rights and thus provide for reduction of vulnerability and addressing hazard risk by enhancing capacities. Article 21 of the Indian Constitution states *"No person shall be deprived of his life or personal liberty except according to procedure established by law"*. The right to life has been employed in a diversified manner in India. Besides the mere right to survive as a species, quality of life, the right to live with dignity and the right to livelihood, etc. are within the purview of Article 21. The Constitution of India provides that all are equal before the law and shall be accorded equal protection of the law. Article 14 states that *"The State shall not deny to any person equality before the law or the equal protection of the laws within the territory of India"*. Article 14 can be used to challenge government sanctions for mining and other activities with high stakes on human rights and environmental impact, where the permissions are arbitrarily granted without adequate consideration of environmental impacts. The Constitution Act of 1976 (Forty Second Amendment) explicitly incorporated environmental protection and improvement as a part of state policy. Article 48-A provides that the state shall endeavour to protect and improve the environment and safeguard the forests and wildlife of the country. Article 51-A (g) imposes a similar responsibility on every citizen to protect and improve the natural environment including forests, lakes, rivers and wildlife, and to have compassion for living creatures. Thus, protection of natural environment and compassion for living creatures was made the positive fundamental duty of every citizen, and relates to actions that reduce disasters by addressing hazards and reducing vulnerability of the land and its people.

13.4.2 Common laws

The term "Common" is derived from Latin Word Lex Communis the body of customary law of England which is based upon the judicial decisions. The Common Law continues to be in force in India under Article 372 of the constitution so far and is not yet altered, modified or repealed by statutory laws. Under the Common Law, an action might lie for causing pollution of environment, viz., air, water, or noise if it would amount to private or public nuisance. The common law remedies against environmental pollution are available under the law of Torts. Tort is a civil wrong other than a breach of trust or contract. The most important tort liabilities for environmental pollution are under the heads of nuisance, trespass, negligence and strict liability.

The Indian Penal Code formulated by the British in 1860, forms the backbone of criminal law in India. The Code of Criminal Procedure, 1973 governs the procedural aspects of the criminal law. Indian Penal Code (IPC), 1860 makes various acts affecting environment as offences (Chapter XIV, section 268 and 294 A). Public health, safety, convenience, decency and morals are dealt under these sections. IPC also cover the negligent handling of poisonous substances, combustive and explosive materials. Criminal Procedure Code, 1973 (CrPC) can also be invoked to prevent to prevent pollution. Chapter X, Part B sections 133 to 143 provides most effective and speedy remedy for preventing and controlling public nuisance. Section 133 can be used against Municipalities and Government bodies. This provision can be extended in the case post disaster debris and waste management as well.

13.4.3 Statutory Laws, Rules and Notifications

The following is list of Acts related to protection and improvement of environment and disaster management, Rules notified and Notifications issued there under.

13.4.3.1 Acts

(i)The Indian Forest Act, 1865 amended in 1878
In 1865, the first Indian Forest Act was passed. It was amended in 1878 when a comprehensive law, The Indian Forest Law Act, came into force. The provision of this Act established a virtual State monopoly over the forests in a legal sense on one hand, and attempted to establish, on the other, that the customary use of the forests by the villagers was not a 'right', but a 'privilege' that could be withdrawn at will. The act of not conservation centric, rather state monopoly and resultant loss of customary rights of the people lead to over exploitation of resources to some extent.

(ii) Indian Forest Act, 1927
The Act was sought to consolidate and reserve the areas having forest cover, or significant wildlife, to regulate movement and transit of forest produce, and impose duty leviable on timber and other forest produces. The Act consolidated and reserved the areas having forest cover or significant wildlife and defined the procedure for declaring forest as reserved, protected and village forest and also levying penalties for not following the provisions of the Act (Section 20, 29 and 28 of The Indian Forest Act, 1927).British India's forest administrators were concerned about the potential long-term environmental effects of deforestation caused by indiscriminate logging. This act is having provisions for conserving the valuable forest resources although the state monopoly continued.

(iv) The Forest (Conservation) Act, 1980.
Objective of the act is to provide for the conservation of forests and for matters connected therewith or ancillary or incidental thereto. The Act restricts the powers of the state in respect of de-reservation of forests and use of forestland for non-forest purposes (the term non-forest purpose includes clearing any forestland for cultivation of cash crops, plantation crops, horticulture or any purpose other than re-afforestation). The Forest (Conservation) Rules were made 1981 in order to exercise the powers conferred by Section 4 of the Act. By restricting the use of forest land for non-forest purpose, the Act serves as an affective legislation for controlling land-use land cover changes and unsustainable development in forest areas which leads to degradation of ecosystems floods and landslides.

(v) Factories Act, 1948

The Factories Act, 1948 was a post-independence statute that explicitly showed concern for the safety and health of workers and environment. The primary aim of the 1948 Act has been to ensure the welfare of workers not only in their working conditions in the factories but also their employment benefits. While ensuring the safety and health of the workers, the Act contributes to environmental protection. The Act contains a comprehensive list of 29 categories of industries involving hazardous processes (Section 2C(b), Factories Act). Factories Act Section 7 and 8 is on the duties of the occupier and manufacturer respectively. These provisions are of great significance in the context process safety and risk reduction (See Section 7A. General duties of the occupier:

(1)Every occupier shall ensure, so far as is reasonably practicable, the health, safety and welfare of all workers while they are at work in the factory (2) Without prejudice to the generality of the provisions of sub-section (1), the matters to which such duty extends, shall include (a) the provision and maintenance of plant and systems of work in the factory that are safe and without risks to health;(b) the arrangement in the factory for ensuring safety and absence of risks to health in connection with the use, handling, storage and transport of articles and substances ;Section 7B. General duties of manufacturers, etc., as regards articles and substances for use in factories.(1) Every person who designs, manufactures, imports or supplies any article for use in any factory shall (a) ensure, so far as is reasonably practicable, that the article is so designed and constructed as to be safe and without risks to the health of the workers when properly used).Section 37.Is about the measures to taken in industries dealing with explosive or inflammable dust, gases.

(vi)   Factories Amendment Act, 1987

Factories Act, 1948 was focuses primarily on the safety and welfare of workers, process safety and onsite emergency issues. However in the Factories Amendment Act, 1987, specific provisions for citing of Industries after examination by a 9 member Site Appraisal Committee comprised of expert members including metrology department, town planning department besides the district authority and factories department. Provision for risk sensitive siting, sharing of information for effective risk communication are few salient features of the Act, which are significant to disaster risk reduction and emergency response. (See Chapter IV Section 41A to 41F; Section 41A. Constitution of Site Appraisal Committees, Section 41B. Compulsory disclosure of information by the occupier, Section 41E. Emergency Standards and Section 41F. Permissible limits of exposure of chemical and toxic substances). Chapter IV of the Act is an excellent example of risk reduction through appropriate land-use planning and risk communication strategy.

(vii)   The Wildlife (Protection) Act, 1972, Amendment 1991.

The Wildlife Protection Act (WPA), 1972, provides for protection to listed species of flora and fauna and establishes a network of ecologically-important protected areas. The WPA empowers the central and state governments to declare any area a wildlife sanctuary, national park or closed area. There is a blanket ban on carrying out any industrial activity inside these protected areas. It provides for authorities to administer and implement the Act; regulate the hunting of wild animals; protect specified plants, sanctuaries, national parks and closed areas; restrict trade or commerce in wild animals or animal articles; and miscellaneous matters. The Act prohibits hunting of animals except with permission of authorized officer when an animal has become dangerous to human life or property or as disabled or diseased as to be beyond recovery (WWF-India, 1999). The near-total prohibition on hunting was made more effective by the Amendment

Act of 1991.Wildlife Protection Act is a powerful law, contribute to the maintenance of ecological diversity and balance and also by limiting the human interference in wild life areas and corridors chances of man-animal conflict will be reduced.

(viii)   The Water (Prevention and Control of Pollution) Act, 1974

Water (Prevention and Control of Pollution) Act, 1974. The objective of the act as envisaged in the preamble is the prevent water pollution and maintain wholesomeness. The Act prohibits the discharge of pollutants into water bodies beyond a given standard and lays down penalties for non compliance. The Act was amended in 1988 to conform closely to the provisions of the EPA, 1986. This helps in assuring clean drinking water to every citizen. Besides, by prohibiting discharge of effluents, the act contributes towards protecting and improving the ecosystems and hence reducing vulnerability.

(ix)   The Air (Prevention and Control of Pollution) Act, 1981

The Act provides means for the control and abatement of air pollution. The Act seeks to combat air pollution by prohibiting the use of polluting fuels and substances, as well as by regulating use of appliances which increases air pollution. Under the Act establishing or operating of any industrial plant in the pollution control area requires consent from state boards. The boards are also expected to test the air in air pollution control areas, inspect pollution control equipment, and manufacturing processes. The Act was amended in 1987 to empower the central and state pollution boards to meet grave emergencies and recover the expenses incurred from the offenders. The act (Section 2(a)is defines air pollutants any solid, liquid or gaseous substance 2[(including noise)] present in the atmosphere in such concentration as may be or tend to be injurious to human beings or other living creatures or plants or property or environment. By controlling air pollution the act serves as an important means to mitigate climate change. The act reduces the vulnerability to health hazards by prohibiting and controlling industrial activities in control areas.

(x)   Environment (Protection) Act, 1986

This Act is an umbrella legislation designed to cover gaps in the areas of major *"environmental hazards"* which were not covered in the other existing laws. Under the section 3 of this Act, the central government is empowered to take measures necessary to protect and improve the quality of the environment by setting standards for emissions and discharges; regulating the location of industries; management of hazardous wastes, and protection of public health and welfare. Section 6 of the Act gives power to The Central Government to make rules to regulate environment pollution and section 8 provides that persons handling hazardous substances are required to comply with procedural safeguards where the discharge of any environmental pollution in excess of prescribed standards occurs or is apprehended to occur due to accident or other unforeseen act or event. From time to time the central government issues notifications under the EPA for the protection of ecologically-sensitive areas or issues guidelines for matters under the EPA. Rules under the EPA Act are of great significance in all phases of disaster management particularly chemical disaster management and protecting and improving vulnerable areas and ecosystems like coastal, mountain, deserts and wetlands (see 4.3.2 and 4.3.4).

(xi)   The Public Liability Insurance Act, 1991

This is one of the most important legislative measures enacted in India to provide immediate relief to victims of accidents which occur while hazardous substances. The Act, for the first time acknowledged the principle of no-fault liability. The Act covers accidents involving hazardous substances and insurance coverage for these. Where death or injury results from an accident, this Act makes the owner liable to provide relief as is specified in the Schedule of the Act. The PLIA was amended in 1992, and the Central Government was authorized to establish the Environmental Relief Fund, for making relief payments. PLIA is an effective measure of risk transfer through insurance and also providing compensation to victims timely although the status of implementation is weak.

(xii)   Biological Diversity Act 2002
The Act provide for the conservation of biological diversity, sustainable use of its component and fair and equitable sharing of the benefits arising from the biological resources and knowledge as well as facilitating access to them in a sustainable manner and through a just process. (See Section 2 (g) "fair and equitable sharing" means sharing of benefits as determined by the National Biodiversity Authority under section 21). The provision of *in-situ* preservation and equitable distribution and decentralisation are some of the salient features of effective governance as envisaged in the Act.

(xiii)   Disaster Management Act 2005
An Act to provide for the effective management of disasters and for matters connected therewith or incidental thereto. Section 2 (d), definition of disaster is worthwhile to note. "Disaster" means a catastrophe, mishap, calamity or grave occurrence in any area, arising from natural or man-made causes, or by accident or negligence which results in substantial loss of life or human suffering or damage to, and destruction of, property, or damage to, or degradation of, environment, and is of such a nature or magnitude as to be beyond the coping capacity of the community of the affected area". Chapter 2, Section 8, the National Executive Committee comprised of Secretaries of ministries including environment is a key step towards mainstreaming DRR.

(xiv)   Forest Rights Act, 2006 (Ministry of Tribal Affairs)
This Act, by recognising forest dwellers' rights, makes conservation more accountable. Provisions of this Act provides for the sustainable use, conservation of biological diversity and ecosystems and strengthening the conservation regime by strengthening the livelihood and food security of forest dwellers.

(xv)   National Green Tribunal Act, 2010
The National Green Tribunal has been established on November 18, 2010 under the National Green Tribunal Act 2010 for effective and expeditious disposal of cases relating to environmental protection and conservation of forests and other natural resources including enforcement of any legal right relating to environment and giving relief and compensation for damages to persons and property and for matters connected

The NGT had fined Rs 1 Lakh each on the Union Environment and Forest Ministry and the Assam Government for letting industries mushroom illegally around the Kaziranga National park over last 15 years. The tribunal ordered the closure of stone crushers, brick kilns, tea factories and others in the no development zone around the biodiversity rich zone and noted that the case was "a clear example of infringement of law to the optimum".
Times of India, Saturday, September 8, 2012

therewith or incidental thereto. It is a specialized body equipped with the necessary expertise to handle environmental disputes involving multi-disciplinary issues. The Tribunal shall not be bound by the procedure laid down under the Code of Civil Procedure, 1908, but shall be guided by principles of natural justice. The Tribunal's dedicated jurisdiction in environmental matters shall provide speedy environmental justice and help reduce the burden of litigation in the higher courts. The Tribunal is mandated to make and endeavour for disposal of applications or appeals finally within 6 months of filing of the same. Initially, the NGT is proposed to be set up at five places of sittings and will follow circuit procedure for making itself more accessible. New Delhi is the Principal Place of Sitting of the Tribunal and Bhopal, Pune, Kolkata and Chennai shall be the other four place of sitting of the Tribunal. The act has specific provisions for compensation and also for protecting ecosystems as evident from the example of Kaziranga where the NGT had fined Rs 1 Lakh each on the Union Environment and Forest Ministry and the Assam Government for letting industries mushroom illegally around the Kaziranga National park over last 15 years.

(xvi)The Mines and Minerals (Development and Regulation) Bill, 2011
The act is to consolidate and amend the law relating to the scientific development and regulation of mines and minerals under the control of the Union. Chapter 4 Section 21 (1) (iv) mandates the collection of baseline information on environmental conditions even before the reconnaissance or prospecting operations. Section 21(1) (v), propose steps to be taken for protection of environment which shall include prevention and control of air and water pollution, progressive reclamation and rehabilitation of the land disturbed by the prospecting operations, a scheme for the plantation of trees, restoration of local flora and water regimes and such other measures for minimizing the adverse effect of prospecting operations on the environment. Such measures are extremely important for reducing the risk of landslides, floods and also for controlling the environmental degradation and damage to ecosystems.

13.4.3.2 Rules

(i) The Manufacture, Storage and Import of Hazardous Chemical Rules, 1989 (Notified under the EP Act, 1986 on November 27, 1989, 20 Rules and 12 Schedules)
In MSIHC rule a Major Accident is defined as an "incident involving loss of life inside or outside the site or ten or more injuries inside and/or one or more injuries outside or release of toxic chemical or explosion or fire of spillage of hazardous chemical resulting in 'on-site' or 'off-site' emergencies or damage to equipment leading to stoppage of process or adverse effects to the environment". This rule covers issues broader than process safety and onsite emergencies. As per the provisions of the rules various pre-disaster measures like Hazard Identification, Risk Assessment, Preparation of onsite plans and offsite plans conducting mock drills etc. are mandatory. Refer to Rules related to the responsibility of occupier, identify major accident hazards. Rule 4(2)(a),take steps to prevent major accidents, and to limit the consequences. Rule4(2)(b)(i),Train persons at site and provide equipment for safety. Rule 4(b)(2)(ii),Notify major accidents within 48 hours. Rule (5), prepare MSDS Rule 17(2), label containers of hazardous chemicals, Rule 17(4), inform import of hazardous material. Rule 18(2).Rule 13 mandates the preparation of on-site emergency plan by the occupier and mock drill (once in every six months). Rule 14 mandates the preparation of off-site emergency plan by the district authority. Rule 15 is about sharing of the information related to the potential hazards to persons

liable to be affected by a Major Accident. This rule primarily looks at Chemical Accidents although the damage to environment is also covered in the definition of major accident.

(ii) The Chemical Accidents Emergency Preparedness and Response Rules (EPPR), 1996

The key objective of this rule is to strengthen the Administrative Response to Hazardous substance accidents and to supplement the MSIHC Rules of 1989. The rule mandated the constitution of Crisis Groups at 4 levels, i.e. Central, State, District and Local Level. It is worthwhile to quote the definition of Major Chemical Accident as given in section 2 (a). "Major Chemical Accident means an occurrence including any major emission, fire or explosion involving one or more hazardous chemicals and resulting from uncontrolled development in the course of Industrial activity or due to

> Schedule 1 has mentioned ten categories of biomedical waste which include human anatomical waste, animal waste, microbiology and biotechnology waste, waste sharps, discarded medicines and cytotoxic drugs, soiled waste, solid waste, liquid waste, incineration ash and chemical waste.

natural events leading to serious effects both immediate or delayed, inside or outside the installation cause substantial loss if life and property including adverse effects on environment". Some of the important provisions are Rule 4: Central government shall set up a functional control room with networking with state and districts; Rule 5: Functions of the Central Crisis Group; Rule 6 and 7: Constitution and Functions of the State Crisis Group respectively; Rule 8, 9: Constitution and Functions of the District Crisis Group respectively; Rule 10: Function of Local Crisis Group – Industrial Pocket level; Rule 13: Providing Information by the CCG, SCG, DCG and LCG.

(iii) Manufacture, Use, Import, Export and Storage of Hazardous Microorganisms, Genetically Engineered Organisms or Cells, 1989

The key objective of the rule is to regulate the manufacture, use, import, export and storage of hazardous microorganisms and genetically engineered cells in industry, hospitals, research institutions and other establishments which handle micro-organisms or which are engaged in genetic engineering. Under Rule 4, a committee of experts is established which recommends various safety regulations and prescribes the procedures for restricting or prohibiting production, sale, import and use of specified organisms. Under Rule 7, handling, manufacture and use of hazardous microorganisms is prohibited except with the approval of Genetic Engineering Approval Committee. A district level committee is required to prepare off-site emergency plans for major accidents caused by the escape of harmful micro-organisms.

(iv) Dumping and disposal of fly ash discharged from coal of lignite based thermal power plants on land, Rules, 1999.

The main objective of the rule is to conserve the topsoil, protect the environment and prevent the dumping and disposal of fly ash discharged from lignite-based power plants. The salient feature of this notification is that no person within a radius of 50 km from a coal-or lignite-based power plant shall manufacture clay bricks or tiles without mixing at least 25 per cent of ash with soil on a weight-to-weight basis. For the thermal power plants the utilisation of the flyash would be as follows:

a) Every coal-or lignite-based power plant shall make available ash for at least ten years from the date of publication of the above notification without any payment or any other consideration, for the purpose of manufacturing ash-based products such as cement, concrete blocks, bricks, panels or any other material or for construction of roads, embankments, dams, dykes or for any other construction activity.

b) Every coal or lignite based thermal power plant commissioned subject to environmental clearance conditions stipulating the submission of an action plan for full utilisation of fly ash shall, within a period of nine years from the publication of this notification, phase out the dumping and disposal of fly ash on land in accordance with the plan.

(v)   Bio-Medical Waste (Management and Handling) Rules, 1998
These Rules have been enacted to regulate through a licensing and reporting system the biomedical waste generated by hospitals, clinics, blood banks and other organisations who generate, collect, receive, store, transport, treat, dispose or handle bio-medical waste in any form (Rule 2). Rule 3(5) defines biomedical waste as any waste which is generated during the diagnosis, treatment, or immunization of human beings or animals or in research activities or in the production or testing of biological and including categories as mentioned in Schedule 1. Rule 4 envisages duty on the occupier to take steps to ensure the wastes are handled without any adverse effect to human health and the environment. Rule 6 talks about the preventive measures and provides that bio-medical waste should not be mixed with any other wastes and shall be segregated at the point of generation prior to storage, transportation, treatment and disposal. No untreated waste shall be kept for more than 48 hours except with the permission of the prescribed authority (State Pollution control Board in states and the Pollution Control Committee in Union Territories, Rule 7) if necessary. Rule 9 puts an obligation on the Government of every state/ Union Territory to constitute an advisory committee to advise the government and prescribed authority about matters related to the implementation of these rules. Rule 12 talks about the accident reporting while handling of bio medical waste to the prescribed authority.

(vi)   The Hazardous Wastes (Management and Handling) Rules,1989
Rule 2 provides that these rules shall apply to the handling of hazardous substances as specified in the schedules to the rules (except waste water and exhaust gases which are covered under Water Act, 1974 and Air Act, 1981; waste discharge from ships covered under Merchant Shipping Act, 1958; and radioactive waste as covered under the provisions of Atomic Energy Act, 1962). Rule 3(x) ensure that hazardous wastes are managed in a manner, which will protect human health and the environment against the adverse effect, which may result from such wastes. Rule 4 envisages responsibility of the occupier and operator of facility for handling of the wastes whereas rule 5 talks about the grant of authorization for handling of hazardous waste from State pollution Control Board. Rule 7 provides that occupier shall ensure that hazardous wastes are packaged, based on the composition in a manner suitable for handling, storage and transport. Packaging, labelling and transport shall be in accordance with the rules made under Motor Vehicles Act, 1988. Rule 8 emphasises to undertake an Environmental Impact Assessment of the disposal sites for hazardous wastes. Rule 11 talks about the trans-boundary movement of the hazardous wastes and provides that export and import of hazardous wastes for dumping and disposal shall not be permitted except as a raw material for recycling or re-use (Rule 12). The Ministry of Environment and Forest will grant the permission for import. According to rule 15 the movement of hazardous waste shall be considered illegal if it is without prior permission of the Central Government, fraud or falsification permission and does not confirm to the shipping details provided in the document.

(vii)   Environment (siting for industrial projects) Rules, 1999

These rules are precautionary in approach where it prohibits the setting of certain industries in the area within the municipal limits and 25km belt around the cities having population more than 1 million, 7 km around the periphery of wetlands as listed in annex 1, 25 km around the periphery of protected areas including national parks, sanctuaries, biosphere reserves and 0.5 km wide strip of either sides of national highways and railway lines (Rule 2). Rule 2(3) poses restriction on the establishments of new units and expansion or modernization of existing units of industries in the Taj Trapezium zone. According to rule 2(4) establishment of new units, as listed in annex III, around the archaeological monuments shall not be allowed within 7 km periphery of the of the important archaeological monuments listed in annex IV of the rule.

(viii)   The   Noise   Pollution   (Regulation   and   Control)   (Amendment)   Rules,   2000

These rules provides that the increasing ambient noise levels in public places from various sources have deleterious effects on human health and the psychological wellbeing of the people. Therefore, it is considered necessary to regulate and control noise pollution with the objective of maintaining ambient air quality standards in respect of noise. Rule 4 provides that the noise levels in any area zone shall not exceed the ambient air quality standards in respect of noise as mentioned in the schedule. Rule 5 imposes restrictions on the use of loudspeakers/public address system except after obtaining written permission from the authority and shall not be used in the night (10.00 pm to 6.00 am). Any person may, if the noise level exceeds the ambient noise standards by 10 dB (A) or more make a complain to the authority (Rule 7) and authority shall act on the complaint and take action against the violator in accordance with the provisions of these rules and any other law in force.

(ix) The Municipal Solid Waste (Management & Handling) Rules, 2000

These rules mandates every municipal authority shall, within the territorial area of the municipality, be responsible for the implementation of the provisions of these rules. The municipal authority shall also be responsible for any infrastructural development, collection, storage, segregation, transportation, processing and disposal of municipal solid waste (Rule 4). According to Rule 7 any municipal solid waste

> According to the MSW Rules, bins for storage of biodegradable wastes shall be painted green, for recyclable wastes shall be painted white and those for storage of other wastes shall be painted black.

generated in a city or town shall be managed and handled in accordance with the compliance criteria and procedure laid down in Schedule II. Rule 7(1) provides for the collection of municipal solid waste from various sources, Rule 7(2) provides for segregation of waste and promote reuse and recycling of segregated materials. Rule 7(3) envisages establishment and maintenance of storage facilities in such a way as they do not create unhygienic and unsanitary conditions around it. Storage facilities shall have easy to operate design for handling, transfer and transportation of waste. Rule 9 provides that when an accident occurs at any municipal solid wastes collection, segregation, storage, processing, treatment and disposal, the municipal authority shall report the accident in the prescribed form to the Secretary In-charge of the Urban Development Department in metropolitan cities and to the District Collector in all other cases. This rule is of great importance since municipal waste management is a challenge in post disaster situations and effective management of municipal wastes in the disaster affected areas and relief camp is import for averting disease outbreaks in the post disaster phase.

(x) Hazardous Waste (Management, Handling and Trans-boundary Movement) Rules, 2008
Hazardous waste is any waste that is toxic, explosive, flammable or corrosive excluding the e-waste, radioactive wastes and municipal waste. Rule 4 describes the responsibility of the occupier for handling hazardous waste. According the Rule 4 (5) (i) and (ii) occupier should take necessary steps for prevention of accidents and mitigating the impacts on human being and environment. The rule 13(i) prohibits import of hazard waste for the purpose of disposal in India. Export of waste subjected to PIC and after ensuring that the country importing waste have environmentally sound techniques for treating and disposing the waste. Rule 24 provides for the accident reporting mechanism and it mandates the occupier and transporter to report to state pollution control board in case of an accident while handling or transporting such wastes. Rule 25 provides for the penalty in-case of an accident occurs while handling or transporting hazardous chemicals. The penalty to be imposed on occupier/transport as per the rules of State Pollution Control Board.

(xi)   Explosives Rules 2008
Explosive Rules covers all regulatory aspects concerning manufacture, possession, use, sale, transport and import & export of explosives. The rules provides for regulations pertaining to grant authorization of explosives, issue license to possess explosives, give permission to import, export and transport of explosives and testing, analysis, monitoring and disposal of explosives. Key provisions are (i) employment of 'Competent Persons' for supervision (Rule 11)(ii) Prescribed precautions to be observed in handling explosives (Rule 12) (iii)Prohibition of smoking, fires, lights and dangerous substances (Rule 14)(iv) Transport (Rule 61-3)- Engage drivers or cleaners whose antecedents are verified by local police (Re-verification in each year) , Transportation only between sunrise and sunset (v) Road van always attended by two armed guards at the expenses of licensee (Rule 67)(vi)Restrictions on transportation of different explosives in the same carriage (Rule 33) (vii) Maximum quantities (Rule 36) (viii) Protection from fire and explosion (Rule 41) and (ix) Safety distances (Rule 137). These rules have several provisions related to safety and prevention of accidents and also addressing the security issues associated with the handling and transport of explosives.

(xii)   Wetlands (Conservation and Management) Rules, 2010
Wetlands are conserved and regulated as per the provisions laid down in Ramsar Convention, 1971 to which India is signatory. As declared by UNESCO, World Heritage site, high land or wetlands are identified by the designated authority. Activities like reclamation of wetland, setting up of industries around the vicinity, solid waste dumping, manufacture or handling of hazardous wastes; discharge of untreated wastes and any construction work or any other activity having adverse impact on environment are strictly restricted within wetlands (Rule 4). Rule 5 envisages the constitution of Central Wetlands Regulatory Authority for conservation, preservation and wise use of wetlands. There are total 25 wetland sites as identified by the Ramsar Convention (Rule 3(i)). Wetlands are the most important part of the hydrological cycle and provide wide range of ecosystem services which includes provisional, recreational, regulatory and supporting services. Wetlands act as natural buffers to control floods and reduce the impact of tides and storm surges and helps in waste assimilation, water purification, erosion control and groundwater recharge. Further wetlands are ecosystems which support wide range of biological diversity and hence a major source food.

(xiii)   Plastic Waste (Management and Handling) (Amendment) Rules, 2011
Salient features of the rules are ban on use of plastic materials in sachets for storing, packing or selling *gutkha,* tobacco and *panmasala*, no food stuffs will be allowed to be packet in recycled plastics or compostable plastics, recycled carry bags to have specific BIS standards, colour to the prescription by the Bureau of Indian Standards (BIS), uniform thickness shall not be less than 40 microns in carry bags etc. One of the major provisions under the rules is the explicit recognition of the rule of waste pickers. The new Rules require the municipal authority to constructively engage agencies or groups working in waste management including these waste pickers. This is the very first time special recognition for the rag pickers has been made. The Municipal authority shall be responsible for setting up, operationalization and coordination of the waste management system and for performing the associated functions, This include to ensure safe collection, storage, segregation, transportation, processing and disposal of plastic waste:, no damage to the environment during this process, setting up of the collection centres for plastic waste involving manufacturers, its channelization to recyclers, to create awareness among all stakeholders about their responsibilities, and to ensure that open burning of plastic waste is not permitted. Plastic wastes are harmful to environment since they contaminate surface and ground water, block the water channels (natural and manmade drainage) and increase floods.

13.4.3.3 Notifications

(i)   Coastal Regulation Zone (CRZ) Notification (revised in 2011)
"Key objective of the notification is to ensure livelihood security to the fisher communities and other local communities, living in the coastal areas and to conserve and protect coastal stretches, its unique environment and to promote development through sustainable manner based on scientific studies and taking into account the dangers of natural hazards in the coastal areas, sea level rise due to global warming etc.". The coastal stretches of the country and the water area upto its territorial water limit, excluding the islands of Andaman and Nicobar and Lakshadweep and the marine areas surrounding these islands upto its territorial limit, declared as Coastal Regulation Zone (hereinafter referred to as the CRZ) and restricts the setting up and expansion of any industry, operations or processes and manufacture or handling or storage or disposal of hazardous substances   as specified in the   Hazardous Substances (Handling, Management andTran boundary Movement) Rules, 2009 in the aforesaid CRZ. Section 5 of the notification mandates the preparation of Coastal Zone Management Plans by respective State and UT in consultation with scientific institutions and various stake holders. Guidelines for preparation of Coastal Zone Management Plans given in Annexure 1. Hazard mapping (Step D) is one of the components as envisaged in the guideline. This notification is one of the key steps towards conserving the marine and coastal resources and improving the livelihood options of the community. Conserving the resources like mangroves, corals and other coastal ecosystems will help in creating natural buffers against tsunami, storm surge and coastal erosion.

(ii)   EIA Notification 1994 (revised 2006)
Section 7 of the notification prescribes the environmental clearance process for new projects comprised of a maximum of four stages, all of which may not apply to particular cases as set forth below in this notification. These four stages in sequential order are (i) Stage (1) Screening (Only

for Category 'B' projects and activities) (ii) Stage (2) Scoping (iii)   Stage (3) Public Consultation and Stage (4) Appraisal. Section 9 of the notifications provides detailed list of factors to be taken into consideration which could lead to adverse environmental effects or the potential for cumulative impacts with other existing or planned activities in the locality. Part III of the Section 9 sets the criteria for "Environmental Sensitivity" taking into consideration of various environmental, cultural and social factors. Areas occupied by vulnerable groups or sensitive land uses/buildings like hospitals, schools, places of worship; community facilities are included in the sensitive zone. Areas susceptible to natural hazard like earthquakes, subsidence, landslides, erosion, flooding or extreme or adverse climatic conditions are also included under these criteria. Check List of Environmental Impacts (Annexure II of the EIA Notification) is comprehensive and address impact of the project on land environment, water environment, vegetation, fauna, air environment, aesthetics, socio-economic aspects, building materials and energy conservation. The notification provides for the formulation of the Environment Management Plan consisting of all mitigation measures for each of the component mentioned above and activity to be undertaken during the construction, operation and the entire life cycle to minimize adverse environmental impacts as a result of the activities of the project. It would also delineate the environmental monitoring plan for compliance of various environmental regulations including the steps to be taken in case of emergency such as accidents at the site including fire. This notification is having provisions for disaster prevention, mitigation, emergency management and integrating environment and disaster management in development planning process.

13.4.4 Policies

(i)   National Forest Policy, 1988
National Forest Policy is a key policy is having several provisions related to DRR. The policy envisages the maintenance of environmental stability through preservation and restoration of the ecological balance that has been adversely disturbed by serious depletion of the forests of the country. The policy recommends checking soil erosion and denudation in the catchment areas of rivers, lakes, reservoirs in the "interest of soil and water conservation, for mitigating floods and droughts and for the retardation of siltation of reservoirs and controlling the extension of sand-dunes in the desert areas of Rajasthan and along the coastal tracts. This policy by conserving the forests helps in reducing the risks of floods, landslides and enhanced the livelihood options of the people depending on forest. Disaster Risk Reduction measures in the policy are evident in the section 2 i.e. the basic objectives.

(ii)   National Water Policy,2002
Objectives of the National Water Policy of India are planning, development and management of water resources. The policy by addressing the need of conserving and effectively managing water resources is paving way to mitigate impact of hydro-meteorological hazards and climate change. Water policy recognised water as a key environmental resource for sustaining all life forms. 'Water is part of a larger ecological system' (1.3). Disaster risks like floods and drought are of concerns in the water policy of India. Socio-economic aspects and issues such as environmental sustainability, appropriate resettlement and rehabilitation of project-affected people and livestock, public health concerns of water impoundment, dam safety etc. are envisaged in the policy while planning and implementation of water resources projects.

In the policy it is recognised that problems of water logging and soil salinity have emerged in irrigation command areas, leading to the degradation of agricultural land. Besides the physical issues the social problems like equity and social justice in regard to water distribution are required to be addressed. Considering the decline in quality and quality of ground water due over exploitation in certain parts of the country the concern and need for judicious and scientific resource management and conservation is envisaged in the policy. The policy recommended water zoning of the country and the development activities including agricultural, industrial and urban development should be guided and regulated in accordance with availability and constraints of water resources. Both in urban and in rural areas adequate safe drinking water facilities should be provided to the entire population. Water environment management with watershed, forestry, soil conservation, and agriculture and river-basin approach is the central philosophy of water policy (3.3 to 3.5). It also prescribes for drainage and irrigation management (6.6), groundwater (7.4), water quality (14), zoning (15), with devoted sections like – floods (17), sea or river erosion (18), drought (19) and use of science-technology and environmental impacts (25).Water Policy is giant leap towards reducing the impact of hazards and vulnerability reduction by assuring drinking and irrigation water supply through structural and non-structural measures. The policy also recommends integration of concerns of water resource management in developmental activities.

(iii) National Agricultural Policy

India's agriculture sector accounts for 18 per cent of GDP, and employs around 60 per cent of the workforce. National Agriculture Policy (NAP) aims to attain an annual growth rate of 4 per cent in the agricultural sector over two decades (2000-2020). NAP is another important policy to address the issues pertaining to protection and management of natural resources and ecosystems in rural areas in particular for hydro-meteorological disasters and climate change issues. The new Agriculture Policy, 2000, of India emphasized on Sustainable agriculture, Food and nutritional security, Technology generation & transfer, risk management and organizational framework. Agro-ecosystems include mainly the man-made ecological production systems – farms, plantations, ponds, etc. also the natural systems used for bio-productivity purposes, covering the purposes of food, dairy, fisheries and other livestock, etc.

(iv) National Environment Policy 2006

The National Environmental Policy, 2006 adopts an integrated approach including coastal zone management, management of wetlands and river systems; conservation and development of mountain ecosystems; land use planning; watershed management and reducing the impacts natural hazards like, flood, landslides, storm surges and climate change. EIA notification envisages for Hazard Mapping, Vulnerability and Risk Assessment Report as a part of environment management plan of the projects.

(v) Urban Sanitation Policy, 2008

Sanitation is a key aspect of disaster management due to its significance in pre-disaster phase in relation to water and garbage related health disasters (epidemics), as well as in the post-disaster risks in context of hydro-meteorological disasters like flood, cyclone and drought, etc. during response and relief phase. While this policy pertains to management of human excreta and associated public health and environmental impacts, it is recognized that integral solutions need to take account of other elements of environmental sanitation, i.e. solid waste management; generation of industrial and other specialized/hazardous wastes; drainage; as also the management of drinking water supply. Key Sanitation Policy Issues like addressing poor and served people, integrated approach, technology choice, occupation and organizational aspects of sanitation. The policy prescribes for a city-sanitation plan and its implementation. Policy also emphasized the need of emergency preparedness and response aspects of sanitation management in cities.

(vi)   National Disaster Management Policy 2009

Definition of Disaster as per the DM Act of India recognises damage to environment as a disaster. Introduction to disaster risks in India (Refer 1.2.1), recognize environmental degradation and climate change as factors increasing people's vulnerability. Objective of the policy (2.4.1) encourage mitigation measures based on technology, traditional wisdom and environmental sustainability. Section on 'Environmentally Sustainable Development' (5.1.6). The policy recognise that environmental considerations and developmental efforts, need to go hand in hand for ensuring sustainability. Restoration of ecological balance in Himalayan regions and raising coastal shelter belt plantations need to be incorporated wherever necessary in disaster management plans. Ecosystems of forests, islands, coastal areas, rivers; agricultural, urban and industrial environment are also to be considered for restoration of ecological balance and sustainable development. Zonal regulations should ensure preservation of natural habitats. Climate Change Adaptation (5.1.7) with focus on glacial reserves, water balance, agriculture, forestry, coastal ecology, biodiversity and health in order to reduce disaster risks and vulnerability.

   (6.3.1) Environmental and hazard data for formulation of alternative land-use plans for different   geographical and administrative   areas with   a holistic approach Institutional arrangements (12.2.1) emphasised the need of close interaction with Central Ministries and Departments of Agriculture, Atomic Energy, Earth Sciences, Environment & Forests, Health, Industry, Science & Technology and Space; and with academic institutions.

13.4.5 International Environment Laws and Implications

Environmental law in India has developed tremendously in the last couple of decades in parallel and complimentary to the development of International Environmental Law. The UN Conference on Human Environment and Development at Stockholm in 1972 is considered to be the Magna Carta of environment protection and Sustainable Development. It recognises the healthy and disaster free environment as an extension to the right to life. The report of the World Commission on Environment and Development (Brundtland Report) in 1987 not only provided impetus to sustainable development but also brought into focus the common concerns of the people and endeavours which we need for peace, security development and environment protection. The UN conference on Environment and Development in 1992 (popularly known as Earth Summit) was the largest UN conference ever held expands further the concept of sustainable development and reaffirms the importance and centrality of Polluter Pays Principle, Precautionary Principle and Environment Impact assessment (for detail please refer Module-1, Environment Legislation for Disaster Risk Management). The major achievement of Rio conference was Rio Declaration on Environment and Development, Agenda 21, Forest principles and two legally binding conventions on Climate Change and Biodiversity that are aimed at preventing global climate change and the eradication of biologically diverse species. In 2002, the United Nations organised the World Summit on Sustainable Development (WSSD) in Johannesburg to reaffirm the commitment among the nations to build a humane, equitable and caring global society cognizant of the need for human dignity for all. WSSD is a step ahead in moving from concepts to actions. The world once again came together at Rio De Janerio in June 2012 (popularly known as Rio +20 Earth Summit) to discuss issues around Sustainable Development but summit did not result in any concrete outcome and failed to recreate the history of 1992. However concern for environment is always there which is evident from various treaties, protocols and conventions time to time. The table below summarises status of India vis-à-vis International Environmental Conventions.

Table 13.1: Status of India Vis-à-vis International Environmental Conventions

Convention	Effective from	Year signed and enforced
International Convention for the Prevention of Pollution of the Sea by Oil (1954)	1974	1974
The Antarctic Treaty (Washington, 1959)	1998	1983
Convention on Wetlands of International Importance, Especially as Waterfowl Habitat (Ramsar, 1971)	1982	October 1, 1981 (ac)
Convention Concerning the Protection of the World Cultural and Natural Heritage (Paris, 1972)	1978	1977
Convention on International Trade in Endangered Species of Wild Fauna and Flora (Washington, 1973)	1976	1974
Convention on the Conservation of Migratory Species of Wild Animals (Bonn, 1979)	1982	1979
Convention on Early Notification of a Nuclear Accident (1986)	1988	1986
United Nations Convention on the Law of the Sea (Montego Bay, 1982)	1995	1982
Protocol on Substances That Deplete the Ozone Layer (Montreal, 1987)	1992	June 19,1992 (ac)
Convention on the Control of Transboundary Movements of Hazardous Wastes and Their Disposal (Basel, 1989)	June 24, 1992	March 5, 1990
Amendments to the Montreal Protocol on Substances That Deplete the Ozone Layer (London, 1990)	1992	June 19,1992 (ac)
Protocol on Environmental Protection to the Antarctica Treaty (Madrid, 1991)	1998	1992, 1996
United Nations Framework Convention on Climate Change (Rio de Janeiro,1992)	1994	November 1,1993
Convention on Biological Diversity (Rio de Janeiro, 1992)	February 18,1994	June 5, 1992
Convention to Combat Desertification in Those Countries Experiencing Serious drought and/or Desertification, Particularly in Africa (Paris, 1994)	December 17, 1996	October 14, 1994
International Tropical Timber Agreement (Geneva, 1994)	1997	October 17, 1996

Convention	Effective from	Year signed and enforced
Rotterdam convention on the Prior informed Consent Procedure for certain Hazardous Chemicals and Pesticides in International Trade,(1998)	2004	2005
Protocol to the United Nations Convention on Climate Change (Kyoto, 1997)	2005	1997
Cartagena Protocol on Biosafety (Nairobi, 2000)	January 23, 2001	January 17,2003
Stockholm Convention on Persistent Organic Pollutants (2001)	2004	2006

Source: Handbook on International Environment Agreements: An Indian Perspective. Accessed at http://awsassets.wwfindia.org/downloads/mea_handbook_cel.pdf

13.5 INTEGRATED ENVIRONMENT DRR FRAMEWORK

The Disaster Management Act 2005 recognises damage to or destruction of environment as disaster. The National Disaster Management Authority, the apex guiding organization on disaster management in India, has developed a number of guidelines on disaster management which prescribe for various environmental approaches in disaster mitigation and post-disaster management covered widely under environmental policies and laws .The 1992 UN Convention on the Protection and Use of Trans-boundary Watercourses and International Lakes calls on each party to define water-quality objectives and to adopt criteria and set guidelines for this purpose. Some bilateral and regional agreements on freshwater and air foresee or mandate water-quality objectives. They significantly address the precursors of the hazards in the river-zones and costal zones known to aggravate the impacts of river or sea erosion, flooding, cyclone. Such private regulations may constrain behaviour of breaching by exercising a moral or practical (sanctioning) influence and litigants may argue that breach of such codes or standards may be an evidence of malpractice or negligence.

13.6 CONCLUSION AND RECOMMENDATIONS

Paradigm shift in approach from relief centric to mitigation and mitigation centric approach was witnessed since mid 1990. Hyogo Framework of Action advocates mainstreaming of Disaster Risk Reduction in developmental planning processes. Implication of the HFA is widely reflected in various international and national legislation and policies pertaining to environment, disaster management, land use, forest, water and natural resources. Laws and policies related to environment and natural resource management in India covers various aspects of the disaster management and risk reduction by protecting and safe guarding environment. Indian Judiciary recognised right to pollution free environment as a fundamental right (Article 21). Right to livelihood, equality, freedom of expression and speech etc are few notable provisions in the constitution. Special emphasis on Ecological Security is evident from various laws where human beings are also considered as part of larger ecosystems. Forest Rights Act and PESA are examples. Environmental laws in 70s and 80s were focussed primarily on the control of pollution as evident from Air Act and Water Act. Indian forest laws are one of the earliest laws to abate the uncontrolled use of timber and other forest products although the laws were not conservation centric till 1980 Forest Conservation Act. During the British period the forest laws were dictated by the colonial interest of reserving forest primary for own vested interests in timber and other forest produces. Ensuring safety measures through law was primarily focussed on industrial (Chemical Mining etc.) safety. Oldest laws on safety can be traced back to Explosive Act 1884 to to regulate the manufacture, possession, use, sale, [transport, import and export of explosives. Insecticide Act 1968, an act to regulate the import, manufacture, sale, transport, distribution and use of insecticides with a view to prevent risk to human beings or animals, and formatters connected therewith. The Factories Act 1948 is having provisions for assuring the safety and health of workers. The Factories (Amendment) Act 1987, (in the aftermath of Bhopal disaster) chapter 4 added and provisions for siting of industries is added (Section 41(a)).Clearing of cases related to environmental issues were always been a challenge to judiciary and need of involving subject experts was felt. Environmental Appellate Authority and Green Tribunal in 2010 to deal with cases related to environmental damages. Compensation and provisions for relief and environmental rehabilitation fund are emerged under the EPA Act and rules there under. Risk transfer mechanism through insurance and principle of absolute liability were introduced through the Public Liability Insurance Act. India is having a well evolved Techno-legal framework, where codes are prescribed for various hazard prone areas, industries, fire safety and so on (techno-legal is beyond the scope of this paper). Long term developmental issues since the beginning of a project is very well address through various provision of EIA Notification, EMPs , Coastal Zone Regulations, Flood Zone Ordinance etc. Risk Sensitive Land-use Planning is also emerged as a recent trend where hazard profile and state of environment is made as part of the regional and master plans. Provisions for monitoring in the form of Environmental and Safety Audit reports are mandated in EPA and Factories Act. Fortunately Indian Environmental Laws and Policies have several provisions to address various natural and human induced disasters.

Few limitations are (i) cases related to environment come under civil liability and not criminal liability (Refer NGTA, 2010). Penalties (financial) are not deterrent in many statutory laws and needs amendment. Incentive approach is to be promoted further encourage green development. Liability is more attached to industrial and chemical disasters and natural disasters are yet considered as "Act of God" where no one is held liable. Liability on the authority's by-passing the regulations like Flood Zone Regulation, Coastal Zone Regulation etc. to be made more stringent. Relief and disaster management codes needs to evolve further to accommodate environmental damages of disasters. Environmentally sustainability mitigation options and the concept of 'greening disaster-response' and 'sustainable-recovery' need to be promoted within the framework of sustainable development by integrating SEA to the developmental planning process. SEA and EIA (already exiting) scope need to necessarily include hazard-risk and vulnerability assessment within the framework.

References

Alex, J.P. (2006). *Disaster Management: Towards a Legal Framework.* New Delhi: Indian Institute of Public Administration and UNDP.

Burton, I., R. W. Katesand G. F. White (1993). *Environmental Hazards.* The Guildford Press, London.

Centre for Environmental Law, WWF-India (2006).*Handbook on International Environment Agreements: An Indian Perspective.* Accessed at http://awsassets.wwfindia.org/downloads/.

Department of Agriculture and Cooperation (2000).*National Agriculture Policy*. Department of Agriculture and Cooperation. Government of India

Diwan, S., and Rosencranz, A. (2001). *Environmental Law and Policy in India*. Mumbai: Tripathi and Tripathi.

GDRC (n.d). *A GDRC Special Feature on Environmental Management and Disaster Risk Reduction*. Retrieved in March 2012 from http://www.gdrc.org/uem/disasters/disenvi/-index.html.

GoI. *The Factories Act, 1948 as amended by the Factories (Amendment) Act, 1987.* www.ilo.org/dyn/natlex/docs/WEBTEXT/32063/64873/E87IND01.htm

Gupta, A.K. (2010).Policies, Strategies and Options for Disaster Risk Reduction interventions in India. In A.K. Gupta, S.S. Nair, S. Chopde and P.K. Singh (eds) Proceedings of International Workshop on Risk to Resilience: Strategic Tools for Disaster Risk Management. NIDM New Delhi and ISET, Colarado, US (with Winrock International, DFID and US-NOAA).

Gupta, A.K., and Nair S.S. (eds) (2011). Environmental Knowledge for Disaster Risk Management – Concept Note in Abstract Book of the International Conference 9-10 May 2011, New Delhi. National Institute of Disaster Management, New Delhi and GIZ Germany.

Gupta, A.K., and Nair, S.S. (eds) (2012). Ecosystem Approach to Disaster Risk Reduction. New Delhi: National Institute of Disaster Management.

Gupta, A.K., and Nair, S.S. (2012). Environmental Legislation for Disaster Risk Management, Module-I. Environmental Knowledge for Disaster Risk Management Project, National Institute of Disaster Management and Deutsche Gesellschaft für Internationale Zusammenarbeit (GIZ) GmbH, New Delhi.

Jaswal, P.S. and Jaswal, N. (2004). Environmental Law: Environment Protection, Sustainable Development and the Law (2[nd]Edn.). mea_handbook_cel.pdf.

Ministry of Law and Justice Government of India (2011).*Constitution of India, Updated upto (Ninety-Seventh Amendment) Act, 2011*. Retrieved from http://lawmin.nic.in/olwing/coi/coi-english/coi-indexenglish.htm

MHA (2005). Disaster Management Act, 2005. Ministry of Home Affairs, Government of India.http://mha.nic.in/pdfs/DM_Act2005.pdf.

MoEF (1988). *National Forest Policy*. Ministry of Environment and Forests, Government of India.

MoEF (2006).*The National Environment Policy*. Ministry of Environment and Forests, Government of India.

MoEF (n.d). Legislations on Environment, Forests, and Wildlife. Ministry of Environment and Forests.http://www.envfor.nic.in/legis/legis.html

MoUD (2008). *National Urban Sanitation Policy*. Ministry of Urban Development, Government of India.

MoWR (2002). *National Water Policy*. Ministry of Water Resources, Government of India.

NDMA (2009). *National Policy on Disaster Management*. National Disaster Management Authority, C.

Pelling & Holloway (2000). *Legislation for Mainstreaming Disaster Risk Reduction*. Middlesex, UK: Tearfund.

PESO. Explosive Act, 1884 and Explosive Rules, 2008. peso.gov.in/PDF/UNEP & UN-ISDR (2010). Environment and Disaster Risk: Emerging Perspective, p.13. United Nations Environment Programme, Post-Conflict and Disaster Management Branch, Geneva, Switzerland, Web: http://postconflict.unep.ch

UNEP, UNISDR-PEDRR (2010). Opportunities in Environmental Management for Disaster Risk Reduction: Recent Progress - A Practice Area Review. In *Contribution to the Global Assessment Report on Disaster Risk Reduction*.Special circulation. Retrieved from http://www.preventionweb.net/english/hyogo/gar/background-papers/documentsChap5/thematic-progress-reviews/UNEP-Environmental-Management-for-DRR.pdf

UN-ISDR (2005).*Hyogo Framework for Action 2005-2015; Building the Resilience of Nations and Communities to Disasters, 2005.*

Effectiveness of Environmental Legal Systems:
A Study of the Tribal Areas of Himachal Pradesh

V.B. Negi and Jai Shree

14.1 INTRODUCTION

The word, 'Paryavarana' is used for the environment, which is a protean term as it readily assumes different forms and characters but it is defined to mean the natural and human made world, excluding economic and social matters of the particular areas. Therefore, the environment includes the ecosystem including biodiversity and natural resources; all areas and structures modified or built by humans; and all factors affecting human health and the quality of human life including cultural heritage and amenity exist within the area. However, the economic and social matters are excluded to confine the concept of the environment to its common-usage. The concepts of environment differ from age to age, since it depends upon the condition, prevalent at that particular time for the preservation of the environmental condition, nature, ecology and ecosystem for the better livelihood of the living creature. The human impacts on the environment are not circular because it simply focuses on regulating that part of the environment that can be controlled by humans; the meanings of the environment have been extensively considered and are well enough understood to need little elaboration. The environment, then, is all around us and humanity is part of it. The human impacts on the environment are regulated and it should not lose sight of the fact that human impacts are themselves part of the environment.

Environment is well-defined in the ancient time particularly in the Vedic period. The knowledge of Vedic seers about the basic elements of environment is highly recognized. The ancient Vedic literature and Upanishads described that the universe consists of five basic elements such as, Earth or Land, Water, Light or lustre, Air, and Ether. A disturbance in percentage of any constituent of the environment beyond certain limits disturbs the natural balance and any change in the natural balance causes lots of problems to the living creatures in the universe. Different constituents of the environment exist with set relationships with one another. The relation of human being with environment is very natural as the living organism cannot live without it. The universe is made on scientific principles, and that's why it is well measured. Therefore, the

Disaster Management and Risk Reduction: Role of Environmental Knowledge; Editors Anil K. Gupta, Sreeja S. Nair, Florian Bemmerlein-Lux and Sandhya Chatterji; Copyright © 2013, Narosa Publishing House, New Delhi

environment is the spatial unity of all materials, forces, situations and living creatures, including humans and their behaviour, which influences the continuance of life and welfare of humans and other living creatures. However, the objective of the definition of environment is consist simply of an identification of the component parts of the environment such as land, air and water; the atmosphere; organic or inorganic matter and living organisms; and human-made or modified structures and areas.

14.2 EFFECTIVENESS OF THE ENVIRONMENTAL LEGAL SYSTEM

Environment is one of the most important features for the survival of human life, which is considered as a basic fundamental human right and a social goal in the present era. It is the prerequisite and an integral part of the development of human being. It is man's most precious possession. It influences all his activities; it shapes destinies of all people. It is the essence of productivity in life. Pollution free environment is a relative feeling of being well in body, mind, spirit and normal functioning of tissues, organs and other parts of the body. It is conducive to harmonious, constructive and qualitative life and surrounding of the area and entire ecosystem.

Environmental regulation system is an integral component of wider government policy cycles, planning processes, management, problem framing, policy framing, policy implementation, policy monitoring and evaluation are the good policy for the protection of the environment and natural resources as the environmental laws have developed around the world for the protection and preservation of the environment. The importance of protecting the environment properly for the survival and quality of human life and life on earth makes the environmental legal systems a vital issue and effectiveness of this systems is important for social and economic reasons because they are often an areas for intense political and social conflict, and a significant constraint on development activities for economic growth of the people and state.

India has taken the serious view over the continuing degradation of the environment and for the better management of its natural resources, the proper preservation and conservation of the environment for wellbeing of future generation, the maintenance of the a proper balance between the economic development and the consequential environmental degradation, the provisions and policies have been framed in addition to the constitutional provisions and regulatory framework. The Constitutional Forty-second Amendment Act, 1976 incorporated Articles 48-A and 51-A (g) in the body of the Constitution, which imposed a duty on the state and the citizens to protect and improve the environment and safeguard the forests and wildlife of the country and the environmental policies received a constitutional sanction. The environment legal system comprises of legal rights, duties, powers and liabilities contained in international treaties, customary international law, domestic legislation, and the common law are a vital tool. The extent and exercise of these laws can depend on legislative and administrative objects, policies and principles. Environmental law includes, but is not- limited to, traditional categories such as environmental protection, conservation, pollution, mining, fisheries, cultural heritage, environmental impact assessment, and planning and development of laws to protect the deterioration of environment of the country.

Environmental legal system seeks to set up the administrative structures, limits and procedures that apply to decision-making by government agencies for the protection of environment and such decisions may take place within a wide range of government activities such as assessment of a proposed new development, approval of land use change, the license approval process of the decision to impose sanctions. Legal mechanisms to protect the environment commonly involve various forms of public participation in the government decision-making. The allocation of environmental responsibility to the government means that environmental law is making has existence across the legal system, rather than in one specific locus of responsibility. This undermines the development of a clear identity for environmental law in India. The important environmental legal system is likely to be found in the environment legislation, articles in a range of statutes, regulations and guidelines. The role of the courts is particularly important for the preservation of environment as the Judiciary has come up with the 'judge driven implementation' of environmental administration in India. For the protection of environment, the courts evolved the Doctrines of Public Trust, Precautionary Principle, Polluter Pays Principle, Absolute Liability Principle and Sustainable Development.

When a judicial decision is handed down, a public definitive pronouncement on the interpretation of words in the text of legislation is provided and the core meaning of rules take shape. Without this process, 'rules' in environmental legislation in India will remain simply as words within a text, subject to conflicting interpretation and possible misapplication. The responsibility of the courts is a heavy one. A rule that the courts refuses to apply will not be part of the legal system, even if it is contained in lawfully enacted legislation. Furthermore, if the courts consistently interpret a rule contrary to its original or literal meaning, their reading of it, not its original sense, becomes law.

The environmental degradation in India is divided in to three factors such as, social, economic and institutional factors. Population, poverty and urbanization are the social factors and non-existent or poorly functioning markets for environmental goods and services; market distortions created by price controls and subsidies; the manufacturing technology adopted by most of the industries which generally is based on intensive resource and energy use; expansion of chemical based industry; growing transport activities are the economic factor and expansion of port and harbour activities and the Institutional factors includes lack of awareness and infrastructure. These three factors are interrelated with each other largely. However, the social factors are affected by the degradation of the environmental condition for achieving the economic growth. The damaged environmental condition of the vulnerable areas of the particular region can be redirected by making an ecological sustainable development by effective implementation of environment legal system and enforcement and implementation of the provisions of environmental laws by the state government. The complexity of the environment and the laws, themselves evaluating the effectiveness of an environmental legal system is a Herculean, multi-disciplinary task requiring the integration of environmental science and law. Without a clear conceptual analytical framework and properly supported from the government this task is practically impossible and the communication of any results of such policy improvement is severely hampered. The most complex and difficult environmental issue currently faced by the society is overexploitation of its natural resources by the unsustainable developmental activities. The environmental legal system is the combination of environmental laws with the courts, government departments and other bodies that administer it within a particular jurisdiction or geographic area. It includes the decision-making processes, policies, practices and constitutional constraints that affect the administration of the law. Evaluating the effectiveness of an environmental legal system considers whether it is likely to achieve its goal of ecological sustainable development.

The significance for the evaluation of the effectiveness of the whole or selected parts of environmental legal system is an integral component of wider government policy cycles and planning processes for the developmental activities. A good environmental policy must be based on a cyclical process with four major stages such as problem framing, policy framing, policy implementation, and policy monitoring and evaluation. The integral part of the wider policy process and planning cycle, evaluating the effectiveness of environmental legal systems makes an essential contribution to the system by constantly evolving and changing in response to new information. The environment, society and life are continuous and constantly changing and this is an ongoing and difficult task with no endpoint or final solution. The reason for the evaluation of the environmental legal systems continues to evolve and it should be expected to continue to do so in the future and deal with complex, difficult policy problems and these are themselves changing the scale in nature and for which there are often large gaps in knowledge and information for the proper

The legislative framework in India has attempted to create several legal spaces for the protection of fragile ecosystems. Many government and non-governmental sectors have used clauses of the Environment (Protection) Act, 1986, and the Environment (Protection) Rule to highlight the sensitivity of a region and thus grant it a special status, "to protect and improve quality of the environment". In the more recent instances, these areas have been called ecologically sensitive areas or ecologically fragile areas. The use of the concept of ecologically sensitive areas/ecologically fragile areas together with the Environment (Protection) Act is gradually gaining recognition as a strategy for the conservation and sustainable development of sensitive and fragile areas. The effective participation of local communities for the planning and management of these areas will prevent the damage of environment and help the government and non-government agency to protect the natural resources of the country.

14.3 DEGRADATION OF ENVIRONMENT IN THE TRIBAL AREAS OF H.P

The abode of eternal snow, Himachal Pradesh, is located in the northern part of the Indian sub-continent in the vicinity of the western Himalayan mountain ranges, snow-capped mountains, complex geological structure and rich temperate flora, fauna and biological natural resources. Himachal Pradesh with its rich biodiversity is a natural tonic wrapped up in the form of a wonderland or paradise for the stressed city folk from all over the world. Within its hectares of heavily forested mountains and lowlands, there are numerous varieties of plants and trees, innumerable species of animals & myriad variety of birds and insects which make the state different. The biodiversity of Himachal Pradesh is not only important for the state for generating revenue, butit also has a strong influence on ecology, climate, soil and other natural flora, fauna, environment and ecosystems. The state of Himachal Pradesh consists of twelve districts and has a population of about 60.8 lakh with an area of 55673 square kilometres and density population of 109 persons per square kilometres. Out of these twelve districts, Kinnaur and Lahaul-Spiti districts in their entirety, and Pangi and Bharmour Sub Divisions of Chamba district constitute the scheduled areas in the state. The state had declared these areas as Scheduled areas under the Fifth Schedule of the Constitution by the notification of the President of India as per the Scheduled Areas (Himachal Pradesh) Order, 1975 (CO 102) dated the November 21, 1975. Total geographical area of these tribal areas of the state continues 23655 sq. m. and the population is 1, 66,402 according to the census of 2001, with the average density population of 7 per sq. km.

The tribal areas of the state command certain uniqueness among other districts of the state from the angle of socio-economic development with the difficult climatic conditions. The main inherent constraints of these areas are: (i) their relative inaccessibility, (ii) their terrain and topography characterized by ruggedness and fragility of mountain sides, (iii) Farming and general economic activities in the whole area remaining restricted over calendar time and in the physical space, and (iv) The performance of all activities always entailing higher fiscal cost and requiring more vigorous effort in view of harsh physical climate conditions. The important conditions characterizing the tribal areas of the state, for operational purposes, separating mountain habitats from other areas, are called mountains as well as hilly specificities such as, inaccessibility, fragility, marginality and diversity or heterogeneity in high vulnerability. Natural suitability of the entire tribal area of the state is fragile, and fragility here refers to the vulnerability of hill resources to rapid degradation through small disturbances caused by the development activities intensification of high risk of disaster like earthquake and landslide etc. With the high altitude and steep slopes in association with geologic, edaphic and biotic factors that limit the farmer's capacity to withstand even a small degree of disturbance in these tribal areas. Their vulnerability to irreversible damages, due to overuse or rapid changes, extends to physical land surface, vegetative resources, and even the delicate economic life-support systems of mountain communities. Consequently, when mountain resources and environment start deteriorating due to any disturbance, it happens at a fast rate and in most cases, the damage is irreversible or reversible only over a long period of time.

The other characteristics of these tribal areas are their biophysical and societal factors, internal diversity or heterogeneity and these areas depend on operational implications. The diversity of mountain habitat generating both range of opportunities and a set of constraints based on difficulties and fragile conditions of these areas with hilly environments are also well known for their range of micro-environmental variations, having risk of various types of disaster. In order to use such an environment to its fullest extent and to reduce the emerging risk of any type of disaster, people of hills have developed various practices that are highly labour intensive, due to its diversity and well recognised for implications.

State of Himachal Pradesh, like rest of the states in the country, also embarked upon the path of economic development for realization of social uplifting and could not escape the damage caused to environment by pursing such strategies. By and large such strategies could neither visualize nor comprehend the impact on local environment which being implemented for the economic benefit of its people, pursuance of economic goals often meant sacrificing the environmental consideration, which has now started making its adverse impact on the state's ecology. Further, population pressure on land and other natural resources also leads to degrade the environment for timber and fuel wood etc., in addition to various development activities including construction of large scale Hydro Power Projects being carried out in the state to achieve the economic growth. Activities related to the economic development and growths in population in these areas are contributing degradation of environment. These include pressure on land, soil degradation, forests, habitat destruction, loss of biodiversity, changing consumption pattern, rising demand for energy, air pollution, global warming, climate change and water scarcity and water pollution etc. The deterioration of natural resources by overexploitation in these areas affects the environment and health of the people. Construction of roads, hydropower projects, irrigation channels etc., are accompanied by over exploitation of land, fresh water and other natural resources and use of fertilizers and pesticides in the horticulture and agriculture products in huge quantity have also increased many folds in the degradation of environment.

The effectiveness of an environmental legal system is a significant for the protection of the environment of the fragile area particularly in context of the entire tribal area of Kinnaur, Lahual Spiti districts and Pangi & Bharmour sub division of Chamba district, which are highly rural topography of the state being high areas specialties. The importance of protecting the environment of these hill areas for the survival and quality of life of humans and life on earth makes the effectiveness of environmental legal systems a vital issue. The effectiveness of the environmental legal system is also important for social and economic reasons to provide quality life to these tribal zones because they are often an arena for intense political and social conflict, and a significant constraint on business activity.

Nature has not only bestowed Himachal Pradesh with an abundance of natural beauty, but also priceless natural resources, that are fully capable of making it a powerful state. The economy of the state is strengthened by the natural resources, water of overflowing and gushing rivers which are also one of the precious natural resources of the state. Himachal Pradesh has been blessed with the vast hydroelectric potential as the five major river systems, which emanate from the Western Himalayas, namely Yamuna, Satluj, Beas, Ravi and Chenab, pass through Himachal Pradesh which provides an ideal situation for optimum exploitation of hydro energy generation. Hydropower is a renewable, economic non-polluting and environmentally benign source of energy, and is perhaps the oldest renewable energy technique known to the mankind for mechanical energy conversion as well as electricity generation. Hydropower represents use of water resources towards inflation free energy due to absence of fuel cost with mature technology characterized by the highest prime moving efficiency and spectacular operational flexibility. The total identified hydropotential of these rivers is estimated to be 20415.62 MW out of which 6418.27 MW potential has been exploited so far to make the Himachal Pradesh as Power State.

The most of the hydroelectric power projects in the state of Himachal Pradesh are located in the tribal areas beside the rivers originating from the western Himalayas. Therefore, during the construction of the hydropower projects in these tracts has led to huge destruction of land mass as a result of landslides, landslips, soil erosion and development of high risks of earthquake. The Himalayas are geologically weak and having fragile rock strata, because of which high intensity blasting and rock cutting to create the expensive infrastructural requirement and construction of roads, tunnels, levelling of project sites and different construction activities for hydropower projects give rise to the problems of landslips, landslides and increase the emerging risk of earthquakes as the rocks are highly fractured and jointed in the entire tribal belt. The hydropower development in these tribal areas has also been drawn in to controversies mostly due to social issues like displacement and rehabilitation of people in the project area and non-involvement of the public at large in decision making. The development activities associated with the hydropower projects carried out in these tribal areas have long term effects on environment and ecology. The environmental degradation and development of risks of earthquake in different ways in these tribal areas and fragile topography have been developed and the need of sustainable development is required for the economic growth. In support of the preservation of the environment and ecosystem being damaged by the development activities, the Supreme Court decision on the Sardar Sarover Dam on Narmada River in the case Narmada Bachao Andolan Vs. Union of India, in which the principle of sustainable development and precautionary principle was developed to ensure remedial measures to preserve the ecological balance and reduce the risk of disasters.

The entire tribal areas of the state of Himachal Pradesh are being disturbed by the development activities and construction of large scale hydro power projects being a seismic zone. The environmental baseline of these areas in term of physical, ecological and socio-economic parameters is being disturbed due to the construction of these development activities. The following are the major components of the environment being disturbed by the overexploitation of natural resources in these fragile zones such as:

(i)   Air Environment (Air quality);

(ii) Water Environment (Water resources, water use, water quality, hydrology);

(iii) Noise Environment (Noise levels);
(iv) Land Environment (Land use, geology, seismology and soils);
 (v) Ecological Environment (Terrestrial and Aquatic ecology); and
(vi) Socio-Economic Environment (Demography, Socio-economic, public health etc.).

These types of environmental degradations as a consequence of resource exploitation such as natural/common property resources is a global phenomenon, which is evident from the increasing levels of deforestation, polluted water, air and land resources leading to disaster. The analysis of the environmental legal system framework for the protection of the environment is an important tool to protect it but non implementations of its provisions are the weaknesses in the formal aspects of environmental law in India. It has been found that the provisions of the constitutional right to a 'good and healthy' environment are the basic constitutional foundation to protect the environment as the state obligation is vaguely expressed.

To analyse the effectiveness of the environmental legal system for the protection of the environment of these tribal areas of the state being damaged by the ongoing large scale hydropower projects and other developmental activities, the perceptions of the people has been obtained and analysed in this paper. To obtain primary data the respondents have been divided into eight categories according to their occupation, i.e., administrative services, academics, hydropower & energy, horticulture, agriculture and forestry, PRIs, legal professionals, banking & financial institutions and others. The views of these respondents with regard to effectiveness of the environment legal system to maintain the balance with the environment and natural resources of these tribal areas and find out the ways to reduce the risk factors of disaster have been analysed as under:

14.3.1 Degradation of Environment by Development Projects and other Activities

Denial of Fundamental Rights under Articles 14, 19, and 21 of Indian Constitution the Constitution of India is already seized of the seriousness of the matter and makes provisions for the protection of ecology and environment to the natural form. Several other legislative provisions and acts are also binding on the state to implement the provisions for sustainable development and sanctity of life of future generations particularly in these tribal areas. In order to preserve tribal interest in the scheduled areas, so as to achieve constitutional goals, the fifth schedule is unique, paramount and this aspect was also described in Samatha Vs. State of Andhra Pradesh, (1997) 8SCC 191, that to preserve the environment of the tribal area and over exploitation of the natural resources the court stated that, "an integral scheme of the Constitution with directions... philosophy in which the anxiety is to protect the tribal from exploitation and to preserve the valuable endowment of their land for their economic empowerment to elongate social and economic democracy with liberty, equality, fraternity and dignity of their person and our political Bharat." Constitution of India has invoked the provisions of the environment protection in part III, popularly known as golden triangle of the constitution in its articles 14, 19 and 21. The right to wholesome environment is treated as a part of life guaranteed by Constitution of India under article 21. The approval of the government for the construction of large-scale development projects having negative impacts on the environmental conditions can be invoked by article 14 balancing the environmental imperative with the right to carry on any occupation covered in article 19 (1) (g). For the socio-economic development of the state, large numbers of hydropower projects are being constructed in the major river basin and its tributaries in the entire tribal zone of the state. These hydropower projects are causing degradation of ecosystem of the area and denial of right to life of the people.

Figure 14.1, reveals that 70% respondents in all categories agreed that the large scale developmental activities including hydropower projects are developing three types of adverse environmental impacts such as inevitable, reducible and avoidable and affecting livelihood of the living organism, the life support system and denial of Wholesome environment. The highest are 89% from academics, 85% from legal professions and 75% from others, whereas 21% respondents disagreed that the development activities including hydropower projects don't deny livelihood to the people of these areas. In this category, the highest are 40% from hydropower and energy, followed by 31% from administrative services and 25% from horticulture, agriculture, and forestry and 24% from PRI's. However, 9% respondents did not respond, may be on account of ignorance about the fundamental rights.

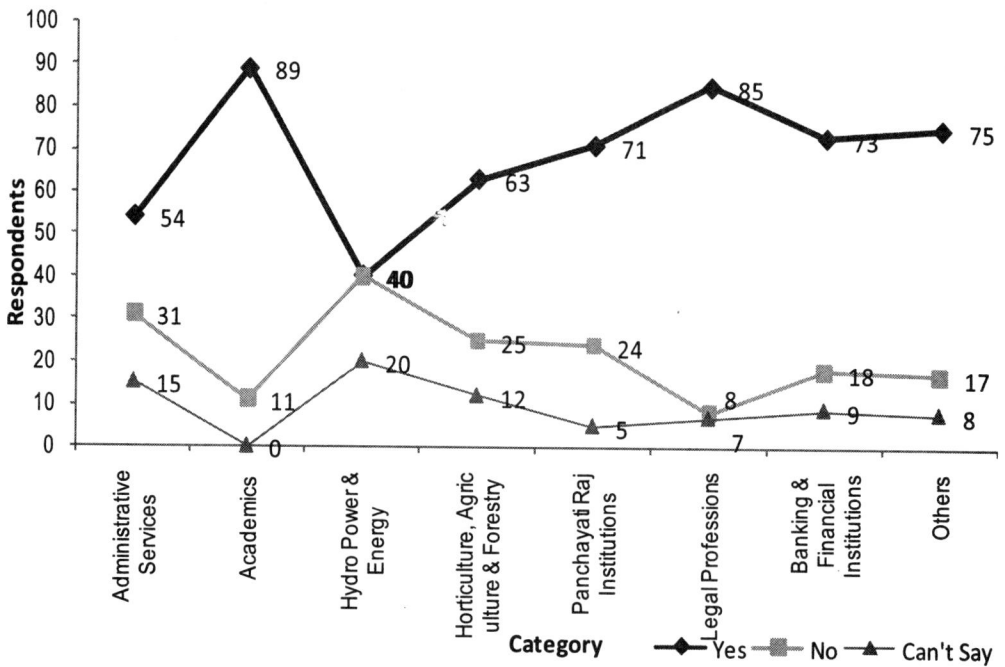

Figure 14.1: Environmental degradation: Denial of fundamental rights

14.3.2 Prevention Strategy through Enforcement of Environment Legal Systems

Sound economic development, which reduces the environmental loads by maintaining healthy and productive environmental conservation activities, encouragement of willingness for environment conservation and environmental education and public awareness clarifying the responsibilities of citizens and local bodies strict enforcement can lead the state to achieve the socio-economic growth in reality. Figure 14.2 reveals that 91% respondents from all categories agree that destruction of natural resources by the developmental activities including hydropower projects and for the prevention of these resources, the Environmental Impact Assessment (EIA) and

Environmental Management Plan (EMP) and strict enforcement of environment legal system is required to be enforced strictly and public awareness for prevention of environmental status is to be raised. In this category, the highest are 95% from PRI's, followed by 92% each from administrative services, others and legal professionals, 91%from banking & financial institutions and 89% from academics. However, 8% respondents disagree. In this category the highest are 20% from hydropower & energy, followed by 12% from horticulture, agriculture and forestry and11% from academics. This may be because of the fact that respondents are in related departments, whereas 1% respondents did not respond.

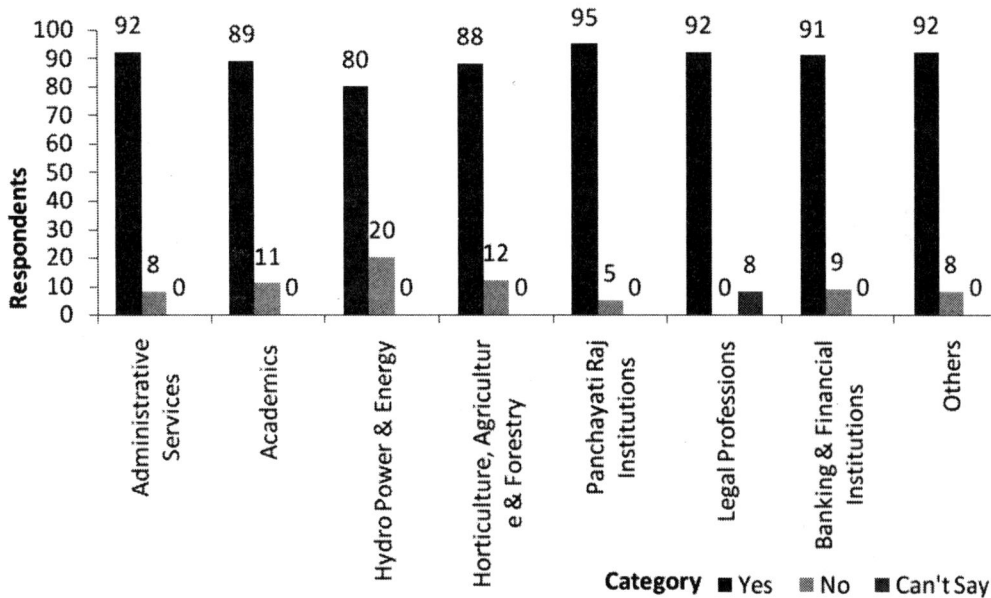

Figure 14.2: Effective prevention strategy

14.3.3 Implementing Environmental Legal System– Challenges

Resistance of political, economic &social forces poses difficulties in implementing environment laws. With the construction of hydropower projects, the direct and indirect impacts on natural vegetation cover occurs at the dam site, areas selected for quarrying and spoil disposal activities, flooded zone and along the access roads on steep slopes. This impact on vegetation includes loss of individual's plants throughout the Right of Way (ROW) of road construction, development of sites and development of quarries and labors camps.

The development activities - construction of roads, buildings, irrigation channels hydropower projects for the overall growth of economy leave social impacts which includes all social, cultural and economic consequences that alter the ways in which people, live, work, relate to one another, organise to meet their needs and generally cope as members of society within the social impacts by these developmental activities with the development of these positive and negative impacts and to achieve economic and social development. Political forces are ignoring the adverse environmental impacts such as deforestation, soil erosion, landslide, over exploitation of natural resources and loss of biodiversity. To achieve the economic growth, the environment aspect is being neglected by the state and implementation of the environmental protection laws and policy as provided in various legislative provisions stand in effective. Figure 14.3 indicate that 83% respondents in all categories out of which, the highest are 89% from academics, followed by 86% from PRI's and 85% from administrative services agree that environment protection and programmes are difficult to implement and harder to enforce, competing as these are with the socio-economic and political forces. According to those, political, economic and social forces, the environmental aspect is less important and development and economic growth is more important to the state, therefore, the implementing agencies do not enforce the programmes and provisions of environment protection laws seriously, however 12% respondents disagree. The highest are 40% from hydropower and energy, followed by 15% from administrative services and 13% from horticulture, agriculture and forestry. This may be because, that these are the agencies responsible for degradation of environment and implementation of the provisions of laws, whereas, 5% respondents do not say anything.

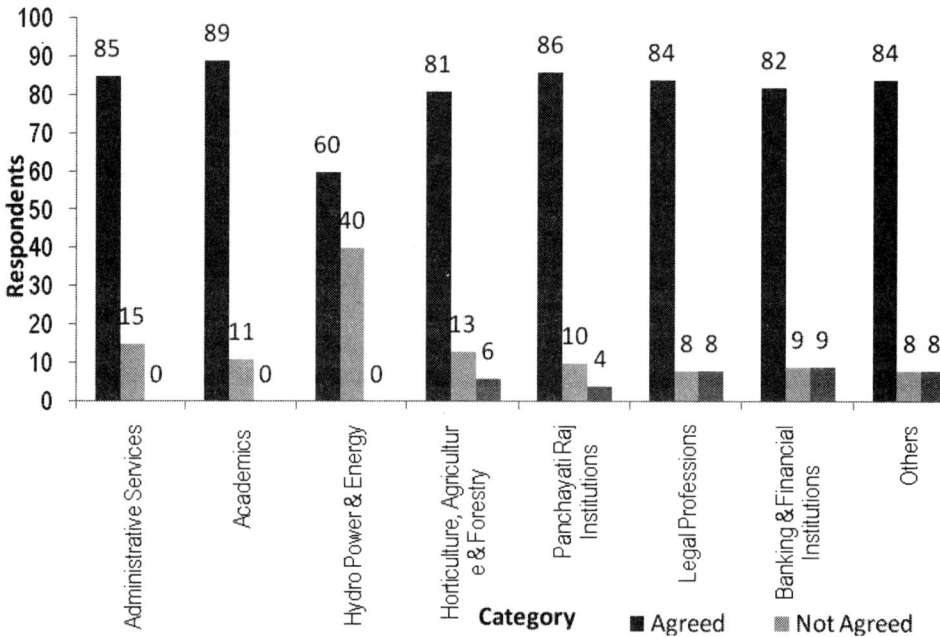

Figure 14.3: Resistance of political, economic &social forces

14.3.4 Environmental Fruits Depend on the Effective Environmental Legal System

Laws cannot affect the natural processes that cause environmental changes but can regulate human behavior in response to natural disasters. Though, humans have been responsible for most of the ecological imbalances and people will protect the environment simply because, it is the cardinal basis for their survival. Laws are essential for effective protection of the environment. It is only through the rule of law that society can protect environment of the area being degraded by the various factors. For this the duty of the every citizen has been provided in article 51-A (g) fundamental duties of the Constitution of India, which states that, "every citizen is having the fundamental duty to protect and improve the natural environment including forest, lakes, rivers wildlife and to have compassion for living creature." It has been seen that the environmental conditions in the entire tribal area of the state of Himachal Pradesh has been degraded and ecological imbalance is seen by the developmental activities being carried out. To maintain the ecological balance and to enjoy healthful environment as well as to achieve the goals of constitutional provisions as stated above, the local residents of these areas are also responsible by making the sincere efforts to protect the environment of these fragile areas.

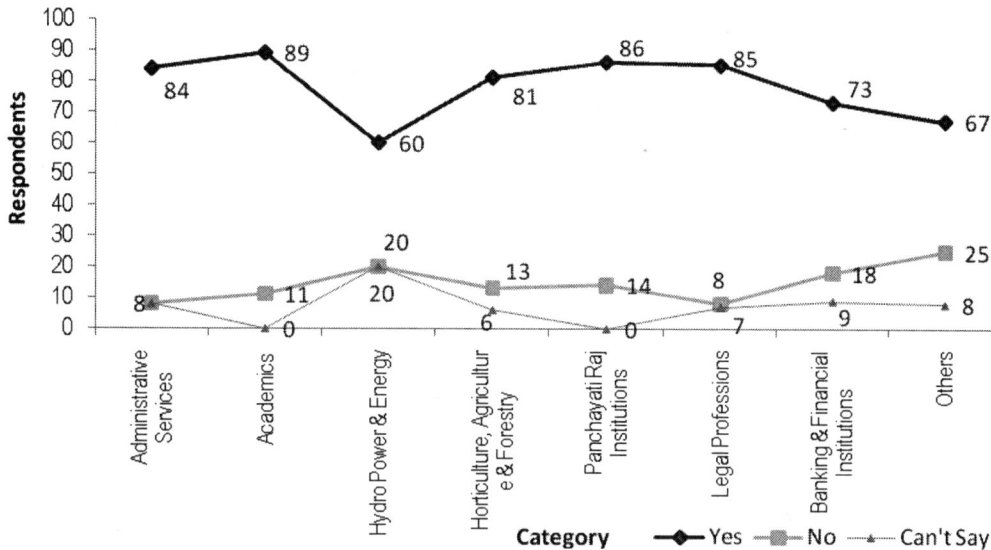

Figure 14.4: Showing ecological imbalances

Figure 14.4 above reveals that 80% respondents in all categories out of which the highest are 89% from academics followed by 86% from PRI's, 85 % from legal professions and 84% from administrative services agree that every person is responsible to maintain the ecological balance if the people of the area do not realize their responsibility, they cannot enjoy a healthy environment as the degradation of the environment is due to over exploitation of natural resources. Whereas, 14% respondents disagree that for the degradation of environment, the general public is responsible and the agencies involved for the development activities are responsible and the state action is required to stop over exploitation of natural resources causing environment degradation. In this category the highest are 25% from other category, followed by 20% from hydropower & energy and 18% from banking and financial institutions, whereas 6% respondents did not respond.

14.3.5 Obligation of Individual and State for the Protection of the Environment

Development activities in the fragile zone of Kinnaur, Lahaul & Spiti districts, Pangi & Bharmour Sub Division of Chamba district of the state require due care and caution. Coupled with the state action, general people of the area are also statutorily obliged to preserve and protect the local environment with the socio-economic development. The Constitution (Forty Second Amendment)

1976, explicitly incorporated the environmental protection and the improvement, by incorporating the article 48-A, in the directive principle of state policy, which provides that the, state shall endeavour to protect and improve the environment and safeguard the forests and wildlife of the country.

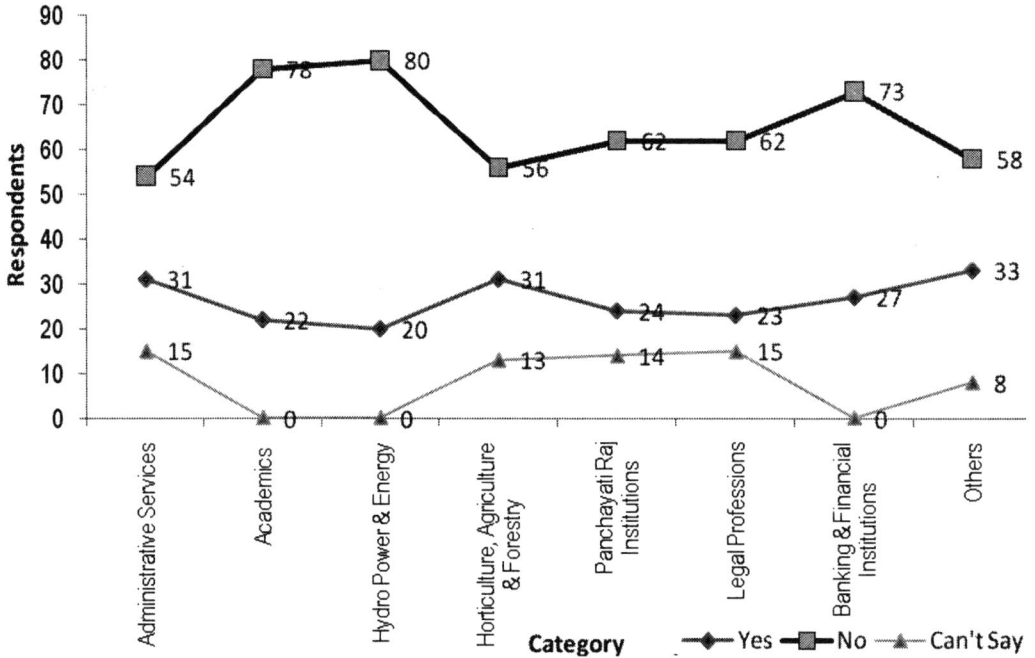

Figure 14.5: Inhabitants protecting the environment

Moreover, article 51A (g) imposes a similar responsibility on every citizen 'to protect and improve the natural environment including forests, lakes, rivers and wildlife, and to have compassion for living creatures. Therefore, protection of natural environment and compassion for living creatures were made the positive fundamental duties of every citizen and both the provisions substantially send the same message.

These provisions for the protection of environment have further correlated the decision of Hon'ble Supreme Court and observed that the livelihood of tribals should be considered in the context of maintaining ecology in the forest of these areas. If there is shrinkage of forest area, the government must take steps to prevent any destruction or damage to the environment, the flora and fauna and wild life under the article 48A and article 51A(g) in Animal and Environment Legal Defence Fund Vs. Union of India: (1997) 3, SCC 549.

Figure 14.5 shows that 27% respondents, in which the highest are 33% from others followed by 31% each from administrative services and horticulture, agriculture and forestry stated that the state government machinery and general people of these areas of the state are doing their obligatory duty to fulfill the objective of constitutional provisions by preserving the environment of these areas including biodiversity and ecosystem, which are being damaged by various developmental activities in these areas, whereas, 63% respondents stated that the developmental activities are damaging the environmental strata of these areas by the over exploitation of natural resources, but the state action and participation of general public under obligations of the article 48-A and 51-A (g) are negligible to preserve and protect the environment. In this category, the highest are 80% from hydropower and energy followed by 78% from academics and 73% from banking and financial institutions. These respondents stated that for the protection of environment in these seismic zones, more action of the state government and general public is required to preserve the natural resources of these areas. 10% respondents did not respond.

The responses of the general public as analyzed above show that the planning strategies for the conservation of natural resources and enforcement of laws for protection of environment are need of the day to encourage sustainable development. The protection and preservation of environment is a pressing issue, as it encompasses not only pollution but also sustainable development, conservation of natural resources; such as land, water vegetation, and ecosystem, which is essential in providing livelihood and environmental security. These respondents further reflected that the increase demographic pressures coupled with developmental activities are causing tremendous pressure on the natural/common property resources leading to various kinds of ecological disasters such as droughts, floods, soil erosion, landsides, deterioration of water bodies, and loss of biodiversity at risk of the earthquake.

The analysis of the data further shows that the development of hydro power projects and other developmental activities in these fragile tribal zone of the state of Himachal Pradesh has given jolts to the physical environment by creating reservoirs, tunnels, widening of roads and construction work, which led the catchment area to get dried and the perennial river is producing hydro-electricity of different magnitudes in the basin are in different stage of their execution. This much disturbance has created a considerable effect on the climatic conditions of the whole tribal areas of the state. The people who are living in the vicinity of these power projects are paying the cost by overexploitation of the natural resources, while some by living in the big and meteors cities are enjoying its fruits. Because of these climatic changes people are facing all sort of ill effects like decrease in crop production, water mills are getting dried, static water of reservoirs is helping to create health hazards and spreading water borne diseases, sources of livelihood have been snatched etc., and the lives of the people have become worse because of these development activities. The affected people of these tribal areas, where the hydro power projects are in development stage are still waiting for the justice along with the people who remained and living in the vicinity reservoirs of these hydropower projects are also facing the problems of climate change. Further it can be asserted on the basis of these citations that not only the affected people in revenue record but all residents who are living in these areas are facing the ill effects by these development activities and power projects in form of changes in the climatic conditions. People of the whole river basin have never experienced the problems of increase in temperature or unusual and unseasonal rain before the installation of hydropower projects but after the installation of hydro power projects, it became routine phenomena in these areas and the agriculture and horticulture crops are also affected largely.

14.4 CONCLUSION

The study reveals that quantitative and non-quantitative impacts by the construction of hydropower projects and other development activities arise in two ways: On the one hand, it is the question of rights of indigenous communities, viz. the model of conservation that alienates them by degradation of their natural environment by damaging the ecology; on the other hand, there is greater commercialization of these inaccessible areas in the name of economic development and increase of high risk of earthquakes and other disasters. Therefore, there is an urgent need for a new model of conservation and sustainable development in which conservation of ecology, development and social equity go hand in hand to reduce the risk of disaster. The present status is evident to prove that the ecology of the tribal areas of the state has been disturbed the environment by these large-scale hydropower projects in the river basins leading to high risks of disaster. The emerging environment scenario of these areas calls for immediate attention for conservation and proper use of natural resources during the course of development intervention in these areas. There is a need to integrate the environmental consequences of developmental activities and planning suitable mitigation measures in order to ensure and achieve the goal of sustainable development through public participation in decision-making to reduce the increasing risk of earthquake and other types of disaster. To achieve these goals, the quality of environment in and around the construction area should be enhanced by proper adoption of mandatory guidelines, method and measures for conservation of natural resources along with strategy for the prevention of adverse environmental and social impacts up to the extent possible on the construction sites, keeping in view the fragile region; for the mitigation of the possible adverse environmental social impacts by means of different policy and scientific method and for avoiding degradation of ecology in the area in the near future by development activities in the name of economic development and these development activities should make pace with the environment to achieve the principle of sustainable development.

The hydro power generation and large development projects involves lot of environment implications and it must be eco-friendly and sustainable in nature and there must be considerable improvement in human welfare. This means a significant advancement of human development, which is economically viable, socially equitable and environmentally sustainable. The local issues must be taken into consideration properly and with true spirit, not in paper and must not be engulfed by the red- tapism. The policies should be framed by visiting local sites and taking local issues into consideration. The hill slopes are prone to landslides, landslips, rockslides and soil creep. These hazardous features have hampered the overall progress of the region as they obstruct the roads and flow of traffic, break communication, block flowing water in stream and create temporary reservoirs and also bring down lost lots of soil cover and thus add enormous silt and gravel to the streams. Development activities including hydro power development along with the population increase the frequency of landslides and land subsidence has increased. Heavy construction work coupled with the lack of planning for water outlet; increase water seepage culminating in the land slides. Huge amount of explosives used in construction works of hydro-power and roads have adversely affected the ecosystem of the region and the stability of stabilized mountain slopes.

Environmental legal system is set of the provisions for the protection of the environment and natural resources of the nation. In its administrative law aspect, environmental legal system needs to guide decision-makers and guarantee legal accountability in government activities for the preservation of the environment and in its regulatory law aspect, environmental legal system must seek to change community attitudes and, where necessary, to provide the means to enforce legal prohibitions and obligations, sometimes against powerful interests in society. The failure to develop a clear concept of the role of legal rules in the administrative and regularity aspects of environmental law has enabled the system to ignore the issue of unbalanced power relations and, hence, has allowed the environmental condition to be exploited by powerful sectors of society. The importance for the preservation of the natural resources, biodiversity conservation and management of biological environment in the hill region of the state particularly in the tribal areas are the most important aspect of the environment legal system of the country. Since the environment and the natural resources belong to all living beings, so it needs proper protection by all for the welfare of all. A set of efforts by the government and non-government agencies including local society to protect environmental viability against pressures for change and/or negative impacts that arise because of developmental works and hydro power generation related activities are the need of the day. The effectiveness of the environment legal system is the foundation to support the life of humans and other living creatures of the nation particularly in the fragile and seismic zones of the hill state of Himachal Pradesh, therefore, the effective enforcement tool and active participation of the public is required to mitigate the ongoing environmental deterioration under the provisions of the constitution and statutory laws of the country.

References

Bakshi, P.M. (1997). *The Constitution of India, Delhi* (Reprint).New Delhi: Universal Book Traders.

Batta, R.N. (2000). *Tourism and the Environment: A Quest for Sustainability*, with special reference to developing countries and policy analysis on Himachal Pradesh. New Delhi: Indus Publishing Company.

Bhuiyan, M., and Hossain, J. (2007). *Application of International Environmental Law Principles By the National Court.* Fifth ISIL International Conference on International Environmental Law, 8-9 December 2007, New Delhi.

Coase, R. (1960). The Problem of Social Cost. *Journal of Law & Economics, 3*(1), 1–44.

Committee on Public Participation and Planning (1969). *People and Planning: Report of the Committee on Public Participation and Planning.* London: HMSO.

FARN (n.d). "*Public Participation and Sustainable Development*" On-Line Module developed by FARN with the cooperation of IISD (Canada); The REC (Hungary) and SEI (Sweden) within the framework of the Sustainable Development Communications Network, http://www.farn.org.ar/docs/p01/publicaciones1f.html.

Government of Himachal Pradesh (2010-11). *Annual Tribal Sub Plan Planning Department.* Shimla: Government of Himachal Pradesh.

Gupta, A., and Asher, M.G. (1998). *Environment and the Developing World: Principle, Policies and Management.* New York: John Wiley and Sons.

Joseph, Hoe Kiat Chun (2002). *A Critical Analysis of the Contribution to Sustainable Development of the Law and Practice on Public Participation in the Authorization of Radioactive Waste Disposal.* Doctoral Thesis, University College of London. Retrieved from http://www.scribd.com/doc/29051051/null.

Kumar, A., Rees, G., and Raghuvanshi, T.K. (n.d.). Small Hydropower Assessment- a Solution through Hydra-HP Software. *GISdevelopment.net.* Retrieved from http://www.gisdevelopment.net/application/Utility/Power/untilityp0009pf.htm

Kumar, N. (1997). *Constitutional Law of India.* New Delhi: Pioneer Publications.

Mamgain, D. (1971). *Himachal Pradesh District Gazetteers: Kinnaur, Lahaul & Spiti, Chamba Districts.* Quick Printer, Ambala Cantt.

McAuslan, P. (1980). *Ideologies of Planning Law* (Urban and regional planning series). London, UK: Pergamon Press

Negi, V.B. (2007). *Hydropower Projects in Kinnaur District of Himachal Pradesh: Issue of Environment Protection and Sustainable Development.* At International Seminar on recent Trends in Teaching and Research in International Law and International Seminar on Law of the Sea. Indian Society of International Law New Delhi.

Negi, V.B. (2008). *Public Participation in Sustainable Development: A Study of Satluj Base Hydro-Power Projects Development in Kinnaur District of Himachal Pradesh.* Paper presented at the International Conference on issues in Public Policy and Sustainable Development, IGNOU, Maidan Garhi, New Delhi,

Negi, V.B. (2008). *Sustainable Agriculture Development and Environment Protection: A Case Study of Tribal Areas of Himachal Pradesh.* Paper presented at Regional Conference on Food Security and Sustainable Agriculture Development, Agartala.

Nirmal, B.C., and Shukla, P.C. (2007). *Public Participation in Decision-Making: International Standards and Indian State Practice.* Paper presented at Fifth ISIL International Conference on International Environmental Law, December 8-9, 2007 New Delhi.

Oluyemisi, A. (2007). *Environmental Protection in the Nigerian Oil Industry: Balancing Sustainable and Economic Development.* At Fifth ISIL International Conference on International Environmental Law December 8-9, 2007 New Delhi.

Ortolano, L. (1997). *Environmental Regulations and Impact Assessment.* New York: John Wiley & Sons, pp. 402-422.

Sabharwal, Y.K. (n.d.). *Human Rights and the Environment.* Accessed from http://www.supremecourtofindia.nic.in/ new_links/ human rights.htm).

Saxena, P. (2006). *Overview of Small Hydropower Development in Indian, Himalayan Small Hydropower Summit.* October 12-13, 2006, Dehradun, Uttarakhand.

Singh, J.S. (2006). Sustainable Development of the Indian Himalayan Region:Linking Ecological and Economic Concerns. *Current Science 90 (6)* March 25, 2006.

UNECE (1998).*Convention on Access to Information, Public Participation in Decision-Making and Access to Justice in Environmental Matters.* Denmark: United Nations Economic Commission for Europe. Retrieved from http://www.unece.org/env/pp

United Nations (1992). *Agenda 21 Programme of Action for Sustainable Development. Report Of The United Nations Conference on Environment and Development. Rio de Janeiro, 3-14 June 1992.*UNDoc A/CONF. 151/26. Retrieved from http://www.un.org/documents/ga/conf151/ aconf15126-1annex1.htm

15

Bhopal Gas Tragedy:
Saga of the Divided World

Vijita Agarwal

15.1 INTRODUCTION

Companies use differing standards in Developed and Developing countries. It might come as a surprise to many that there is nothing illegal, as per United States law, for a U.S. Corporation to employ child labour in its overseas operations whereas employing a child for its home operation is a big crime. The various acts from the Fair Labour Standards Act (FLSA), to the Occupation Health and Safety Act (OSHA), to the Family Leave and Medical Act of 1993, and so on, seem to be inapplicable beyond the borders of the United States, and the same is being practiced by many multinationals of USA such as Bridas Corporation, Petroecuador, EBX Group and others that are following different rules in their home country and host country. In the case of Union Carbide Corporation (UCC), it even did not follow the "Right to Know Act" of USA in which industries need to disclose the volume of certain chemicals released annually by them in environment. This act makes the employees of the company to access the environmental information and this act is being followed by UCC in USA but was not adopted in India. This makes clear that they were also having differential standards for US and India (Jasanoff & Sheila, 2007).

It has been observed that the developed countries have much stricter safety norms and better implementation practices compared to developing countries. In some cases, the wastes from hazardous industries are exported, whereas in other cases, the whole industries are shifted to third-world countries. For example, owing to shortage of suitable disposal sites and incineration facilities for hazardous wastes in the US, the companies started exporting the chemical wastes to other countries. Publicity about pending chemical waste export deals has led to enraged public reactions in West Africa, Mexico, Haiti, and the Dominican Republic. Yet, the export of toxic wastes is increasing.

15.2 BHOPAL GAS TRAGEDY

The Bhopal disaster is the world's worst industrial massacre that happened in India. This disaster happened on the night of December 2, 1984 at the Union Carbide India Limited's (UCIL)

Disaster Management and Risk Reduction: Role of Environmental Knowledge; Editors Anil K. Gupta, Sreeja S. Nair, Florian Bemmerlein-Lux and Sandhya Chatterji; Copyright © 2013, Narosa Publishing House, New Delhi

pesticide plant in Bhopal, Madhya Pradesh, India due to a leak of Methyl Isocyanate gas and other chemicals from the plant which resulted in deaths of thousands of people. The estimates vary on the death toll. The reports by the government of Madhya Pradesh confirmed a total of 3,787 deaths related to the gas release. A government report in 2006 stated the leak caused 558,125 injuries (Dubey, June 21, 2010).

The cases are still pending in the district court of United States, and also in the District Court of Bhopal, India, on the employees, including Warren Anderson who was CEO of UCC at the time of the disaster. Seven ex-employees, including the former UCIL chairman, were convicted in Bhopal of causing death by negligence and sentenced to two years imprisonment and a fine of about $2,000 each, the maximum punishment allowed by law in June 2010. Before judgment was passed an eighth former employee who was also convicted died (Chouhan et al., and others 1994, 2004).

Moreover, many observers, such as those writing in the Trade Environmental Database (TED) Case Studies as part of the Mandala Project from American University, have stressed on the "lack of communication gap and gaps between Union Carbide's administration and its operation in India", described by "the parent companies hands-off approach to its overseas operation" and "cross-cultural barriers" (Hanna et al., 2005). Some studies also reveal that the personnel management policy led to an exodus of skilled personnel to better and safer jobs (Cassels, 1993).

MIC in Bhopal but did not equip the plant with the safety mechanisms to deal with accidents. UCC was aware that some of the technology it transferred was not proven, 140 and entailed operational and safety risks. UCC did not export the same standards of safety in design or operations to Bhopal as it had in place in the USA. In particular, UCC failed to set up any comprehensive emergency plan or system in Bhopal to warn local communities about leaks, even though it had such a plan in place in the USA. As early as 1982, UCC was aware that there were major safety concerns regarding the Bhopal plant. Months before the accident, UCC was warned of the possibility of a reaction similar to the one that caused the eventual leak in Bhopal.

15.3 SAFETY MEASURES IN UCC PLANTS IN THE USA AND INDIA

UCC had claimed the Bhopal plant to be a model facility using latest technology. Its manager, when well-versed of the MIC leak, said, "The gas leak just can't belong to my plant. Our technology just can't go incorrect" (Bogard, 1987). UCC used to work on a alike plant in Institute, West Virginia and claimed that both plants are same. However, UCC plant in West Virginia was better equipped compared to the plant in Bhopal. There were a number of critical differences in levels of design and operations of the Bhopal and Institute plants. This clearly indicates that the company was following two different approaches in developed and developing country.

Table 15.1: Differences in levels of design and operations of the Bhopal and Institute plants

	Institute, West Virginia, USA	Bhopal, Madhya Pradesh, India
Capacity	High production of MIC matched with high processing capacity. MIC not stored for long periods of time.	High production capacity of MIC but low processing capacity. MIC stored in large quantities for long periods of time.
Emergency	MIC storage tank equipped with	No emergency caustic scrubber to

	Institute, West Virginia, USA	Bhopal, Madhya Pradesh, India
scrubbers	emergency scrubbers (to neutralize any escaping MIC) designed to operate under emergency conditions.	neutralize any MIC leak.
Computerized monitoring	Computerized monitoring of instruments (gauges, alarms, etc.) and processes to support visual observation.	No computerized monitoring of instruments and processes. Relied solely on manual observation.
Cooling system	MIC field storage tanks used a cooling system based on chloroform (inert and nonreactive with MIC).	MIC tanks used a cooling system based on brine (highly reactive with MIC).
Refrigeration unit	Refrigeration unit to control temperature in the tanks was never turned off.	Refrigeration unit had been turned off since June 1984.
Nitrogen pressure	MIC was always maintained under nitrogen pressure.	MIC tanks had not been under nitrogen pressure since October 1984.
Emergency plan	An elaborate four-stage emergency plan to deal with toxic releases, fires, etc., including a general public alert linked to community police, river and rail traffic and local radio stations. Various emergency broadcast systems in place to alert and disseminate appropriate information to the public.	No system to inform public authorities or the people living adjacent to the plant. No emergency plan shared with communities living adjacent to the plant; no system to disseminate information regarding emergency to the public with the exception of a loud siren.
Maintenance programme	A maintenance programme to determine and evaluate replacement frequency for valves and instrumentation and alarm systems. Weekly review of safety valves and reviews and maintenance recorded extensively.	No evidence of an effective instrument maintenance programme. Safety valve testing programme largely ineffective and no proper records maintained of reviews of instruments, valves and alarm systems, etc.
Lab analysis	A lab analysis of MIC was conducted to test quality and check for contamination prior to storage, processing or distribution.	No lab analysis of quality was undertaken. MIC stored for long periods without testing for contamination.
Training	Extensive employee training programme to ensure high level of training and information among all employees of normal and emergency procedures.	Operators put in charge without sufficient training.
Protective equipment	Extensive provision of appropriate personal protective equipment to employees including protective clothing, air respirators, etc.	Personal protective gear and breathing air equipment not easily accessible, inadequate and of poor quality.

(Source: "Cloud of Injustice – Bhopal Disaster 20 Years On", Amnesty International, 2004)

15.4 SAFETY COMPROMISE AT BHOPAL PLANT

Due to monetary problems, UCC Bhopal had started lowering the necessary worker amenities. Pipelines were repaired by means of epoxy resins and seals with permission of the administration (Pareek, 1998). An unbeneficial plant should have been shut down. However, UCC Bhopal tried to make it gainful or cut losses by closing down the safety systems. Storage of a huge amount of MIC was not warranted by the process. UCC Bhopal had not needed it but the headquarters in the US had overruled the opposition. Additionally, an interchange production route was presented that did not manufacture MIC as an intermediate (Lees, 1996). UCC either did not know of it or determined that the MIC route was not as unsafe as it really turned out to be.

Serious operational and maintenance failures were major causes of Bhopal gas tragedy. Had the plant been properly managed and maintained as the plants are being managed in developed countries but. Different regulation system adopted by the companies is the major cause of such accidents. As discussed above many multinationals of USA have different regulations of operating in the home country and in the host country.

There are differential rules and regulations for the companies operating in the US and US companies operating abroad. The ease of transferring Carbide's technology of production from West to East, India's attempts to transfer the legal sanctions for failure back to the risk-exporting country proved much less easy. This gives strength for the argument that multinationals, by virtue of their global purpose, organization, and resources, should be treated as single, monolithic agents, rather than as a network of discrete, non-interdependent units.

15.5 SIMILAR ACCIDENTS AT OTHER SITES

After the Bhopal tragedy, both the UCC and US Occupational Safety and Health Administration (OSHA) announced that the similar type of accident could not happen at the Institute, West Virginia, plant as of the plant's better equipment, superior personnel and America's generally "superior level of technological culture" (Perrow, 1986). There were almost 60 leaks of MIC in the W. Virginia plant between 1980 and 1984. As per EPA investigation many of them went unreported (Lagadel, 1990). Yet, only 8 months afterwards, a like accident happened there leading to brief hospitalisation of about 100 people.

As in Bhopal, the warning signal was late and the company was sluggish in making the information available to the community (Lagadel, 1990). OSHA fined UCC US$1.4 million charging "steady, willful, and obvious violations" at the plant and a common atmosphere and attitude that "a small number of accidents here and there are the price of production" (Perrow, 1986). In the UCC, Institute, WV, case, a Federal organisation (OSHA) gave a green signal to the WV plant subsequent to the Bhopal accident, and 8 months afterwards a severe toxic leak happened there.

The report on the Esso Australia disaster states that the government there had reduced both the power and control of the OH&S Authority by making it placed first under the department of business and employment and then amalgamating it with the workers' payment insurance agency (TCE, 1999). At the Texaco refinery, the report states that the corporation had not learned anything from precedent experience on same plants.

15.6 OWING RESPONSIBILITY OF THE ACCIDENT

At the UCC Bhopal plant, the company did not tell the government authorities what the leaking chemical was and did not advise any remedy for it. The doctors treating the suffering community were in the dark for many days and this badly affected the treatment protocol used. Likewise, at another UCC facility in the US, a leak of toxic gas harmed a shopping centre. Emergency treatment was administered to a number of people. For two days doctors were not familiar with what the toxic chemical was or where it came from because UCC denied of the leak's existence (Perrow, 1986). Even at the local government level, the local government did not act specifically and purposefully on previous accidents and ignored newspaper articles predicting disaster.

15.7 LEGAL ACTION AGAINST UNION CARBIDE

Legal proceedings were started against UCC, the United States and Indian governments, local Bhopal authorities, and the disaster victims started immediately after the disaster. Legal proceedings leading to the settlement started from December 14, 1984, when the CEO of Union Carbide, Warren Anderson, addressed the US Congress, emphasizing on the company's "commitment to safety" and ensuring that a similar accident "will never happen again" (Mitchel & James, 1996). An act was passed by the Indian Government i.e. the Bhopal Gas Leak Act in March 1985, making the Government of India to act as the legal representative for victims of the disaster that helped in leading to the beginning of legal proceedings.

The US government called for inquiry into the Bhopal disaster, in 1985, by Henry Waxman, a Californian Democrat, which resulted in US legislation related to the accidental release of toxic chemicals in the United States.

In March 1986 saw Union Carbide offer a settlement figure, providing a fund for Bhopal victims of between $500–600 million over a period of 20 years. In May, the litigation was transferred to Indian courts from US by US District Court Judge as UCIL was a "separate entity, owned, managed and operated exclusively by Indian citizens in India". UCC's forum request was granted by the judge in US thus making the case to move to India. From this, it is evident that under US federal law, the company had to submit to Indian jurisdiction.

During 1988, litigation continued in India. US$ 350 million was claimed by The Government of India from UCC. In November 1988 the Indian Supreme Court told both sides to come to an agreement and "start with a clean slate". In 1989, Union Carbide agreed to pay US$ 470 million for damages caused in the Bhopal disaster, (15% of the original $3 billion claimed in the lawsuit). By the end of October 2003, as by the Bhopal Gas Tragedy Relief and Rehabilitation Department, compensation had been awarded to 554,895 people for injuries received and 15,310 survivors of those killed. On an average $2,200 was given to the families of the dead (Eckerman & Ingrid, 2005).

The Indian Supreme Court heard appeals that were made against the settlement throughout 1990. The Supreme Court upheld $470 million, dismissing any other outstanding petitions that challenged the original decision in October 1991. It was ordered by the Court that Indian government "to purchase, (out of settlement fund), a group medical insurance policy to cover 100,000 persons who may later develop symptoms" and cover any shortfall in the settlement fund. It also requested UCC and its subsidiary to "voluntarily" fund a hospital in Bhopal, at an estimated $17 million, to basically treat victims of the Bhopal disaster. To this the company also agreed.

The various charges were put against Warren Anderson and others. On December 7, 1984, UCC CEO Warren Anderson was arrested and released on bail by the Madhya Pradesh Police in Bhopal. The arrest, which took place at the airport, ensured Anderson would not meet any harm by the community of Bhopal. Anderson was taken to UCC's house after which he was released six hours later on $2,100 bail and flown out on a government plane. In 1987, the Indian government summoned Anderson, eight other executives and two company affiliates with homicide charges to appear in Indian court. Union Carbide baulked, saying the company is not under Indian jurisdiction. The local authorities of Bhopal charged Anderson, who had retired in 1986, with, a crime that carries a 10 years in prison that was the maximum penalty in 1991. On February 1, 1992, he failed to appear at the court hearings in a culpable homicide case in which he was named as the chief defendant. So he was declared a fugitive from justice by the Chief Judicial Magistrate of Bhopal. Orders were passed to the United States from the Government of India to press for an extradition. In October 1993, the victims of the Bhopal disaster could not seek damages in a US court as Supreme Court refused to hear an appeal against the decision of the lower federal courts.

During 2004, the remaining settlement funds to victims were released to Indian government as ordered by the Indian Supreme Court. In September 2006, the Welfare Commission for Bhopal Gas Victims made an announcement that all original compensation claims and revised petitions had been made. In 2006, in New York, the Second Circuit Court of Appeals upheld the dismissal of claims that were remaining in the case of Bano v/s Union Carbide Corporation. This made a move of blocked plaintiffs' motions for class certification and claims for damages caused to property and remediation. In 1999 Haseena Bi and several organizations representing the residents of Bhopal raised the issues in the class action complaint first filed against Union Carbide.

In June 2010, seven ex-employees of the Union Carbide subsidiary, (all were the nationals of India and many in their 70s), were charged of causing death by negligence and each sentenced to two years imprisonment and fined Rs.1 lakh. All were released on bail shortly. The names of those who were convicted were: Keshub Mahindra, former non-executive chairman, Gokhale, managing director, Kishore Kamdar, vice-president; J. Mukund, works manager; S.P. Chowdhury, production manager; K.V. Shetty, plant superintendent; and S.I. Qureshi, production assistant. Federal class action litigation, Sahu v/s Union Carbide et al., is still pending presently, on appeal before the Second Circuit Court of Appeals in New York. The litigation seeks damages for individual injury, medical monitoring and providing relief in the form of clean-up of the drinking water supplies for residential areas near the Bhopal plant. In the Southern District of New York, a similar complaint seeking similar relief for property damage claimants is stayed pending the outcome of the Sahu appeal before the federal district court. This shows that cases that are from developing countries are still pending whereas the cases from developed countries get justice very soon. For example a case of Sahu v/s UCC related to Bhopal gas tragedy of India is still pending in the New York court whereas a similar disaster (BP disaster of US) where justice have been provided to all sufferers and the case is closed now as the countries i.e. the governments are very strict in following the legal laws for the companies operation in their home country rather than its overseas operation.

15.8 CONCLUSION

The above discussion points out that the major disasters can occur in any country in spite of the development level. In this paper apart from the Bhopal gas tragedy, the rest of the incidents occurred in the developed countries i.e. Australia, Canada, UK, and USA. The reason behind choosing such disasters that have occurred after the Bhopal tragedy (except the TMI) in developed countries is to examine, if the latter has had a salutary effect, at least in the developed countries.

However, from the discussion, it can be concluded that a Bhopal type massacre entailing the extent of mortality and morbidity that happened in Bhopal could not have happened in 1984 in any of the developed country. All the disasters that occurred in developed countries were neither due to the managerial operations nor due to the safety regulations which were the biggest component in the Bhopal gas tragedy as the management gave only lip service to process and personnel safety and the governments did not ensure compliance with the regulations. The case of Bhopal gas tragedy is still pending and suffered people are still searching for justice and those disasters that have happened in developed countries such as BP disaster in US in 2010, people have got the fair justice.

The condition since the early nineties has transformed overall for the better. Based upon expansion in the CPI and not a balanced increase in the number of losses and insurance claims for fire and explosion, it has been concluded that several hundred lives and over a billion dollars in damages have been saved (Kletz, 1998 (a.b); Sutton, 1999). (The cost of add-on safety measures has not been factored in.) Though, major incidence still do happen which could have been prevented.

It should be pleasing that the improvements in method of safety due to new legislation, stricter enforcement and personnel training have, more or less, reached a limit. These will, in future, create only small incremental improvements in safety. Legislation is evolving after each new type of accident. Learning from accidents has its own limits and is very costly indeed in terms of cost as well as human misery. If it is true that technology up gradation takes place as much by overcoming failures as it does by achieving successes, then the price of betterment will always include heartbreaks (Kemp, 1986). Elementary R&D is needed in the causes of accidents and in industrialized processes used in the chemical process industries.

The challenge is scary but there is no alternative way out to make further significant advancement in process safety. Fundamental research to study the causes of accidents in other disciplines such as boiler explosion have, in the past, led to major developments in plan as well as in manufacturing new and better materials. This is what is expected in the field of process safety. The current move in activity related to intrinsically safer approaches will also play a major role in getting better safety and the public image of Chemical Process Industries (Kletz, 1998).

References

Blazier, A., and Skilling, J. (1995). Potential source of data for use in human factor studies In *Major Hazards Onshore and Offshore II. I. Chemical Engineers Symposium Series No. 139.* Rugby, UK: Institution of Chemical Engineers.

Bogard, W. (1987). *The Bhopal Tragedy.* Boulder, Colarado, US: Westview Press.

Browning, J. (1993*). Union Carbide at Bhopal Crisis Response: Inside Stories on Managing Image under Siege.* Michigan, US: Visible Ink Press.

Cassels, J. (1993). *The Uncertain Promise of Law: Lessons from Bhopal.* Toronto, Canada: University of Toronto Press.

Chouhan, T. R., Jaisingh, I., and Claude Alphonso, A. (1994). *Bhopal: The Inside Story—Carbide Workers Speak Out on the World's Worst Industrial Disaster.* Goa, India: Other India Press.

De Grazia, A. (1985). *A Cloud over Bhopal.* Bombay: Popular Prakashan.

Dubey, A. K. (21 June 2010). First 14 News. Archived from the original on June 26, 2010.

Eckerman, I. (2005). *The Bhopal Saga—Causes and Consequences of the World's Largest Industrial Disaster.* India: University Press.

Hanna, B., Morehouse, W., and Sarangi, S. (2005).*The Bhopal Reader. Remembering Twenty Years of the World's Worst Industrial Disaster.* US: The Apex Press.

Hidden, A. (1990). *Investigation into the Clapham Junction Railway Accident.* London: HMSO.

Jasanoff, S. (2007). Bhopal's Trials of Knowledge and Ignorance. *Chicago Journals, 98* (2), 344-350

Jungk, R. (1958). *Brighter than a Thousand Suns.* London: V. Gollancz.

Keenan, J. G. (1979). *Report of the President's Commission on Three Mile Island.* Washington, DC: US Government Accounting Office.

Kemp, E. (1986). Calamities of Technology. *Science Digest, 7,* 50–59.

Kletz, T. (1998a). *Process Plants: A handbook of inherently safer designs.* London: Taylor & Francis.

Kletz, T. (1998b). Review of 'The explosion and fire at the Texaco refinery, Milford Haven, July 24, 1994'. *Chemical Engineering Progress, 94* (4), 86.

Lagadel, P. (1990). *States of Emergency.* London: Butterworth–Heinemann.

Lees, F. P. (1996). *Loss prevention in the Process Industries* (2nd ed.). Oxford: Butterworth–Heinemann.

Leveson, N.G. (1995). *Safeware System Safety and Computers.* Reading, MA: Addison-Wesley.

Mitchel, J. (1996). *The Long Road to Recovery: Community responses to industrial disaster.* New York: United Nations University Press.

Pareek, K. (1999).The Management did not adhere to Safety Norms. *Down to Earth, 8*(1),56.

Pareek, S.K. (1998). Personal Communication.

Perrow, C. (1982). The President's Commission and the normal accident. In D. L. Sills, *Accident at Three Mile Island: the human dimensions.* Boulder, CO: Westview Press.

Perrow, C. (1986). The habit of courting disaster. *The Nation, October.*

Spooner, P. (1995). Disasters: A family group's view. In *Major hazards onshore and offshore I. I. Chemical Engineers Symposium Series No. 139.* Rugby, UK: Institution of Engineers.

Sutton, I. (1999). Engineering Process Safety. *Chemical Engineering, 106* (5), 114–121.

Willey, R. J., Hendershot, D. C. and Berger, S. (2006). *The Accident in Bhopal: Observations 20 Years Later.* Orlando, Florida, USA.

Natural Resource Management Policy Implications on DRM Practices:
Insights from North-East Cambodia

Kathlyn Kissy H. Sumaylo

16.1 INTRODUCTION

Located in the fertile Mekong River Basin, Cambodia is one of the most disaster-prone areas in South-East Asia (IOM, 2010; NCDM PDNA, 2010). The kingdom, considered a Least Developing Country, has very high vulnerability levels in terms of human development indices (NCDM and WFP, 2003a; MoP and WFP, 2003b) and low adaptive capacity of institutions, livelihoods, communities, and ecosystems to natural disasters and climate change (Helmers and Jegillos, 2004; Yusuf and Francisco, 2009).

Cambodia suffered from deep social and cultural intergenerational trauma, economic stagnation, large population displacements into, and institutional fragmentation as a result of nearly twenty years of civil war from 1975 to 1991. The signing of the Paris Peace Accords in October 1991, gave Cambodia its first comprehensive political settlement, and the United Nations Transitional Authority in Cambodia (UNTAC), was formed under UN Security Resolution No. 745 to "lead peace restoration efforts, to hold free and fair elections leading to a new constitution and to spearhead the rehabilitation of the country" (UN website). The holding of Cambodia's first democratic elections under the auspices of the UNTAC in 1993 also paved the way for the foundations of decentralized governance. With the completion of the post-conflict repatriation of displaced Cambodians, the multi-donor integrated development project phases, Cambodia Area Rehabilitation and Regeneration Project (CARERE) 1 and 2 and the Partnerships for Local Governance, shifted their focus from emergency post-war recovery and rehabilitation to a developmental approach. This new approach addresses Cambodia's long-term challenges and needs for institutional governance, planning systems, infrastructure and capacities delivered under the government program SEILA. Natural resource management was a key output under the program's objective on poverty alleviation, alongside delivery of rural infrastructures and de-mining. Community-based natural resource management was the centrepiece of the SEILA/CARERE's NRM program in Ratanakiri Province to address communal lands, forests and

Disaster Management and Risk Reduction: Role of Environmental Knowledge; Editors Anil K. Gupta, Sreeja S. Nair, Florian Bemmerlein-Lux and Sandhya Chatterji; Copyright © 2013, Narosa Publishing House, New Delhi

water resources management (UNDP CARERE website). The program, which was piloted in Ratanakiri Province in the north-eastern region and in two north-western provinces, produced trained Department of Environment staff in forest protection; 12 community forest pilots; formation of two Provincial Land Use Planning committees; and issuance of 7,600 land titles. At the height of the SEILA program implementation, the first policy stipulation for disaster risk management- the National Committee for Disaster Management Policy Paper- was also released in 1995 which underwent three revisions and was not approved till 2001.

From 2000, national administrative and legal bases for environmental and natural resource management began to take shape. Alongside, decentralization experiments in national administrative and local development planning were posting development gains. From 2001 onward, significant NRM and environmental policies and frameworks were put in place on four areas: land tenure and management; forestry protection; protected areas and wildlife conservation; and water resources management. These policies intersect with national and local policy developments on disaster risk reduction and management (DRRM) within a decentralized governance system, which is broadly described below as they impact on the DRRM and NRM local policies and practices in North-East Cambodia. Decentralization work in Cambodia was initiated in 2001 and with the elected representatives in 2002 and DM policy in 2002.

16.2 LAND TENURE AND MANAGEMENT

The Land Law of 2001 forms the basis of Cambodia's present land use, tenure and management, setting the legal regime for the ownership of immovable properties in the country as determined by nature, by purpose and by law (Articles 1 and 2, Land Law, 2001). Legally-recognized properties in Cambodia are distinguished into three types: State Public Property, State Private Property and Private Property. The distinction is significant for natural resource management and environmental conservation as it determines the mode of land tenure, use and management. For instance, state public properties, having public interest as its primary purpose "may not be sold or transferred to other legal entities, though subject to occupancy and use rights that are strictly temporary in nature, such as logging concession in the forest reserve" (Land Law, 2001; Oberndorf, 2006, p.3). Meanwhile, state private property, or lands owned by the State and public legal entities that are idle or excess, "may be sold or transferred to other legal entities, such as to land recipients under the Social Land Concession framework" (Obendorf, 2006, p.4). Article 10 of Land Law of 2001 defines immovable property by nature as all natural grounds such as forest land, cleared land, land that is cultivated, fallow or uncultivated, land submerged by stagnant or running waters and constructions or improvements firmly affixed to a specific place created by man and not likely to be moved whole, while immovable property by purpose are those fixed to the ground or incorporated into the constructions and which cannot be separated there from without damaging them or altering them, such as trees. Article 14 on state private property strongly stipulates that that no transformation of a land concession, unless in response to social purposes determined by the State, can be converted into a right of ownership (Land Law, 2001, p.8).

The same law also provides for land concessions, bestowed as a legal right to qualified persons or entities for economic or social reasons. Economic land concessions (ELCs) in Cambodia are characterized by large scale plantations such as rubber, and for agro-industry such as corn, cashew nuts and soybeans, for a maximum period of 99 years (Sub-Decree No. 146, 2005). Sub-decree No. 146 on Economic Land Concession was issued in December 27, 2005, in compliance with the provision of the Land Law of 2001. The SD defines ELC as a mechanism to grant private state land through a specific economic land concession contract to a concessionaire to use for agricultural and industrial-agricultural exploitation. This land use arrangement is a particularly contentious form of natural resource exploitation because of its associated socio-economic and political implications on land access and tenure especially on traditionally held agricultural lands, and environmental damages brought about by excessive land clearing and related activities. While an environmental impact assessment is required before an approval of a concession activity, many reported ELC commence their operations before the EIA approval and beyond the land area limit set by law. Social land concessions on the other hand are a land allocation mechanism that may be granted to repatriated persons, poor homeless households for farming or residential purpose, demobilized soldiers; to those displaced by natural disasters or by infrastructure development and needing resettlement; and to workers of land for ELCs (Sub-Decree No. 19 on Social Land Concessions, 2003).

16.3 FORESTRY PROTECTION AND MANAGEMENT

The Forestry Law (2002) defines the framework for the sustainable management, harvesting, use and development of forests in the Cambodia. The permanent forest reserves of the country consist of production forests, protection forests, and conversion forestland. Production and protection forest sub-classification are particularly significant for indigenous people and local communities who have traditionally used the forest and harvested its by-products. It also provides for the establishment of community forests as a mechanism to balance traditional use of forest areas for livelihoods, residence and practice of customary beliefs in the Permanent Forest Reserves with preserving natural balance. The authority to allocate such forest areas for traditional user rights rests with the Ministry of Agriculture, Forestry and Fisheries. The arrangement is enforced through a Community Forest Agreement (CFA) between the local community and the Forestry Administration cantonment for a maximum period of 15 years. The sub-decree defines community forest is a forest planted under the state's public property, the rights of which are given to a local community living in or near the forest to manage and utilize it in a sustainable manner. The implementing guidelines are laid out on the Sub-decree on Community Forestry Management issued in 2003, which includes the formation of community forestry management committees and the crafting of a forest management plan, scope of use and harvesting within the designated area, and the roles and responsibilities of the implementing institutions (Forestry Administration and the Department of Agriculture, as well as local communities).

Wildlife conservation is another important contribution of the law, which reiterates stipulations in past sub-decrees and policies on the designation of protected areas in the country, a number of which are located in the north-eastern provinces of Cambodia. However, the law also prohibits certain forest activities that may lead to pollution, damage, over harvesting, excessive natural resource exploitation. Shifting cultivation and harvesting of high-value forest products such as resin or rare species for instance are not allowed within the permanent forest reserves (Protected Areas Law, 2008).

An equally significant dimension of the law with regard to forest management is the forest concession management in production forest not under use, which could be entered by the Royal Government with any investors or legal entity. Forest concessions must be entered through a bidding process and shall have an Environmental and Social Impact Assessment (ESIA) in their Forest Management Plan. They must also ensure that concession operations do not interfere with customary practices on registered land property held by indigenous peoples, as mandated by the Land Law of 2001, and customary access and user rights within and close to the concession areas (Forestry Law, 2002).

16.4 PROTECTED AREAS AND WILDLIFE CONSERVATION

The mandate on protected areas is governed by the Protected Area (PA) Law (2008) which establishes and modifies the scope of protected areas into eight classifications from the four as provided in the Royal Decree on the Protection of Natural Areas in 1993. The PA Law identifies four management zones for each of the eight protected areas: (1) Core zone, or areas of high conservation values containing threatened and critically endangered species and access to the zone is allowed only for researchers and officials of the Nature Conservation and Protection Administration for scientific studies and observation for conservation and protection of biological resources; (2) Conservation zone, also of high conservation value, consists of natural resources, ecosystems, watershed areas and natural landscape which are adjacent to the core zone. Limited access to and strict control of the zone is allowed for small scale community uses of non-timber forest products to support local ethnic livelihoods; (3) Sustainable use zone are areas allocated for national and local development, and conservation of protected areas for local communities and indigenous ethnic minorities. These include areas for national cultural heritage, eco-tourism, wildlife, preservation and recreation, and infrastructure development (irrigation, reservoir, hydro-power), mining and sustainable resin exploitation; and (4) Community zones, which are reserved for the socio-economic development of local communities and ethnic minorities as well as areas with existing residences, paddy fields and upland fields. Criteria for zoning shall be based on area management objectives, potential values of the area, socio-economic implications, and geographical location. The law encourages the participation of local communities in the management of the area.

16.5 WATER RESOURCES MANAGEMENT

Cambodia sits in the Mekong River Basin which spans six countries in Asia, including Thailand, Lao PDR, Vietnam, China and Myanmar. The Agreement on the Cooperation for the Sustainable Development of the Mekong Basin signed in 1995 forms the basis of the river basin management trans-boundary cooperation framework by the four countries in the Lower Mekong Basin (Lao PDR, Cambodia, Thailand and Vietnam), while China and Myanmar sit as Dialogue Partners (MRC website). The Lower Mekong Basin is home to 60 million people. Its rivers and wetlands provide sources of food, livelihoods, power, transport and trade to many communities and industries, The entire river basin has one of the most productive inland fisheries in the world. The Agreement aims to "optimize the multiple use and multiple benefits" of water resources and to "minimize the harmful effects that might result from natural occurrences and man-made activities." To further operationalise the plan, the governments of Cambodia and Vietnam have further signed a Memorandum of Understanding (MoU) and Action Plan to promote cooperation on cross-border water resource management (IOM, 2010, p. 29).

The country has been noted for its unique hydrological system wherein the Mekong River and the Great Tonle Sap Lake are joined by the Tonle Sap River, which reverses its flow twice a year. Water profile of Cambodia indicates that in early November, when the level of the Mekong decreases, the Tonle Sap River reverses its flow, and water flows from Lake Tonle Sap to the Mekong River and thence to the Mekong Delta.

16.6 INTERFACE OF DRRM AND NRM: INSIGHTS FROM NORTH-EAST CAMBODIA

Cambodia has suffered from the devastating effects of flood and drought, as well as from storms and storm surges and strong winds. Between 1987 and 2007, 12 floods killed 1,125 people and caused approximately $300 million in damages (SNAP, 2008). In 2009, Cambodia was among the countries in Southeast Asia heavily hit by Typhoon Ketsana, which swept through 14 of its 24 provinces and severely affected 50,000 families. The Ketsana Comprehensive Post Disaster Needs Assessment report (2010) led by the National Committee for Disaster Management reported that agriculture and fisheries were among those severely affected, which cover 80 per cent of the country's still largely rural population that are heavily dependent on climate-sensitive livelihoods and natural resources.

Institutional mechanism for disaster risk management was first laid out in a policy paper in 1995 creating the National Committee for Disaster Management to be the lead agency in coordinating disaster management activities in the country. As an inter-ministerial body, the NCDM is composed of 22 ministries, representatives of the Cambodian Armed Forces, Cambodian Red Cross, and Civil Aviation Authority. Royal Decree No. 0202/040 of 2002 mandated the NCDM to play a leading role in facilitation and coordination, including the exercise of a comprehensive approach to disaster management; the analysis of existing and potential hazards and taking measures to mitigate against hazards; and provide and formulate effective disaster preparedness and response plans.

Structures at the sub-national level are present at the provincial, district, commune and village levels, in line with the decentralization policy of the government. Since the creation of the NCDM, major policy developments on DRR have been accomplished nationally and internationally. International policies and declarations that have translated to national action plans include the Hyogo Framework of Action which led to the signing and crafting of the Strategic National Action Plan (SNAP) 2008-2013 on six key areas: institution-building; local and community based disaster risk management, early warning; IEC and knowledge management to promote culture of safety and resilience; mainstreaming of DRR and disaster preparedness at all levels. Another important development was the crafting of the National Adaptation Programme of Action to Climate Change (2006) in compliance with Cambodia's commitment to the United Nations Framework Convention on Climate Change (UNFCCC). Decentralized mechanisms have been the main channels through which DRRM and NRM policies are synergized, and harmony is established between sectoral agendas. The Decentralization and Deconcentration (D&D) Strategic Framework (2005) emphasizes the adherence of governmental strategies and programs to the pillars of representation, participation, accountability, effectiveness, and poverty reduction: all rooted in bringing services and decision-making processes closer to citizens. The new law on the administrative management of the Provincial, Capital, District and Khan councils (2009), also called the Organic Law, spells out the roles and responsibilities of sub-national councils, including tasks in agriculture, forestry, natural resource management and environment, land use, electricity production and distribution, and water management. Emphasis is placed on allocating the use of resources within the framework of a larger poverty reduction scheme that considers the needs of vulnerable groups; the law also recognizes the need for a disaster management plan along with the mandated local development and investment plans. Guidelines on provincial/capital development plans are being drafted and contain sections on natural resource, environment and disaster risk analysis for natural disasters and climate change. A network of tributaries called the 3-S Rivers (Sesan, SrePok and Sekong rivers) flows through the three provinces from the bordering countries of Vietnam and Lao PDR before they empty into the Mekong River. In the last ten years, the region has opened up to economic development, inviting economic boom from land development and natural resource extraction as well as cross-border trade and immigration. Just as the region is endowed with natural resources, it is also subject to intense natural resource exploitation and environmental degradation. The three provinces also have very high poverty rates estimated from 41 to 100 per cent according to a recent poverty survey for the Cambodian Millennium Development Goals (MoP and RGC, 2009, p.6). IOM vulnerability assessments in Ratanakiri, Mondulkiri (2009) and Stung Treng (2010) identified the region as having high disaster risk levels to flood, drought and insect infestation, although they were not included in the priority areas under the Strategic National Action Plan (SNAP) for Disaster Risk Reduction 2008-2015 (SNAP, 2008). These provinces are seasonally exposed to slow onset and flash floods caused by a combination of above average precipitation during the monsoon, overflow of rivers and streams, and water releases from the upstream hydropower dams; and to agricultural and hydrological droughts resulting in greater water stress for competing local uses, lower agricultural production, and insecure clean water sources. Other hazards affecting local populations include insect infestation, animal diseases, emerging and recurrent human diseases, in combination with the environmental hazard risks from deforestation, rapid changes in hydrological regime of the Mekong tributaries with contributions from water pollution and upstream hydropower dams, which further raise the risk levels of communities to extreme weather events.

More than 70 per cent of the communities in all surveyed villages were found to have medium to high disaster risk levels to flood and drought, with secondary hazard impacts such as animal disease outbreaks and recurring and emerging human diseases. Climate-induced hazards are emerging hazards that is increasingly being faced by the region. Among the manifestations of rapid climactic changes include longer duration and intensity of droughts, highly irregular precipitation levels, increase in average temperatures, and smaller cycles and frequency of floods which disrupt traditional agricultural farming practices and contribute to worsening existing health, socio-economic, environmental and material vulnerabilities of the communities studied (IOM, 2009a, 2009b, 2010).

The vulnerability assessments further indicate that the key features of social/organizational vulnerability in the North-East are the 1) lack of vertical linkages between national and sub-national committees for disaster management, and sub-national authorities and local communities on disaster preparedness and response; 2) the absence of coordination mechanisms for timely receipt and dissemination of reliable early warning information to affected populations; and 3) the low priority given to disaster risk reduction as a development issue, with PCDM and its line departments convening only during the onset of disaster events

Based on IOM's field assessments, the interactions between environment and natural resource management policy and program implementation and DRRM in the context of the north-east region of Cambodia can be summarized in the following key points below:

(i) *The need to build on local and indigenous practices and knowledge of their natural environments for effective disaster preparedness, response to and recovery from seasonal natural and environmental hazards.*

Traditional knowledge and warning signals are present and considered to be reliable means for predicting the onset and behaviour of hazards. Rootedness to land and natural resources as central part of the local identity may serve as an impetus for basic but sustained community disaster risk management initiatives. However, the increasing unpredictability of seasonal and climactic patterns, and prolonged and more frequent periods of drought, renders this knowledge unreliable to changing conditions (IOM, 2009b, p. 3). Traditional knowledge needs to be complemented by access to safe areas and facilities, timely receipt and actions based on early warning signs both traditional and modern, and alternative livelihood strategies. A broader environment and natural resource management campaign for local communities is important in raising awareness on the linkage between disaster risk management and natural resource management.

 (ii) *Changes to livelihood, food and social securities driven by environmental degradation may increase the impact of natural disasters and further exacerbate the vulnerabilities of affected communities.*

Food insecurity and insecure access to water are central features of vulnerability to natural disasters in the region. While the region is abundant in water sources, the lack of appropriate technology and storage to tap and deliver water to households and paddy fields is a challenge. Availability of new crop fields is decreasing due to restrictions placed by conservation policies, economic land concessions as well as increasing demand from growing population in the assessed villages. For instance, in Ratanakiri Province, some of the assessed villages expressed concern over changes in access to natural resources following the passage of the Protected Area Law. Efforts are underway to ensure community participation in the co-development of a community protected areas (CPA) management plan. While management zones are provided by law, there are yet no defined boundaries between protected area zones. Some areas already identified as community protected areas cover resin and malva nuts which are harvested by indigenous communities and old crop fields, but prohibit the clearing of new areas (IOM, 2009b, p. 26). In Stung Treng Province, land conversion for agriculture, cash crops and resettlement expansion in response to population growth and migration, and infrastructure development are identified as main threats to management of the Ramsar site. Conservation for international bird areas (IBA) in the same province is threatened by extreme exploitation of riverine birds for commercial and local uses, and habit loss due to flash floods and forest clearance for agriculture and logging. In Mondulkiri Province, two wildlife sanctuaries face management challenges such as unmonitored timber poaching, harvesting of wood for fuel, charcoal production and wildlife hunting and trading, and habit changes due to forest land conversion for settlement and agro-industries. Across the three provinces, forest areas that were affected include communal forests that are under the traditional use of indigenous people for resin tapping, grazing lands, wild game, and for spiritual purposes (IOM, 2010, p. 25).

16.6.1 Environmental protection, disaster preparedness and population stabilization

One potentially significant consequence of their high vulnerability to natural disasters and climate-induced hazards and environmental degradation is on local community's mobility patterns. Although no large movements have been reported in the North-East, both temporary and permanent movements that had been documented in the last ten years (IOM 2009a, 2009b, 2010; 3SPN, 2007) which indicate movements as part of local coping strategies to environmental and natural hazards, including flood and drought, and to socio-economic factors such as search of new lands for cultivation and settlement. Indigenous communities, having a tradition of moving within village boundaries, have noted the difficulty of moving their rice fields or houses in response to these hazards or in the practice of their traditional beliefs. Even when temporary movements serve as a coping strategy, there are no water and sanitation facilities and shelters available in safe areas identified by communities. Migration to safer areas is also limited by the lack of financial resources, the lack of available lands within the village, and the limitations posed by conservation policies in forest and protected areas. There is a need to understand how migration can be maximized as a coping strategy and as part of community-based preparedness and response to floods and drought (i.e., identification of safe areas and safe area management, and inclusion of migration/population displacements in pre and post disaster risk assessment).

As road and border access to the region improves, local economies experience growth as well as population growth that stems from seasonal labour migration, permanent in-migration, and cross-border trade. The combination of these factors makes unplanned population movements unsustainable. The survey and development of land use plans down to the commune level is still at its early stage, and when finalized these plans can better inform local officials and communities in their local land use. The experience of Mondulkiri Province demonstrates the importance of government, NGO and community cooperation in natural management that is crucial for CBDRM wherein the Forestry Administration led the participatory land use planning (PLUP) in partnership with NGOs like the Wildlife Conservation Society for the Seima Biodiversity Conservation Area. Communities were engaged in identifying different land uses through a participatory approach resulting in a PLUP agreement with the community. Under the agreement, each family will be allocated five hectares to use for their residence and farming activities (IOM, 2009a, p. 23). In Ratanakiri, the Forestry Administration in cooperation with local NGOs and indigenous communities has identified 31 areas to be allocated for community forestry. In Stung Treng, local community projects have been set up by NGOs in three districts, but have yet to be officially endorsed by the Forestry Administration (IOM, 2010, p.25).

In line with the mandate of the Land Law of 2001 as well as the Sub-Decree on Procedures of Registration of Land of Indigenous Communities, communally held lands by indigenous communities can be recognized subject to the establishment of their legal identity as a community and subsequent legal registration as defined by the Ministry of Interior. In Ratanakiri Province, two legal identities and in Mondulkiri and one legal identity for a Phnong indigenous village have been registered (IOM, 2009a, p. 23; IOM, 2009b, p.24).

"Institutional capacities for convergence of DRRM and NRM are crucial in effective development planning and rationalized resource mobilization and community engagement and in addressing the emerging impacts of climate-induced hazards".

Among local communities surveyed particularly in areas where chronic poverty is high, disaster preparedness was limited to household-level preparations—such as preparing food stock and evacuating livestock to safety during floods or storage of water during drought—and some village-based mechanisms such as sharing of boats and identification of safe areas during evacuation. These disaster preparedness activities proved sufficient in the past during small scale floods and drought; however, recent trends indicate unpredictable flooding and precipitation patterns that have increased in intensity, duration and scale are rendering these mechanisms insufficient. Overall disaster response capacity is also very limited owing to the lack of technical and human resource capacity, as well as financial and material resources to undertake emergency response, relief and recovery. Often, Provincial Committee for Disaster Management (PCDM) officials meet only during onsets of disasters to mobilize resources from different line departments and from other funding sources, indicating a very high dependence on external assistance. At the district, commune and village levels, there are no dedicated technical and material resources to first lines of response at the onset of disasters nor are there trained local human resources that can be tapped to undertake systematic disaster response and evacuation, and assessment of damages and displacements and assistance to be given to affected populations.

Institutional capacities by the Provincial Committee for Disaster Management (PCDM) are limited by both technical and financial resources as well as low levels of appreciation of disaster risk reduction beyond emergency response. Past disaster management efforts had focused on remedial measures for drought and emergency response during and after floods and there is very low inter-departmental and administrative coordination on disaster management. DRR is not identified in the provincial development plan as prioritized development issue and has yet to be fully mainstreamed into natural resource management, environment and health programs in the provinces. Hence, there is no human, technical and financial resources tied to DRR-related activities. Awareness on climate change is at its earliest stage at the institutional and community-levels through the Provincial Department of Environment and some local non-governmental organizations.

The low level of awareness on climate change makes it challenging for the interaction of DRR and climate change at field level. Both concepts are new and their application in local development planning is limited as yet. IOM assessments had shown that inter-sectoral departmental coordination by key provincial departments such as the agriculture, forestry and fisheries, water resources and meteorology, and environment, for DRR, NRM and climate change has not yet taken place due to the lack of guidance for the operationalisation of these mandates. The lack of coordination and convergence of sectoral planning and focus at lower levels of government more importantly reflects the difficulty of such convergence at the national level, where each responsible ministry under the Strategic National Action Plan (SNAP) for DRR has defined responsibilities. The shift to a developmental focus of DRR is gradually taking place but this eventual shift has yet to be supported by national and provincial wide capacity building and administrative reform process. Clear definition of roles and responsibilities and budgetary allocation at provincial, sectoral department levels, and other sub-national levels must be reinforced. Field experience from piloting of the Village Disaster Management Teams (VDMT) in two provinces had shown that community-based disaster risk management could only be effective when informal social mechanisms are linked to formal mechanisms such as through the commune investment and development planning mechanisms. It has also demonstrated how the impacts of climate change and environmental degradation need to be considered in community action planning and risk assessment. Present mechanisms have not yet been fully maximized to allow for this convergence at the local level. For instance, natural resource management priorities supported under the DANIDA CBNRM funding stream could be channelled as a major DRR measure in the annual commune investment planning. Ongoing administrative streamlining following the newly passed national program on sub-national democratic development for 2010-2019 provides for a set of planning guidelines currently on draft at sub-national levels that include DRR and climate change in the risk assessment and planning priorities.

16.7 CONCLUSION

Policy developments in the last two decades in Cambodia have placed environment and natural resource management as pillars to its development strategy and long-term poverty reduction targets. The increasing complexity of Cambodia's socio-economic development challenges calls for more integrated policy and institutional convergence and solutions on key issues such as environmental degradation, impacts of natural disasters and climate change on food security and human vulnerability, and natural resource exploitation. Decentralized mechanisms had shown promising gains in bringing these convergences and mandates to sub-national government and local communities.

Historically exposed and devastated by floods and drought and in recent years, by extreme weather events such as the Typhoon Ketsana in 2009, Cambodia has low adaptive capacity to both natural and climate hazard risks. Two of three north-eastern provinces, Ratanakiri and Mondulkiri, rank high in vulnerability levels and low in adaptive capacity in a recent study on climate vulnerability in Asia (Yusuf and Francisco, 2009).

Research assessments from the provinces of Ratanakiri, Mondulkiri and Stung Treng in North-East Cambodia confirm the high vulnerability levels of local and indigenous communities as well as ecosystems and institutions to the compounding risks from natural hazards, environmental degradation driven by such factors as natural resource exploitation, rapid economic development and in-migration, and climate induced hazards. Institutional capacities for disaster management are low and synergies between environment and natural resource managements are weak. Despite these challenges, pockets of optimism abound for the contribution of natural resource management and environmental preservation to disaster risk reduction and management practices in the region. Field experiences demonstrate the clear benefits of community-based disaster risk management (CBDRM) and community-based natural resource management (CBNRM) as complementary approaches to help address local vulnerabilities to natural hazards. Recent policy approval on decentralization and poverty reduction recognizes the development linkages between addressing vulnerability reduction to natural disasters, increasing ecosystem and human resilience, and poverty reduction. The development of local institutional, technical and human resource capacities for NRM and DRRM is crucial in developing adaptive capacity of local communities and ecosystems to recurrent risks of natural hazard and extreme weather events, environment hazards as well as climate-induced hazards.

References

Food and Agriculture Organization (September 30, 2008). Water Profile of Cambodia. Retrieved from http://www.eoearth.org/article/Water_profile_of_Cambodia.

Helmers, K., and Jegillos, S. (2004). *Linkages between flood and drought disasters & Cambodian rural livelihoods and food security*. Phnom Penh: International Federation of Red Cross and Red Crescent Societies and the Cambodia Red Cross Society.

International Organization for Migration (2009a). *Mapping Vulnerability to Natural Hazards in Mondulkiri*. Phnom Penh: International Organization for Migration.

International Organization for Migration (2009b). *Mapping Vulnerability to Natural Hazards in Ratanakiri*. Phnom Penh: International Organization for Migration.

International Organization for Migration (2010). *Mapping Vulnerability to Natural Hazards in Stung Treng*. Phnom Penh: International Organization for Migration.

Information Technology Section/Department of Public Information (DPI) (2003). United Nations Transitional Authority in Cambodia. Retrieved from http://www.un.org/en/peacekeeping/missions/past/untac.htm

Mekong River Commission (April 20, 2011). About the Mekong. Retrieved from http://www.mrcmekong.org/about_mekong/about_mekong.htm.

Mekong River Commission (1995). *Agreement for the Cooperation on the Sustainable Develop-ment of the Mekong River Basin.* Chang Rai: Thailand, Mekong River Commission.

Ministry of Planning and United Nations World Food Programme (2003). *Poverty and Vulnerability Analysis Mapping in Cambodia: Summary Report.* Phnom Penh: Royal Government of Cambodia.

Ministry of Planning (2009). *Preliminary Results of the CDB-based Research and Analysis Project. Poverty and Select CMDG Maps and Poverty Charts 2003-2008.* Phnom Penh: Royal Government of Cambodia.

National Committee for Disaster Management (1995). *National Committee for Disaster Management Policy Document.* Phnom Penh: Royal Government of Cambodia.

National Committee for Disaster Management, United Nations World Food Programme (2003). *Mapping Vulnerability to Natural Hazards in Cambodia.* Phnom Penh: National Committee for Disaster Management and the United Nations World Food Programme.

National Committee for Disaster Management & Ministry of Planning (2008). *Strategic National Action Plan for Disaster Risk Reduction (SNAP) 2008-2013.* Phnom Penh: Royal Government of Cambodia.

National Committee for Disaster Management (2010). Comprehensive Post- disaster *Needs.* Royal Government of Cambodia. (2003b)

Obendorf, R. (2006). *Legal Analysis of Forest and Land Laws in Cambodia.* Phnom Penh: Community Forestry International.

Royal Government of Cambodia (1993). *Royal Decree in the Protection of Natural Areas.* [Document prepared by the Ministry of Land Management, Urban Planning and Construction]. Phnom Penh, Cambodia.

Royal Government of Cambodia (2001). *Land Law of 2001.* [English translation by the Ministry of Land Management, Urban Planning, and Construction]. Phnom Penh: Royal Government of Cambodia.

Royal Government of Cambodia (2003a). Sub-Decree on Community Forestry Management [Translation provided by GTZ Cambodian-German Forestry Project]. Phnom Penh: Royal Government of Cambodia.

Royal Government of Cambodia (2003b). *Sub-Decree No. 19 on Social Land Concessions.* [Document prepared by the Ministry of Land Management, Urban Planning and Construction] Phnom Penh: Cambodia.

Royal Government of Cambodia (2005a). *Strategic Framework for Decentralization and Deconcentration Reforms.* Phnom Penh: Royal Government of Cambodia.

Royal Government of Cambodia (2005b). *Sub-Decree No. 146 on Economic Land Concessions.* Phnom Penh: Royal Government of Cambodia.

Ministry of Environment (2006). *National Adaptation Programme of Action to Climate Change.* Phnom Penh: Royal Government of Cambodia.

Royal Government of Cambodia (2008). *Protected Area Law.* Phnom Penh; Royal Government of Cambodia.

Royal Government of Cambodia (2008).*The Law on Elections of Capital Council, Provincial Council, Municipal Council, District Council and Khan Council (Royal KramNo.NS/RKM/0508/018).* Phnom Penh: Royal Government of Cambodia.

Royal Government of Cambodia (2009). *Sub-Decree on Procedures of Registration of Land of Indigenous Communities.* [Approved by the Council of Ministers, Unofficial English Translation as of August08, 2009]. Phnom Penh: Royal Government of Cambodia.

Stephens, A. and Brown, G. (2005). *Indigenous Communities and Development in Ratanakiri.* Phnom Penh: Community Forestry International.

3S Rivers Protection Network (3SPN) (2007). *Abandoned Villages Along The Sesan River in Ratanakiri Province, Northeastern Cambodia.* Banlung, Ratanakiri: 3S Rivers Protection Network.

United Nations Development Programme (June 28, 2001). *The SEILA Programme and the Cambodia Area Rehabilitation and Regeneration Project.* Retrieved from http://mirror.undp.org/carere/index.html.

Bamboo Green Belts:
An Innovative Option for Alternative Livelihood and Sustainable Coastal Protection

K.G. Thara

17.1 COASTAL DYNAMICS IN INDIA

The Indian sub-continent has a long coastline of about 7516.60 km including Daman, Diu, Lakshadweep and Andaman & Nicobar Islands. About 23% (1450 km) of the shoreline along the Indian mainland is affected by erosion, of which 700 km is reasonably well protected by construction of seawalls, groynes, etc. About 8% of the area of the country is prone to cyclone-related disasters such as storm surge, heavy winds and very heavy rainfall. Between 1891 and 2006, 308 cyclones were estimated to have crossed the east coast, out of which 103 cyclones were severe. The west coast however witnessed lesser activity, with 48 cyclones, out of which 24 were of severe intensity. Evaluation of cyclone data for over a century (1891-2000) for assessing the vulnerability of the east coast of India (Mascarenhan, 2004), reveals that Andhra Pradesh received 32 % of the cyclones formed in the Bay of Bengal. Odisha was hit by 27 %, Tamil Nadu by 26 % and West Bengal by 15 % of cyclones. Though the frequency of occurrence of tropical cyclone is relatively low in the east coast of India and in Bangladesh, these regions are highly risk prone in terms of human population at risk.

At the global level, in the last 30 years, the strength of tropical cyclones has more than doubled in the Atlantic region and nearly doubled in the West Pacific region. This is attributed to the increased concentrations of greenhouse gases and concomitant increase of sea surface temperatures (Emanuel, 2005). This change and increased sea activity is most pronounced in the North Pacific, Indian and Southwest Pacific oceans, while the least change is noticed in the North Atlantic Ocean.

17.2 COASTAL VULNERABILITY OF KERALA & NEED FOR PROTECTION

The 587 km long coastal zone of Kerala is vulnerable not only to the perpetual threat of coastal erosion but also to the storm surges caused by cyclones. This coastal belt is one of the most

---

Disaster Management and Risk Reduction: Role of Environmental Knowledge; Editors Anil K. Gupta, Sreeja S. Nair, Florian Bemmerlein-Lux and Sandhya Chatterji; Copyright © 2013, Narosa Publishing House, New Delhi

densely populated regions in the country, which increases its vulnerability to a multitude of disasters such as beach erosion, cyclone, tsunami, storm surge, sea level rise etc. Kerala constitutes only 1.18 % of the total area of Indian Union but accounts for about 3.1 %of Indian population. Almost 16 % of Kerala's population lives along the vulnerable coastal belt.

The state is also susceptible to strong cyclonic winds. Almost the entire state (96.9 % of the total area) lies in the 140.4km/h wind zone (moderate damage risk zone) and the rest of the area lies in 118.8km/h wind zone. The probable maximum storm surge height in the State is 3.5m and minimum is 2.3m. If the Storm surge occurs during high tide, the maximum surge height may reach 4.2 m and minimum storm height may reach up to 3 meters (Figure 17.1) (BMPTC vulnerability Atlas of India, 2006).

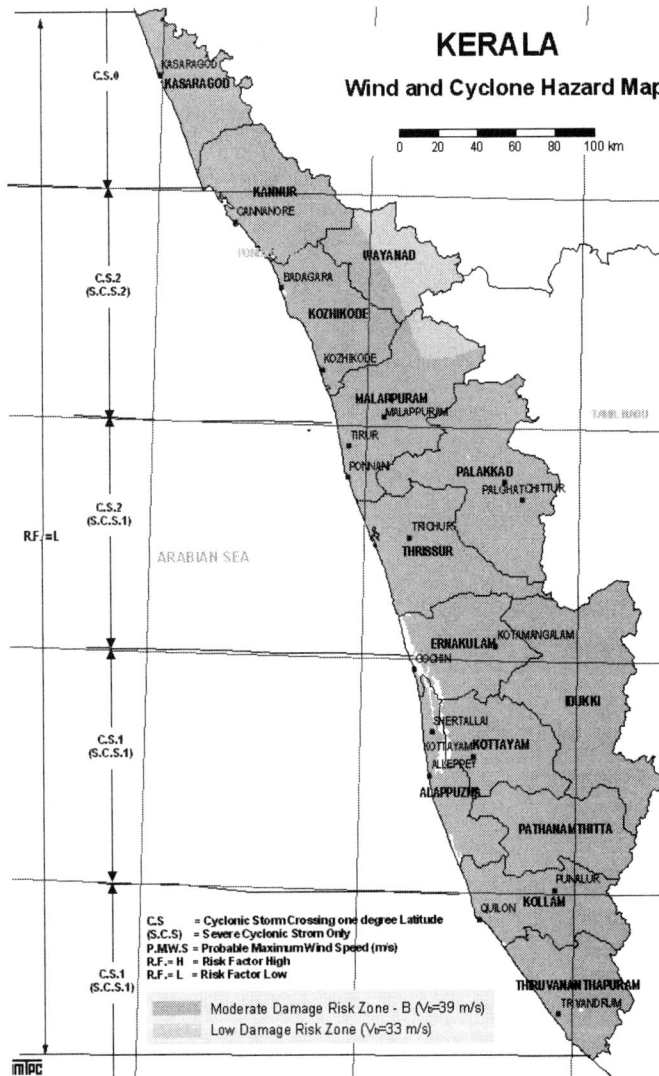

Figure 17.1: Wind and Cyclone Hazard Map of Kerala

Since the density of population is 819 persons/km^2, the third highest density in the country (Economic review, 2010), a very large number of people are vulnerable to damage caused by cyclonic winds.

Topographically, 16.40 % of the total area constituting coastal plains and lagoons is only 10 m from the mean sea level. 54.17 % of the area is low lands falling within an altitude of 10-300 m above the sea level and the remaining 29.43 % lies above an altitude of 300 m above the sea level. Thus, these factors make the estimated 8, 43, 587 fishermen, around 3 % of the total population of the state extremely vulnerable to hydro-meteorological hazards. For around 2.2 lakhs active fishermen, who venture into the ocean on a daily basis, fishing is the main source of income, and they are particularly vulnerable to the hazards mentioned above. The physiographical peculiarities of Kerala the large coastline with the adjoining low lying area and high density of population make the coastal population especially sensitive to sea level rise, high tides, tsunami and coastal erosion. The rapid increase of greenhouse gases in the atmosphere, land degradation, increasing floods and droughts, deforestation, loss of biodiversity and productivity are leading to an ecological crisis that is affecting livelihood options, and leading to poverty, pollution and unsustainable development. As coastal areas are among the most vulnerable locations, it is imperative that scientific management of disasters and augmentation of resilience through natural interventions is given the highest priority so as to protect the poor- those who directly bear the brunt of these adversaries. This suggests, inter alia, a policy to actively encourage and support the plantation of green belts to overcome these constraints.

The rapid increase of greenhouse gases in the atmosphere, land degradation, increasing floods and droughts, deforestation, loss of biodiversity and productivity are leading to an ecological crisis that is affecting livelihood options, and leading to poverty, pollution and unsustainable development. As coastal areas are among the most vulnerable locations, it is imperative that scientific management of disasters and augmentation of resilience through natural interventions is given the highest priority so as to protect the poor- those who directly bear the brunt of these adversaries. This suggests, inter alia, a policy to actively encourage and support the plantation of green belts to overcome these constraints.

In India, interventions for prevention of coastal erosion have primarily been centred on structural measures such as the construction of seawalls, revetment, groynes, off-shore break waters etc. However, with the increasing realization of the adverse effects observed of all structural options, especially in the downstream side, the need to shift to non-structural or soft measures such as vegetative covers, beach nourishment etc., is increasingly being recognized by the policy makers and the local communities. Plantation of trees and shrubs which act as bio-shields are increasingly becoming an option for coastal protection because of their multifold benefits.

Some coastal plants commonly used as green belts are Nyamplung (Calophylluminophyllum), CemaraLaut (Causarinaequisetifolia), Ketapang (Terminaliacattapa), WaruLaut (Hibiscus tillaceus), PutatLaut (Barringtoniaasiatica), Bintaro (Cerberamanghas).Since the choice of species for shelterbelts is dictated largely by the local climate, soil conditions, physical and chemical properties of the substratum and the tolerance to high salt conditions, mangroves are generally the most favoured natural species in tropical coastal regions. However, these are sensitive to many ecological factors, including changes in water flow and salinity. Moreover, mangroves over large offshore regions may limit fish-farming opportunities, and establishment of shelterbelts that are contiguous onshore, may limit the types of agriculture or forestry options that are available in the coastal area.

Figure 17.2: Plants commonly used as green belts (a) Nyamplung (Calophylluminophyllum), (b) CemaraLaut (Causarinaequisetifolia), (c) Ketapang (Terminaliacattapa), (d) WaruLaut (Hibiscus tillaceus), (e) PutatLaut (Barringtoniaasiatica), (f) Bintaro (Cerberamanghas)

It is important to note that a shelterbelt protects an area up to its own height on the windward side and up to twenty times its height on the leeward side, depending on the strength of the wind (Eugene et al., 2007).

In regions where the combination of exposure to tropical cyclones and soil conditions does not favour the growth of mangroves, tall trees such as some varieties of coconut palms and rubber trees are already in use, especially in locations beyond the inundation. Examples of such coastal protective belts include a 3000 kilometre shelterbelt built mainly with *C. equisetifolia* along China's southern coast and use of coconut palms and rubber trees on the west coast of Indonesia (Smith, 2006). *Casuarina equisetifolia* is also used as a shelterbelt species in tropical coastal areas because of its height (up to 30 metres), rapid growth rate and high tolerance to salty environments as well as to high rainfall, although in some regions it is considered unpopular owing to its invasive habit and lack of support for an accompanying diverse ecosystem.

17.3 BAMBOO AS AN OPTION

India ranks second in the world in bamboo reserve and diversity, with 136 species and an estimated 8.96 million ha coverage in the forest area. Though a National Bamboo Mission was formed to promote growing bamboo, the country exploits only one–tenth of its total potential. The history of using bamboos in day to day life in India dates back to more than 5000 years. Bamboo was and is used for construction of houses and buildings, poultry sheds, cowsheds, goat farms, for making furniture and also used as fencing for compounds and protection of agricultural land, construction of small bridges, scaffoldings, etc.

Figure 17.3: Bambusaarundinacea (Retz.) Roxb. (Thorny bamboo)

Bamboo is increasingly being recognized as a healthy alternative in modern sustainable architecture and there is tremendous scope for widening its utility both in urban and rural areas. However, the wide scope for product diversification and use of bamboo in different type of construction such as load bearing structures, thatching, partition walls, staircase reeling's, flooring, frame in minor RCC work etc., is not fully utilized. Bamboo thrives in warm climates where the earth is kept moist by frequent monsoon rains. Bamboo cultivation has scope for promoting the skill of artisans and through value addition to products, strengthening the marketing options. The ability of a bamboo to mature to a height of 20 meters or more in a mere four years and an extensive root system which remains intact, allowing for rapid regeneration , are qualities which make bamboo an ideal plant for areas threatened with the potentially devastating ecological effects of soil erosion. Apart from bamboo baskets, woven boats made of bamboo splits are also used for rescue operations in disaster situations in countries such as Vietnam.

Kerala state has approximately 3600 hectares of area under shelterbelt plantations and another 1800 hectares is planned to be brought under the green belt to address the escalating risk factors in the coastal regions. *Bambusa Arundinacea* (Retz.) Roxb. (Thorny bamboo) is one such species identified by the M. S. Swaminathan Research Foundation (MSSRF) of Chennai that can be used for coastal forests and shelterbelts in India. Apart from reducing the state's vulnerability to cyclones and storms, green belts are expected to stabilize the coast and prevent the wind erosion of sand, protection of life and property from cyclonic storms and floods and ensure the sustainability of coastal zones.

17.5 SOCIAL AND ECONOMIC BENEFITS OF BAMBOO AS A GREEN BELT

Introducing bamboo in the shelterbelt and coastal forests is proposed as an environmentally attractive, innovative and sustainable option, given its status as an ideal species capable of achieving conservation of soil and moisture, repairing degraded land, providing ecological, nutritional, livelihood and economic security, apart from its manifold uses and industrial applications. Most species of bamboo produce mature fibre in about three years and grows much faster than any other tree species. Some species grow up to one metre a day, with the majority reaching a height of 30 metres or more. Despite the National Bamboo Mission actively involved in the promotion, research and marketing of bamboo and bamboo based handicrafts, its use as a coastal protection belt is yet to be explored in a large scale. Bamboo can also be used in many building products, including bamboo-glass fibre composites, plywood substitutes, and laminated flooring that are ideal for disaster resistant construction. By virtue of its structural advantages such as strength and light weight, bamboo buildings are inherently resistant to wind and earthquake forces. These properties can be effectively exploited through promotional plantations and construction initiatives along the coast.

17.6 BAMBOO AS AN ALTERNATIVE SOURCE OF LIVELIHOOD AND INCOME GENERATION

In recent years, the demand for bamboo has increased within the country and abroad as a raw material for furniture making, as panel boards substituting wood, as agricultural implements, house/construction related uses and also as a vegetable. It can also act as the 'Poor Man's Steel' in disaster resistant construction techniques. Bamboo plantations are expected to (i) increase employment opportunities (ii) provide an alternative source for employment and (iii) to produce materials that promotes a healthy human environment as a resource that can be harvested with no harm to the natural habitat. It is pertinent to note that 54 % of fishermen in Kerala, have an income of less than thousand rupees per month and are below the poverty line. 64% of fishing households in Kerala are in debt because of financial loans availed for purchase of fishing implements. Disasters in these coastal areas simply strip these populations off their only means of livelihood, spiralling them into more and more debts and deprivation. The tsunami of 2004 stand a testimony to the severe damage and destruction they can wreak in infrastructure, assets, services and economic activities. Bamboo cultivation would help the fishermen and fisherwomen to become more self-reliant by ensuring a minimum wage, through a simultaneous promotion of bamboo-based handicrafts and innovative construction technologies. This is significant especially in a country which has a 7516 km stretch of sea coast.

17.7 CONCLUSION

The dynamics of the coastal area of the country with its high risk to the resident communities from the multitudes of hazards such as cyclone, storm surge, beach erosion, sea level rise, tsunami, high tides etc., makes it imperative to take up measures to protect life and property of these populated areas. The adverse effects observed in the downstream side of all structural options necessitates the promotion of soft structural measures for reducing or preventing impacts of cyclone, sea surge and beach erosion. The introduction of bamboo as an innovative option in green belts will not only ensure environmental conservation and eco development, but also help our economy by generating alternative income and employment opportunities for those communities whose livelihood options are threatened during any disasters. Qualities such as ability to mature to a height of 20 meters or more in a mere four years and an extensive root system which remains intact make bamboo an ideal plant for areas threatened by coastal hazards and its structural advantage can also be exploited in the construction of disaster resistant construction.

References

BMTPC (2006). *Vulnerability Atlas of India. (2^{nd}Edn)*. New Delhi: Building Materials & Technology Promotion Council (BMTPC).

Economic Review (2004). State Planning Board, Government of Kerala.

Economic Review (2010). State Planning Board, Government of Kerala.

Emanuel, K.A. (2005). Increasing destructiveness of tropical cyclones over the past 30 years. *Nature,436* (7051), 686–688.doi:10.1038/nature03906.

Eugene, S., Chen Takle, T. C., and Xiaoqing, Wu (2007). *Protective Functions of coastal forests and trees against wind and salt spray*. FAO Corporate document repository.

Mascarenhan, A. (2004). Oceanographic validity of buffer zones for the east coast of India: A hydro-meteorological perspective. *Current Science 867*: 399–407.

Selvam, V., Ravishankar, T., Karunagaran, V.M., Ramasubramaniar, R., Eganathan, P., and Parida, A. K. (2005). *Toolkit for Establishing Coastal Bioshield*. Centre for Research on Sustainable Agriculture and Rural Development. M.S. Swaminathan Research Foundation.

Smith, M. (2006).The right way to rebuild Asia's coastal barrier Mangrove forests can act as 'bioshields' against storms. *Science Development Network*. January12.

18

Ecosystem Approach to Disaster Risk Reduction:
A Case Study of Majuli Island, Assam

Sunanda Dey

18.1 INTRODUCTION

The increasing frequency and severity of natural and technological disasters particularly, but not exclusively, in the developing world place them in the centre of debates on human-environment relations and issues of development and sustainability. Disasters occur at the interface of society, technology and environment and are fundamentally the outcomes of the interaction of these features (Smith and Hoffman, 2002). In very graphic ways, disasters signal the failure of a society to adapt successfully to certain features of its natural and socially constructed environment in a sustainable fashion. Since the last three decades, a new perspective has emerged that views hazards as basic elements of environments and as constructed features of human systems rather than as extreme and unpredictable events, as they were traditionally perceived. When hazards and disasters are viewed as integral parts of environmental and human systems, they become a formidable test of societal adaptation and sustainability. In effect, if a society cannot withstand without major damage and disruption, a predictable feature of its environment, that society has not developed in a sustainable way. When disasters strike, whether it is the slow onset of drought, exposure to hidden toxic waste, or the sudden impact of an earthquake or chemical leak, it tends to be a totalizing event or process, affecting eventually most aspects of community life. The traditional communities with their indigenous environmental knowledge which develop on ecosystem approach come to rescue in such situations. The native environmental knowledge and their culture-woven practices make sure that the local communities are safe from disasters which strike in their vicinity. The present article throws light on one such local community i.e. the Mishing tribe of Assam who face annual floods on Majuli Island. Indeed disasters have variously been considered a 'natural laboratory or a criserevelatrice', as the fundamental features of society and culture are laid bare in stark relief by the reduction of priorities to basic social, cultural and material necessities. In that sense, then, there is a fundamental congruence between the analytical requirements posed by disaster studies and the distinctive approach of cultural and social

Disaster Management and Risk Reduction: Role of Environmental Knowledge; Editors Anil K. Gupta, Sreeja S. Nair, Florian Bemmerlein-Lux and Sandhya Chatterji; Copyright © 2013, Narosa Publishing House, New Delhi

anthropology. The holistic, developmental and comparative perspectives of anthropological research placing specifics against larger societal wholes and concerned with issues of social change and evolution are particularly congruent with the totalizing nature of disasters. Anthropological disaster research has taken predominantly outside the Euro-American context, which has been the site of the most disaster research by the other social sciences. The numbers of high-impact technological and natural events are increasing much more rapidly now in the non-Euro-American context, where anthropologists have traditionally worked. Aim of the present study is to understand the indigenous disaster risk reduction measures of Mishing tribe of Majuli Island, Assam in India and study the mainstreaming these practices into the development planning.

18.2 THE ECOSYSTEM APPROACH

The Ecosystem Approach places human needs at the centre of biodiversity management. It aims to manage the ecosystem, based on the multiple functions that ecosystems perform and the multiple uses that are made of these functions. The ecosystem approach does not aim for short-term economic gains, but aims to optimize the use of an ecosystem without damaging it.

It was endorsed at the fifth Conference of the Parties to the Convention on Biological Diversity (CoP 5 in Nairobi, Kenya; May 2000/Decision V/6) as the primary framework for action under the Convention. It comprises 12 Principles. Decision VII/11 of the 7th Conference of the Parties to the CBD supports the application and implementation of the Ecosystem Approach, and welcomes additional guidelines to this effect (Hammerlynck, 2012).

In applying the 12 principles of the ecosystem approach, the following five points are proposed as operational guidance:

1. Focus on the relationships and processes within ecosystem

The many components of biodiversity control the stores and flows of energy, water and nutrients within ecosystems, and provide resistance to major perturbations. A much better knowledge of ecosystem functions and structure, and the roles of the components of biological diversity in ecosystems, is required, especially to understand:

    (i)   Ecosystem resilience and the effects to biodiversity loss (species and genetic levels) and habitat fragmentation; and

    (ii) Underlying causes of biodiversity loss

    (ii) Determinants of local biological diversity in management decisions

Functional biodiversity in ecosystems provides many goods and services of economic and social importance. While there is a need to accelerate efforts to gain new knowledge about functional biodiversity, ecosystem management has to be carried out even in the absence of such knowledge. The ecosystem approach can facilitate practical management by ecosystem managers (whether local communities or national policy makers).

2. Enhance benefit-sharing

Benefits that flow from the array of functions provided by biological diversity at the ecosystem level provide the basis of human environmental security and sustainability. The ecosystem approach seeks that the benefits derived from these functions are maintained or restored. In particular, these functions should benefit the stakeholders responsible for their production and management. This requires, inter alia: capacity building, especially at the level of local communities managing biological diversity in ecosystems; the proper valuation of ecosystem goods and services; the removal of perverse incentives that devalue ecosystem goods and services; and, consistent with the provisions of the Convention on Biological Diversity, where appropriate, their replacement with local incentives for good management practices.

3. Use adaptive management practices

Ecosystem processes and functions are complex and variable. Their level of uncertainty is increased by the interaction with social constructs, which need to be better understood. Therefore, ecosystem management must involve a learning process, which helps to adapt methodologies and practices to the ways in which these systems are being managed and monitored. Implementation programmes should be designed to adjust to the unexpected, rather than to act on the basis of a belief in certainties. Ecosystem management needs to recognize the diversity of social and cultural factors affecting natural-resource use. Similarly, there is a need for flexibility in policy-making and implementation. Long-term, inflexible decisions are likely to be inadequate or even destructive. Ecosystem management should be envisaged as a long-term experiment that builds on its results as it progresses. This "learning-by-doing" will also serve as an important source of information to gain knowledge of how best to monitor the results of management and evaluate whether established goals are being attained. In this respect, it would be desirable to establish or strengthen capacities of Parties for monitoring.

4. Carry out management actions at the scale appropriate for the issue being addressed, with decentralization to lowest level, as appropriate

As noted in the description of the ecosystem approach, an ecosystem is a functioning unit that can operate at any scale, depending upon the problem or issue being addressed. This understanding should define the appropriate level for management decisions and actions. Often, this approach will imply decentralization to the level of local communities. Effective decentralization requires proper empowerment, which implies that the stakeholder both has the opportunity to assume responsibility and the capacity to carry out the appropriate action, and needs to be supported by enabling policy and legislative frameworks. Where common property resources are involved, the most appropriate scale for management decisions and actions would necessarily be large enough to encompass the effects of practices by all relevant stakeholders. Appropriate institutions would be required for such decision-making and, where necessary, for conflict resolution. Some problems and issues may require action at still higher levels, through, for example transboundary cooperation, or even cooperation at global levels.

5. Ensure intersectoral cooperation

As the primary framework of action to be taken under the Convention, the ecosystem approach should be fully taken into account in developing and reviewing national biodiversity strategies and action plans. There is also a need to integrate the ecosystem approach into agriculture, fisheries, forestry and other production systems that have an effect on biodiversity. Management of natural resources, according to the ecosystem approach, calls for increased intersectoral communication and cooperation at a range of levels (government ministries, management agencies, etc.). This might be promoted through, for example, the formation of inter-ministerial bodies within the Government or the creation of networks for sharing information and experience (UNEP, 2012).

The ecosystem approach when studied in context of disaster risk reduction in local communities we can see that the above mentioned principles are in consonance with the practices adopted by them. The Mishing community illustrates how the measures taken by them uses environmental friendly methods, which are symbiotic in nature, help them build their resilience and manage their lifestyle in a way which is culturally acceptable to them. The practices which are followed by generations help them keep their culture intact while adapting to the fury of nature which is in the form of floods in their case.

18.3 INDIGENOUS KNOWLEDGE FOR DISASTER RISK REDUCTION

Traditional knowledge has been used around the world in some or the other form by different indigenous communities. In Burkina Faso, the Mossi farmers use their Indigenous Knowledge to abate problems of drought. They build lines of stones (bunds) on their cultivated land to construct terraces and pits that conserve water. They also fill the bunds with organic material to increase soil fertility. The semi-permeable bunds allow for a gradual seeping in of the water and prevent run-off caused by the scarce but highly intensive rains, thus reducing the risk of crop failure and soil erosion. In the disastrous drought years of 1983 and 1984, crops grew on land with bunds, while on adjoining fields nothing could grow. In India, rural communities play an important role in mitigating the effects of drought by using traditional drought-coping methods such as construction of ponds and dams (anicuts) to save the rain water, which would otherwise be lost due to surface run-off, thus mitigating the effect of drought.

Indigenous knowledge has also been used in mitigating the impacts of earthquakes and cyclones in India. After the disaster of January 26, 2001, Manav Sadhna, an NGO based at Gandhi Ashram in the city of Ahmadabad, Gujarat state, became engaged in earthquake rehabilitation efforts in the village of Ludiya. It implemented a recovery project aimed at reducing impacts of disasters. Remodeled traditional circular houses (bhungas) were constructed to replace the thousands of destroyed homes. The traditional circular homes are known to be resistant to earthquake and are also considered to be cyclone proof. They are constructed with local materials such as sun-dried bricks and straw branches (khip) of the babool tree (UNEP, 2007).

Construction practices in the Himalayan region are still on the lines of the traditional methods. Most of the construction taking place in the region makes use of traditional materials like stone masonry or burnt bricks. Such buildings are highly vulnerable in the high seismic zones. With improvement in the economic condition RC framed buildings are also being constructed. Although these buildings make use of reinforcements little attention is paid to incorporate seismic resistant features in the building design. The buildings in Kangra, Himachal Pradesh had several seismic resistant features inherent in the structure. These features had helped these buildings survive the various earthquakes that the valley has encountered. Buttresses at corners, lateral systems, wooden bands, corner reinforcements, through stones, small openings and buttress-like projections along gable walls were the common earthquake resistant features found in mostly all the buildings (SEEDS, 2007) .

18.4 MAJULI ISLAND AND MISHING COMMUNITY

Majuli, in Assam, has supposedly derived its name from its location between the two streams- the Luit and the KherkatiaSuti. Acknowledged as one of the largest inhabited river islands of the world, Majuli, earlier known as Majali, is situated between 26°45'N to 27°15'N latitude.; 93°45'E to 94°30'E longitude (Figure 18.1). Towards its north lie the north Lakhimpur and Dhemaji districts, Sivsagar and Jorhat lie towards its east, Golaghat is to the south of Majuli while Nagaon and Sonitpur districts lie towards its west. Majuli is situated in the upper reaches of the

Brahmaputra, 630km. Upstream of the Indo-Bangladesh border and 1000km. from its mouth. Prior to 1950, the total area of Majuli was 1,256 sq. km. Continuous erosion since has resulted in depletion of its size. To the north of Majuli flows the old streams of Brahmaputra- the Luit and the KherkatiaSuti- and on the south lies the Brahmaputra. The extreme ends, the east and the west, have been marked by the bifurcation and amalgamation of the two channels of the mighty river mighty. Its elevation from the mean sea level is 84.50 meters.

18.4.1 Cultural Heritage

Majuli is the nerve-centre of the neo-vaishnavite religion, art and culture. In fact, it is considered to be the Vatican of neo-vaishnavism. It was way back in the sixteenth century that MahapurushSrimantaSankardeva along with his chief disciple, Madhabdeva, laid the foundation of the Satra culture in Majuli, which ushered in an era of distinctive religio-cultural heritage. Sankardeva, the great social reformer, founded a new cult of vaishnavism, known as Ek Saran Naam Dharma, meaning Lord Vishnu is at the root of all gods and goddesses and He is to be worshiped only through NaamPrasnga- prayers. The ChoidhyaPrasanga or the fourteen prayers, Gayan Bayan or the playing of drums and cymbals, Borgeet or devotional songs and Bhaona or theatrical performances are some of the instruments used to chant the glory of god. Thesatras- Vaishnavite monasteries situated in Majuli and their influence in religious, cultural and social life of the people have made Majuli, the principal seat of pilgrimage for all people in general and vashnavites in particular. These satras resemble the Buddhist monasteries, where a number of inmates, Bhakats, dedicate themselves wholeheartedly to religious activities. In each satra, there is a Satradhikar- the head of the satra- and the DekaSatradhikar- the deputy head of the satra. Both reside within the satra campus. A model satra has four Hatis- rows of cottages for the inmates or bhakats- in rectangular or square forms, a NaamGharprayer hall and the Monikut, or the sanctum sanctorum. Majuli provided the ideal backdrop for the historic MoniKanchanSanjog or a unique combination – of Assam's pioneer Vaishnavite saint, SrimantaSankardeva and his disciple, Madhabdeva, in sixteenth century, the first satra to be established in Majuli, was named after the place where the two great saints met for the first time. Ever since that meeting of the great minds and the subsequent establishment of the Satras, Majuli emerged as the crowning glory of the Vaishnavite culture of Assam. The followers of Sankardeva, Madhabdeva and Damodardeva established 64 satras in Majuli during the sixteenth and seventeenth century; of which only 22 remain at present. The rest have either been eroded away or had to be shifted to other places.

18.4.2 Diverse Ethnicity

Majuli boasts of a multiplicity of ethnic tribes, which have contributed immensely to its rich and colorful cultural heritage. Mishings- With a population of 63, 572 constituting about 42 per cent of the total population of Majuli (according to the 2001 census), the Mishings who belong to the Burmese branch of Mongoloids, are the most important tribal community of Majuli. Deuris- are believed to be the priestly class of the Chutiyas, who originated from the Tibeto-Burmese branch of the Mongoloids, have a population of 3,498 in Majuli, constituting 3 percent of the total population of Majuli. SonowalKacharis- According to the 2001 census, 1071 people of this community are in Majuli. Most of them reside in the SonowalKachari village, which is about 3 km. to the east of Natun Bazaar in upper Majuli. They are of Tibeto-Burmese branch of the

Mongoloids. This community was engaged in the collection of gold out of the sands of the Subansiri and other rivers of Assam during the reign of the Chutiyas and Ahom King. Mataks-They belong to the historic Moamaria community. In Assamese, mat means decision and ek means one, thus the word matak means unanimity in decision making, which was exhibited bythis community during the Ahom rule and hence they came to be called Mataks. Shri ShriAniruddha Deva is their spiritual leader. In Majuli people of this community are mainly found in the villages of Dekasensowa and located 3km. to the north of Rawanapar. The Mataks are also famous for their exposition of the Mridanga, a musical instrument. The population of Majuli is a medley of the Brahmins, Kalitas, Kochs, Naths, Kaibartas, Mishings, Deuris, SonowalKacharis, Ahoms, Chutiyas, Suts, Nepalis, Bengalis, Mataks and a sprinkling of Marwaris and Muslims.

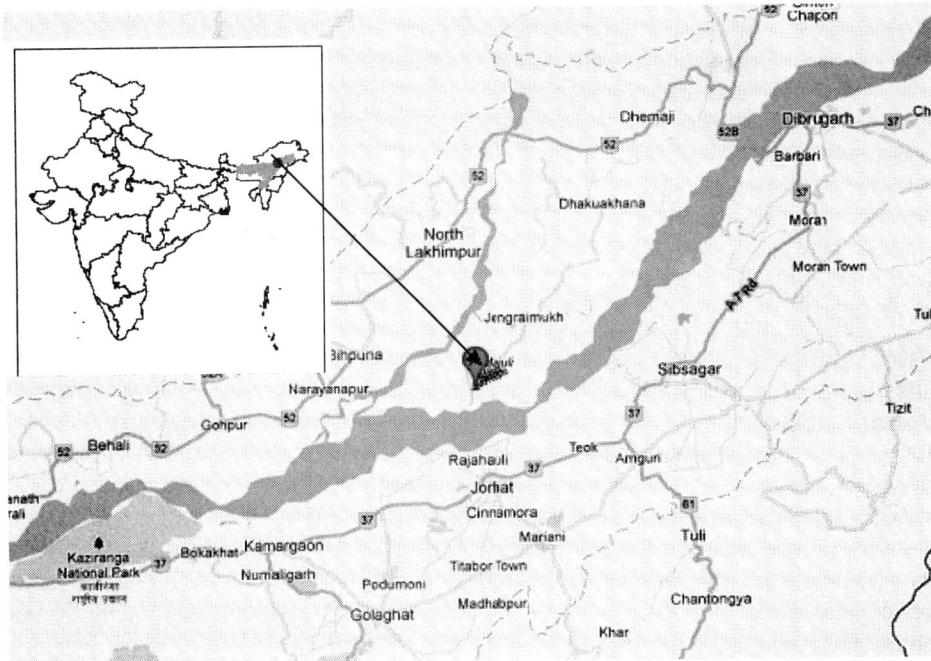

Figure 18.1: Location of Majuli Island in Jorhat district, Assam.
Source: google maps

18.4.3 Problem of Erosion in Majuli Island

Several studies have indicated that the Brahmaputra River changed its course abnormally after the Great Assam Earthquake of 1950 with a magnitude of 8.6, and the attendant historic flood. There was a balance between sediment supply and transport up to 1950 AD, and this balance was disrupted by the great earthquake which produced severe landslides within hilly tracts, and suddenly provided a large quantum of additional sediment.

Descending into the plains, the extra sediment choked the river channel gradually and initiated bank erosion causing channel-widening. Moreover, there has been a gradual increase in channel slope since 1920. The riverbed of Brahmaputra has also shoaled following heavy siltation due to the construction of flood embankments, deforestation, etc. Many other towns besides Majuli, on the banks of the Brahmaputra River, are also under threat due to abnormal changes in the river course.

Figure 18.2: A typical Mishing Chang Ghar, Majuli, Assam

Ground observations on the overall erosion activity around Majuli Island exhibited two types of features. The bowl-shaped shear failure activated by the flow of sediments is more common around GajeraGaon, the easternmost corner of the island. During the high stage of the river, water forces into the massive sand bodies, providing additional support to the bank materials and acting as a continuous system. However, with the fall in water level, the pressure diminishes rapidly and water from the pore spaces of sand bodies tends to flow to the main channel. This flow of water from the sand bodies has caused liquefaction of sediments and the bank materials have been subjected to different degrees of flow. This type of failure is more common around Kumargaon, Batiamarinagar and Bechamara. However, around Salmora where the bank material comprises mainly cohesive clay material, the slope is almost 90° and causes significant oversteepening towards initiation of bank failure. Moreover, the cohesive clay material, on the other hand, produces a comparatively stable land mass that is not prone to erosion thereby helps to generate a node point there. The node point, in turn, also offers significant resistance to the connected flow regime. As the bank materials are relatively stable in this area, the river scours deeper to accommodate the flood discharge, thereby increasing the wash load of the river temporarily. Therefore, below the node point, the river tends to be wide and shoaled. Here the current velocities diminish when large quantities of sediments are deposited and mid-channel bars or chars are formed. Once formed, the chars locally decrease the cross-sectional area and they must cut the bank laterally to maintain a proper cross-sectional area that is in equilibrium with discharge. This might be the most possible explanation about the enhanced rate of erosion activity around Nematighat, Kamalabari and Kaziranga, which has already taken a devastating proportion of land area.

The 2008 floods were the most recent devastating floods when the assessment of the flood situation was done and some base line data was collected. In the following table the number of families which got permanently displaced has been depicted year wise starting 1969 based on district data.

Table 18.1: Number of families affected by problem of erosion till 2007

Year	No of families' permanently displaced	Total area eroded (in Ha)
1969-1999	6497	-----
2000	567	9124
2001	63	14
2002	204	200
2003	100	68
2004	401	667
2005	471	257
2006	150	94
2007	564	230
Total	9017	10650
(Source- District Administration, Jorhat)		

18.4.4 The Mishings

The Mishings are a tribe settled mostly in the districts of Dibrugarh, Lakhimpur, Sibsagar, Jorhat, Darrang, and Goalpara of Assam. Ethnically, the Mishings are mongoloid and belong to its Indo-Tibetan group. They belong to the tribal groups of Miyongs, Padams and Nishis of Arunachal Pradesh. It is said that the Mishings migrated from the hills of Arunachal to the plains of Assam about eight centuries ago and it continued up to the first part of the nineteenth century. Although they identify themselves as Mishing and regard it to be correct name of their community, they have been identified by the term Miri by the people of Assam. The term Miri has been used for the Mishing by the Assamese ever since they came in contact with them in the remote past. Owing to its widespread use the term Miri eclipsed the original name 'Mishing' and was officially listed as Miri in the Scheduled Tribes. The Mishings are divided into two broad divisions, viz., Barogam and Dohgam.

The Mishing originally lived in the hills situated on the north of the district Lakhimpur of upper Assam. This hill tract has been referred to as Miri Hills and is situated between what was once referred to as the Aka-Dafla and the Abor Hills. The region covers the entire territory to the north of the Brahmaputra valley in the upper Assam- roughly the area lying between gorge of the Subansiri in the west to the Dibong river in the east. At present, this region forms the major portion of the subansiri and Siang districts of Arunachal Pradesh.

Tracing the history of Mishings is really difficult because as said by one of the key informants, the history of Mishings was written by their oldest ancestor on the hide of the deer skin. One day there was nothing for the ancestor to eat to survive, so he ate the hide and their history and origin got lost.

18.4.4.1 Religion

The people practice traditional pattern of worship which is a mixture of tribal rituals and Hinduism. It is known among them as Kewalia or Kalhanhati or Nishamaliya. The villagers indulge in various forms of traditional worship where pigs and fowls are sacrificed. They believe in creator (Abutani), heaven (Regi-Regam) and other deities. The main deity of the Mishings is the sun and the moon (Dnoynee-Po:lo). This deity is worshiped in every puja and festival. Traditional pujas and festivals are numerous. Some of them are Po:rag, Ali-ai-ligang, TalengUyu (pujas concern mainly with agricultural activites), Dobur, AshiUyu, YumrangUyu (pujas for diseases), DodgangApin (pujas for ancestors or ancestor worship). These pujas and festivals can be divided into three divisions viz., pujas performed by group or group-puja (e.g. Po:rag, Dobur, Ali-ai-ligang), self or personal puja (e.g. UromApin, Yumrang), pujas due to causes (e.g. TalengUyu, Yalo). Naamghar, the worship place or nowadays used for meeting purposes is found in every village now. Most probably the Mishings converted into Hinduism or Kalhanhati religion during or after the rise of the new Vaishnava Movement in Assam in the fifteenth and sixteenth century.

18.4.4.2 Food Habit

Mishings eat pig meat, foul, mithun regularly and more so particularly during festivities. The consumption of rice beer called as Apong, made from fermentation of rice is quite popular amongst them. They serve Apong as a mark of respect to anyone who comes to their house as guest. It is served in small brass wide mouth katoris. Apart from this, rice is their staple food. They have meals thrice a day in which fish is also included as they are fish lovers and experts in fish catching too. The apong, rice beer is of two kinds with different degree of alcoholic flavour in it. The white or simple apong is milder one as compared to black one which is stronger in flavour and is served during special festivals.

18.4.4.3 Language

The Mishing people speak Mishing dialect. Their language belongs to the northern Assam branch of the tibeta0-burman languages. Assamese language is as popular among them as their own dialect though pronunciation and expression in Assamese is not sound. The Mishings came into contact with the Ahom in the early part of the sixteenth century. Most probably they know Assamese language since then. In the later period Assamese language was the medium of the Gosains and Brahmins who converted them into Hinduism. Assamese is the official language of working in Assam so in Majuli Island. The villagers can also understand Hindi to some extent and Bengali also. But the proportion of the villagers is less comparatively. The Mishing script is roman based.

18.4.4.4 Dress

The dress of the people is simple. The dress of the men is similar to the dress of the non-tribal Assamese. A dhoti (Ugon), and a shirt (Galuk) is the dress of the men. On ceremonial occasions they wear turban (Dumer). The women wear two pieces of garments, a skirt (Mosanam-Age) usually of various colours covering the part from waist to the knee. The upper piece (Gaseng or Gero) fastened above the breast falling along with the skirt. A blouse is sometimes worn to cover the upper part. Old ladies avoid that in summers. Except in the winter season, the upper part of the body remains uncovered. The shawls are made of eri (castor) silk and are woven by the womenfolk of the household. Women are fond of ornaments. The Mishing women are famous for their weaving. Their products are (e.g. Miri Jim, TapumGastor, Dumer) popular among the non-tribal Assamese. Nowadays machine made clothes are worn by youngsters. The middle aged and older generation wear traditional cloths. The school going children prefer non-tribal clothing more like Assamese. It consists of mekhla, blouse and a piece of cloth that is tied around the hips and tucked over the shoulders.

The Mishing community is aware of the annual floods which occur in the region. The Mishings have been coping with floods for a long time now. They have been living intelligently with floods. Each household has a boat and the people live in raised houses built on stilts or "Chang Ghars". The house on stilts (Chang Ghar) is a big hall with a central kitchen for a large joint family. A lower part of the house is used to provide to shelter to animals that every household rears and in river side villages where people lived in flood plains animals were provided shelter on the raised home with people. On the other hand these houses were also made to provide protection from wild animals. Apart from the main house there is a traditional granary on raised platform. It is said by the elders of the Mishing tribes that the present river banks of Brahmaputra were tall grassland and also had very thick vegetation of reeds leading to favourite game area for wild elephants. They say that elephants do not attack houses on stilts and therefore not destroy even the granaries; the grains are also protected from moisture, rodents and floods.

North East including Assam is rich in sylvan resources and most of its forests are richly stocked with bamboos and canes of various species. Bamboo is a raw material of great versatility and forms an integral part of the lifestyle and economy of Assam. The major components of these houses are bamboo, cane and palm leaves for roofing. Bamboo is widely used for pillar, linter, floor, roof, door etc. Bamboos are aptly called the poor man's timber and are found in great abundance in the North Eastern region of India. Their strength, straightness and lightness combined with extraordinary hardness, range in sizes, abundance, easy propagation and the short period in which they attain maturity make them suitable for the purpose (Mipun, 1993, Bhandhari, 1992).

18.5 ECOSYSTEM APPROACH TO DISASTER PREPAREDNESS

The Mishings have their own preparedness plans for the annual floods. The house structure, management of food storage and rescue boats is vital to Mishings. All these practices and measures are adopted keeping an ecosystem approach. The surrounding eco-friendly material (bamboo, dried straws) is used for making and preparing for annual floods. The traditional Chang Ghars are built by them where the impacts of flood are least observed, as it saves their belongings during high floods. The bamboo used houses on slits are their cultural identity and is an indigenous method of protection and adaptation to the local environment.

NGOs also help them in reconstruction of their houses if any damaged during floods, and also provide economic relief through national schemes like Indira Awas Yojna to build their houses. Though some educated Mishings have started to build Assamese style house in some part of Majuli Island- Jengraimukh. They are now more oriented towards modern living, though their number is very less now.

Nowadays due to more of NGOs penetration into the lives of Mishing, they are well prepared to face the wrath of floods as compared to earlier times. There are posters in Assamese language through which they are taught about cleanliness and hygiene to be maintained during flood times and otherwise. Mishing people are provided sanitation kits, soaps and foe ladies- sanitary pads to be used in general life and specially during disasters. They are told by posters about -rescue operation methods, in times of emergency evacuation, personal hygiene to be maintained during floods and in case someone falls ills- the symptoms to detect the disease. The international and national level NGOs are very actively involved in these awareness programmes. Beside all this, they provide mosquito nets and basic medicines to protect them from malaria and other water borne diseases. As the Mishing community is very restrictive in using the things provided by the ngos, the local volunteers help them use them showing their importance.

Figure 18.3: Modern day Mishing Chang Ghar

Figure 18.4: Meeting of Gram Panchayat along with NGO people at Jengraimukh Village

Some modern day help is also taken for preparing for disasters. For the protection of the food items during flood times the NGOs help villagers of a village or a group of villages in constructing granaries. These granaries are built on strong foundation instead of bamboo slits which are their traditional types. The granary's base pillars are covered by steel sheets so that it gets protected from rodents and other small animals attack. The construction is traditional in design but the parts constituting it are modern, so that the sentiments of the villagers are intact and also the purpose of building strong, protective granary is attained. For this ngos sponsor, sometimes also villagers contribute to their capacity.

Ham radio sets are provided to locals before the onset of floods to communicate. Boats are also provided which are used to rescue people during floods. Otherwise bamboo constructed temporary boat kind planks are provided. Besides, every villager knows swimming. They also sow deep water rice which is their specialty. This kind of rice can withstand floods and after the floods can be harvested without worrying about the result. These are sown in low lying areas of the island. In high line areas they grow saali paddy which is another kind of rice.

18.6 EARLY WARNING SYSTEMS FOR FLOODS

As the people living in harmony with nature and sustaining ecosystem over the years learn the various ways of survival from different kinds of hazards of environment. Ecosystem approach helps the community people to detect early warning signs and get alerted about any impending disaster. There are some signals from nature which it provides before any major event that is about to happen. The people living in consonance with nature are able to detect and identify them and are able to protect themselves in prior time. Tribal people are masters in this. Mishing community has their own set of early warning signs which help them to prepare and know about the upcoming natural event. Related to floods, they have some signs and symptoms. The nearest river to Mishing people in Majuli is Brahmaputra and its tributary Luit. When they observe the soil sediments coming downstream, they can get an idea of how heavy the rains would be.

Figure 18.5: Modern day granary of Mishing Community

If soil sediments are flowing in the river before the onset of monsoon, it signals that flood will come. That is the rains would be heavy leading to floods. Another warning system that works for them is observing the rain pattern. If during the beginning of the rainy season, it rains heavily for first week or more than ten days continuously, it predicts heavy flooding. During such times

Mishing community and other community people are evacuated to higher grounds. Another one is when wild animals start to retract to higher grounds, it shows floods are approaching.

Besides this, the local information system is quite active during flood season or the rainy season. Regular updates are circulated by radio and ham radios to the villagers about the approaching floods or any unpleasant natural event. Some localized radio stations are opened up or local body operation centre is set up which updates them on hourly basis or as the situation demands. Nowadays television has penetrated into the lives of Mishing community, so live updates help people to assess the situation properly and act accordingly.

18.7 CONCLUSION

Disaster management and indigenous risk reduction measures are two terms which have found a new relation in recent past. Studies done in the South Pacific Small Island Developing States (SIDS) indicate that despite the influx of outside aid, indigenous communities still show considerable resilience to natural disasters. This is because of adopting ecosystem approach. Strong levels of intra-community cooperation exist and many indigenous groups still utilize traditional building and food preservation techniques to help them escape the ravages of disasters in the region. The natural resource management of Majuli Island by the Mishing community has kept the island alive for so long. The Mishing community is economically not sound but traditionally very rich in cultural practices. The management of their surroundings which they are engaged in has been able to make them sustain and their environment symbiotically. The disaster risk reduction practices which they adopt to avoid the impacts of annual floods are noteworthy and their house structures can give engineers a good challenge for future housing designs in that area. The agricultural practices along with other cultural rituals which help them sustain during floods on Majuli Island are also adopting ecosystem approach which balances the natural resource management of the area and helps the local community survive in difficult environment. The knowledge is acknowledged even by development plan makers and they incorporate their design and daily lifestyle in making any plans for the region. The ecosystem approach has been adopted to prevent the disaster impacts and thus indigenous environment conscious disaster risk reduction measures uphold.

References

Bhandari, J. S. (1992). *Kinship, Affinity and Domestic Group: A Study among the Mishing of the Brahmaputra Valley*. Delhi: Gyan Publishing House.

Collins, A. E. (2009). *Disaster and Development*. London: Routledge.

Census of India, (2001). District Primary Census Abstract of Total Population, (2001), *Assam*.

Hammerlynck, O. (2012). The Ecosystem Approach. Commission on Ecosystem Management, IUCN. Retrieved from http://www.iucn.org/about/union/commissions/cem/-cem_work/cem_ea/.

Jorhat District Administration (2008). Majuli (Assam) Flood and need assessment Report. 2008.

Mipun, J. (1993). *The Mishings (Miris) of Assam: Development of a new Lifestyle*. Delhi: Gyan Publishing House.

SEEDS (2007). Pan-Himalayan Study on Indigenous Technology of Earthquake Resistant Construction of Historic Buildings.

Smith, A.O., and Hoffman, S.M. (1999). *The Angry Earth: Disaster in Anthropological Perspective.* New York: Routledge.

Smith, A.O., and Hoffman, S.M. (2002). *Catastrophe & Culture: The Anthropology of Disaster.* (School of American Research Advanced Seminar Series). School of American Research Press: USA.

UNEP (2007). *Environmental Emergencies News,* Issue April 6, 2007 p. 3-4.

UNEP (2012). Retrieved from http://www.cbd.int/ecosystem/operational.shtml.

19

Traditional Water Management Systems and Need for its Revival:
A Study of Ahar-pynesystem in South Bihar, India

Swati Singh

19.1 INTRODUCTION

Water is an important element and it is associated with social, economic and cultural aspect of human life in many ways. The utilization of water at micro level is very important for social and livelihood requirement of rural society. Each community incorporates water resource into its cultural pattern in its own way (Pandey, 2006). The Hindu Survey of Indian Agriculture, 2002, observed that micro irrigation method is not merely an irrigation technology; it is an integrated management tool in the hands of the farmer. According to Agarwal et al., 2001, over the last one hundred years or so, the world and India, too, have seen two major shifts in water management. One is that state has taken role of individuals and communities of water management. The second is that the simple technology of using rainwater has declined and at its place exploitation of rivers and groundwater through dams and tube wells has become the key source of water. As water in rivers and aquifers is only a small portion of the total rainwater, there is inevitable and growing stress on use of water from rivers and groundwater.

History says that traditional water harvesting systems have met domestic and irrigation needs of the people. Traditional systems have evolved as specific responses to ecology and culture of the people. Not only they stood the test of time but also they have satisfied certain local needs in an environment friendly manner. These systems emphasize ecological conservation in contrast to environmental overuse of modern systems. Traditional systems have benefited from collective human experience since time immemorial and their biggest strength lies in that. India economy agricultural based, where 80% of the cultivation is rainfed. Despite of several irrigation projects only less than 50% of the runoff water that is use for various purposes including irrigation could be addressed. There is water crisis today. But the crisis is not having too little water to satisfy our needs. It is a crisis of managing water so badly that millions of people-and the environment suffer badly.

Disaster Management and Risk Reduction: Role of Environmental Knowledge; Editors Anil K. Gupta, Sreeja S. Nair, Florian Bemmerlein-Lux and Sandhya Chatterji; Copyright © 2013, Narosa Publishing House, New Delhi

19.2 ABOUT AHAR-PYNE SYSTEM

Ahar-Pyne is a traditional floodwater harvesting system which is indigenous to South-Bihar. The meaning of Ahar as found from the local people is to hold water ("Aa" means to come and "Har" means to capture). The physical features of Bihar divides it into three regions-north Bihar plain, the south Bihar plain and the Bihar plateau, also known as the Chotanagpur Plateau. In South-Bihar, the slope is roughly at the rate of 1 m per km and using this terrain, an Ahar was built by erecting an embankment of one meter or two in height. An Ahar resembled a rectangular catchment basin with embankments on three sides. Ahar were sometimes built at the end of small rivulets or artificial channels called Pynes to ensure the supply of water. Pynes were channels constructed to utilize water flowing through hilly rivers intersecting the country. In South Bihar, rivers are generally dry for most of the year but swell during the monsoon season. Given the slope and the sandy soil, the water is either carried away rapidly or it percolates down through the sand. Hence, a system of Pynes was usually developed to lead off water from these rivers to agricultural fields. Some of the biggest Pynes, which were 20-30km long, fed a number of distributaries and irrigated over 100 villages. Ahars and Pynes were collectively used by farmers and they had to synchronize their operations. Pant (2004) posited that Ahar-pyne is historically the most important source of irrigation in South Bihar and even today provides a shining example of participatory irrigation management.

Figure 19.1: Ahar-Pyne to capture excess flood water *(Source: Farmers Forum, October 2010)*

Apart from the irrigation facilities, there is another utility of the system which has rarely been studied. Being a region situated in between the Chotanagpur Plateau and the Gangetic valley, South Bihar is very prone to floods. But the abundance of storage works like Ahars and large scale dispersion of torrential floodwater into pynes minimized the rush and speed of floodwaters passing through south Bihar. Some of the small rivers of South Bihar could never reach any of the main rivers like the Ganga or the Punpun because their water was completely dispersed through several pynes. Ahars and Pynes were extensively used in Gaya district. In Bihar, The Irrigation Commission of 1901-03 noted that these systems irrigated about 6,76,113 ha. In 1949, the flood

Advisory committee of Gaya district in 1949 found that the large scale occurrence of flood in the district is because of the deterioration of the traditional system of irrigation.

19.3 PRESENT STATUS OF AHAR- PYNE SYSTEM

Ahar-pyne system of indigenous irrigation is historically the most important source of irrigation in South Bihar and even today provides a shining example of participatory irrigation management. Ahars, with sides that are more than a km. long, irrigating more than 400 ha are not rare, though smaller ones are more common. However, the average area irrigated per ahar during the early twentieth century was said to be 57.12 ha (Sengupta, 1993). According to O'Malley (1919), this indigenous system is the outcome of the natural conditions and physical configuration of the country, and has been evolved to meet the obstacles which they place in the way of cultivation. However, with the passage of time, the collective institutions of management of the ahar-pyne system have declined. Area irrigated by ahar-pynes is on the decline, accounting for only about 12% of the total irrigated area in Bihar (Table 19.1).

Figure 19.2: BarkaAhar of village Angra in Patan block

The ahar-pyne system of irrigation was overwhelmingly more important in South Bihar, where it was irrigating about 35% of 2.5 mha of cropped land during the first two decades of twentieth century. Compared to it, the irrigation in North Bihar was a mere 3% of 3 mha cropped area (Pant, 2004). During this period, of the 0.98 mha area irrigated by ahar-pyne, 0.88 mha area was irrigated in South Bihar, while only 0.1 mha was irrigated in North Bihar (Tanner 1919). The area irrigated by this indigenous source has witnessed a constant decline.

The extent of decline can be gauged by the fact from 0.94 mha in 1930s in South Bihar, the area declined to 0.64 mha in 1971 and to 0.55 mha by 1975-76. As per the Government figures, the area irrigated by ahar-pyne system in whole of Bihar came down to about 0.53 mha constituting about 12% of all irrigated sources in the year 1997 compared to about 18% in South and North Bihar alone during the first two decades of twentieth Century.

Table 19.1: Area irrigated by ahar-pyne system

Year	Area irrigated (mha)	Region Covered
1930	0.94	South Bihar
1971	0.64	South Bihar
1976	0.55	South Bihar
1997	0.53	Whole of Bihar

Source: Pant (2004)

Figure 19.3: Pyne in dilapated condition due to erosion

19.4 REASONS FOR SUCCESS OF AHAR-PYNE SYSTEM IN THE PAST

The farmers indigenous knowledge of utilisation of water for irrigating their paddy fields was based on great understanding of the local topography, flow of water and positioning of the fields. The major factors that led to the traditional ahar-pyne being so much prevalent in the region are enumerated as below:

1. Fragmented land holdings and equity in water distribution: An interesting pattern noticeable in each of the ahar-pyne areas and in general the whole of South Bihar and adjoining areas is that the land-holdings of the farmers in general are small, fragmented and scattered. As a result, every landholder in the command of a pyne had some land at the head, some in the middle and some at the tail of the irrigation channel. So all farmers have their plots both in advantageous as well as disadvantageous locations - head, middle and tail. Therefore, to optimize their irrigation, they would have to take active participation in all kinds of situations. To safeguard the interest of their tail-end farm, they would work with others so that the water reaches at the tail also. In this way, ahar-pynes seem to overcome the problem of headreach/ tailender conflicts that are a common feature of irrigated commands of major and medium projects. Ahar-pynes ensured equitable distribution of irrigation water in the command (Sengupta, 1985). Further, several irrigation

commands get benefit from the same ahar or pyne and several ahars may get water from the same pyne. Since cultivators have unconsolidated holdings, they are not left with any choice other than to work collectively for a common good.

2. Cheap Source of Irrigation: In the past, ahar-pyne used to be the cheapest and easiest source of irrigation in the region which only needed a collective effort from the villagers. Although presently, ground water through diesel based borings and electric motors are available but the cost of irrigation is very high compared to the ahar-pyne system. In case of ahar-pyne, all major repairs are done by the government and farmers do not have to pay any water charges. Hence, cultivators do not mind working collectively for small maintenance or to meet emergencies like breach in pyne or embankment etc.

3. Uniformity in Cropping: All farmers grow the same crop (paddy) all over the irrigation command around the same dates. As a result, agricultural operations undertaken by all cultivators are similar throughout the irrigation command. Such uniformity of operations is essential when cultivators are utilizing the same irrigation channel. Since ahars and pynes have to be used collectively, all farmers have to synchronize their operations. In such a scheme of things, there is no scope for crop diversity in the same irrigation command. Uniform cropping also facilitates collective action when irrigation system is in the danger of non-functioning (Pant, 2004).

4. Collective action: Communal action for irrigation operation and maintenance referred to as goam consists of large groups. Ahar-pynes have been constructed by the extraordinary concerted effort of the human beings against the oddities of nature. Although in South Bihar also, like rest of India, a rigid caste hierarchy obtains, this does not deter different caste groups, including scheduled castes to come together for a common good and a common concern. All cultivators, who take water from the same pyne or the same ahar, irrespective of the location of their villages and irrespective of their castes, come together for collective action whenever their irrigation is affected or is likely to be affected. According to Pant (2004), cultivators had their vested interests to participate actively in collective actions. This was particularly true in respect of goam to meet the emergencies such as breaches in embankments, diversion in river and pyne routes etc. Hundreds and even thousands of people still come forward for goam even today in South Bihar.

19.5 INSTITUTIONAL AND MANAGEMENT ISSUES

Such a large system of irrigation which would sometime be spread over many villages could not have existed with a strong institutional mechanisms and proper management. Although no written rules existed in most of the cases but there were certain issues that were dealt upon by the people in their own indigenous manner.

1. Equity in allocation and distribution of water: The Ahar-pyne system had well worked-out institutional mechanisms for sharing of water between farmers. Synchronization of the agricultural operations over the year was achieved by earmarking each 14-day period on the lunar cycle for each agricultural operation (Table 19.2). Buchanan (1939) noted that landlords appointed proper persons to divide the water among the tenantry. According to O'Malley (1919), the parabandi System was used to distribute water among the villages from a common source (usually a pyne). Parabandi derived from the term para (turn) and bandi (fixation) meant fixation of turn. Each village had its fixed turns of so many days and hours to avail the water. These turns were assigned by mutual agreements or ancient customs. A detailed register called lalbahi (red register) maintained in some systems specified the irrigation rights of each village. Usually

parabandi arrangements began in the month of Aswin (mid-September), when the demand was acute and supply limited. At other times, all branches of pynes were left open (CSE, 1997). The reliability and timeliness of ahar irrigation is ensured because water is stored in the reservoir and is utilized when pynes do not have any water left and rains are not forthcoming. This is the likely scenario during the hathia period, when water is critically needed by paddy (Pant, 2004).

Equity in water allocation was not a granted right but it was in-built in the system. The total landholding of each individual in a command was highly fragmented. In consequence, every major landholder who could influence the allocation had interests both at the head and the tail regions of the distributary (Sengupta 1993). If water available is not sufficient and does not reach the tail end, a part of the command area remains unirrigated, but everyone suffers. Pynes feed several ahars and several distributaries originate from each ahar.

Table 19.2: Timing of agricultural operations in ahar-pyne system

	Period	**Operation**
1	June 20 to July 5	Seed Bed Sowing
2	July 18 to August 15	Transplanting
3	September 12 to September 25	Field water drained out
4	September 26 to October 7	Fields filled again
5	October 8 to October 20	Standing water in fields
6	October 21 to November 3	Field water drained out
7	November 4 to November 15	Harvesting

Source: Aggarwal and Narain (1997)

The primary level irrigation organizations correspond to the irrigators benefiting from a distributary (ayacut). The peculiar land holding pattern as that every cultivator owns a fragment of upper , middle and lower levels of ayacut have also been noted for irrigation systems in Sri Lanka (Leach, 1961) and Philippines (Siy, 1982). In addition they also were also close to neighbours though they did not belong to a single caste. The crop growing in the ayacut area is paddy, the same for every cultivator. Thus, all of them require irrigation at approximately the same time. Because of the characteristic distribution of the plots no one is deprived of water. Once the fair allocation of water is assured individual cultivators do not lack motivation to join in community works for irrigation (*goam*). However the system is not entirely free of disputes. One village often tries to get more water than it should, or else when rainfall is scarce, lower reach villages seek to get water before their proper turn.

2. Community participation and distribution of responsibilities: In the past community participation was extensive in traditional irrigation management. Community labour for repair, called *kudimarammath* in south India and *goam* in Bihar was an established custom. Ahar-pynes work, particularly the one relating to maintenance and overseeing of water distribution was looked after by three functionaries. These were headman, Barahill (supervisor) and Gudait

(watchman). A unique feature of ahar-pyne management system in was that some posts were associated with particulars castes. For instance job of the watchman, the drum- call for *goam* (Collective physical action) used to be made by beating of drums and the drum beatings used to be done by dafalis (Pant, 2004). Some of these indigenous irrigation systems (pynes) were so large that their water conveyance system ran over 30 km, covering hundreds of villages and irrigating thousands of acres of land. Since the construction of such irrigation works required huge capital investment, only big landlords could do it. In fact, sometimes it required the cooperation of two or more landlords. In such occasions, each cooperating landlord used to appoint his team of officers to look after his interest on the negotiating table during the construction phase. Usually the cost involved in the construction of pynes was much higher than the one involved in constructing ahars. The construction of pynes, particularly the large ones, involved excavation of pynes running several km. In addition, it also involved construction of dams across the river to divert the water to the pynes. Large pynes were mostly constructed several years ago when larger areas were under the control of the single zamindars (landlords) and their authority to enforce their orders and wishes was more absolute (O'Malley, 1919)

3. Repair and Maintenance: The repair and upkeep of the most ahar and its water conveyance system is of two types. The one involves major repairs and the other deals with the minor routine upkeep to make the system work. The responsibility of ahar-pyne construction as well as major repairs was of landlords (Buchanan, 1939; O'Malley, 1919). The amount spent by the estate was later realised from the farmers under the Gilandazi (improvement of irrigation works). Today, minor repairs are not done by the farmers and the repairs are done by the Minor Irrigation Department. In the past, farmers had to pay for the repairs as well as for the irrigation, while today they do not pay for any of these two things. The routine upkeep work involves cleaning and desilting of ahar and pyne and maintaining the water conveyance network, while the system is in operation. As a result, ordinary maintenance such as the periodic clearance of silt, the repair of small branches of the ahars and field channels is done by the cultivators themselves under *goam* system and it starts before the onset of monsoon. Apart from the routine activities, an important task is to keep constant vigil, particularly during monsoon against sudden damage of protective works which may occur due to natural cause or due to man-made reasons. The operational works include cutting and closing embankments for diversion, erection of bandhs or garandis across the pynes, opening and closing of outlets and at times even resorting to manual water lifts to irrigate uplands. *Goam* was and still is very effective in meeting the emergencies. The call for *goam* was made by beating of drums. *Goam* occurs even today every year in hundreds of villages of South Bihar.

4. Central control: Steward (1949) and Wittfogel (1957) opine that irrigation management requires a high degree of discipline and that in turn implied central control and an all-powerful bureaucracy. In the ahar-pynes of South Bihar, it is found that a centralised authority in the form of the landlord played an important role in respect of construction of ahar-pynes, their major repairs and allocation of water to different villages. However, landlords did not play any role in determining the mechanism relating to how water was distributed among different individuals in each micro irrigation command and how they maintained the micro water conveyances structures. Further, Buchanan (1939) mentions that there existed some indigenous irrigation works in South Bihar which were constructed and maintained by tenants and that the landlords had no claims of rents against such works. Even where findings do indicate a centralised management in certain

matters, it is difficult to assume that high level of participation of cultivators in the irrigation management was a natural consequence of the centralised authority (Pant, 2004).

5. Collective choice arrangement for Ahar management: The Ahar system has been managed collectively by the villagers, as the terrain acted as driving force. This system could only facilitate establishment of irrigation system, due to its physiological and climatic factor. The nature of agriculture practiced needed a regular supply of water for rice cultivation and Ahar system collectively provide the solution. Therefore, the collective choice arrangement for improving the livelihood resulted in managing Ahar as common pool resources.

19.6 REASONS FOR DECLINE

As discussed earlier, Ahar-pyne systems were the lifeline of the farmers cultivating Paddy in this region. However, there has been a gradual decline in the area irrigated by the ahars because of the following reasons:

1. Abolition of the Zamindari system: One of the reasons cited for the decline of ahar-pynes is the abolition of the Zamindari system. During the British period all cultivated lands belonged to Zamindars (feudal landlords) who paid a fixed revenue to the British Government. After independence in 1952 this system was abolished and the land was distributed among the erstwhile tenants-Zamindars (Land-lords) regularly organized maintenance and desilting of ahar-pynes before independence. Till the abolition of Zamindari system the Zamindars used to maintain these systems because they had the capital resources and had a vested interest in doing so. Tenants were required to pay Gilandazi (improvement of irrigation works) charges. Gilandazi is an excellent form of investment as the capital spent on it returns a dividend of 40 to 50 % in the first year itself, in some cases 100 percent if the landlord even received only half of the produce of the land irrigated by these works, they would get a very good return on their capital outlay (O'Malley 1919). After the zamindari abolition there are no regular budgeted funds for the repair of these systems.

2. Development of new Irrigation sources: Development of new irrigation sources, notably canals and tubewells leading to easy availability of water made people lose interest in ahar-pynes, which needed community effort for upkeep and maintenance. Many indigenous works were directly suppressed by extension of modern methods. This has been aided by high doses of government subsidies in case of private tubewells. Even in 1970-71, the area irrigated by tube-wells in Bihar was about 17%, this reached above 48% in 1994-95 (Government of Bihar 1972 and 1997). There are such examples from UP and Punjab where shrinkage in traditional well irrigation took place from the extension of canals and the modern groundwater exploitation techniques. Pandey (1979) also reported how traditional ahar-pyne irrigation was suppressed in many different villages by introduction of canal irrigation project in that area.

3. Lack of convergence between old systems and new schemes: Non-integration of the indigenous systems in the new diversion schemes undertaken by the Irrigation Department was also a major reason for the decline. Many authors have noted that the irrigation departments did not have adequate understanding of the value of this system. Hence, often the new irrigation schemes were at variance with the existing ahar-pyne system and no attempt was made to integrate the two. At present, some initiative is restrained because of expectation of assistance from external agencies like the Government and the reduced interest of those who have acquired pump sets.

4. Lack of centralized authority: Literature suggests that centralized authority in the form of the landlord did play an important role in respect of construction of ahar-pynes, their major repairs and allocation of water to different villages. Today, there is no coercive authority of the landlord or anyone to force them to contribute community labour for irrigation management which has led to lack of interest of people.

5. Roads on pynes and ahars: Pynes, as already been mentioned, is the artificial channels that connect one Ahar with the other or Ahar with the river system. It was designed in such manner so that inundated water can be collected from one Ahar to another without mere wastage of precious water. It was in the year 1989, Jawahar Rojgar Yojna (JRY) was started under the leadership of our then Prime Minister Late Mr. Rajeev Gandhi. Though it was a very noble step to connect each and every village of the country with roads for their development but on the other hand it ruined the traditional system of irrigation. Most of the pynes have been converted into roads.

6. Heavy siltation: Heavy siltation is one of the major causes of declination of ahar-pyne. Silt of around 6 to 7 ft has been accumulated in the catchment areas that have reduced the storage capacity of Ahars. Aftermath of siltation is that water gets dry very early and not enough water is left for paddy cultivation. Moreover, siltation has given another cause for the declination of Ahars as most of the silted Ahars beds have been converted into agriculture field.

7. Conversion of Ahars into Agriculture field: Previously, water was collected in Ahars in the month of July – August at the time of heavy rain. Bhao or outlet remained closed at that time because water for irrigation was fulfilled from monsoon. In September outlet was opened and closed according to the requirement. After Hathiyanakshtra (after October) excess water was drained to grow Rabi crops in the Ahar bed. But today mind-set of people has changed and most of the Ahars have been converted into agriculture field and water is drained before October.

8. Small Landholdings: After the Abolition of Zamindari System lands were given to landless people and even Ahars were divided. Today in one Ahar there is land of 10-15 or even more farmers. Small landholdings also limit the average production.

9. Heading towards commercial farming: Farmers have become more commercial in nature. They now take garlic cultivation on large scale because it is more revenue fetching. Farmers having lands in Ahars drain water in September to grow garlic, ginger and many other commercial crops/cash crops. This practice has not only caused the decline of Ahars but has reduced the average production of paddy from 15 quintals per acre to 5 quintals per acre. This has also increased fallow land.

10. Dependence on Government schemes: In spite of going for the revival of Ahars and other traditional system of irrigation people are asking for well or check dam or ponds from the government. Government schemes are such that they always go for new structures but never look into the existing structures. Because of this people have lost interest in the traditional system.

11. Social problem: Social stratification of village has affected the collective action, which was earlier provided under the leadership of zamindars.

12. Increase in population: Rise in population is also contributing towards its decline though indirectly as this is one of the reasons for small landholdings. Conflict for property is a common phenomenon everywhere and due to this reason big land is divided into small chunks.

13. Deforestation: Forest resources are dwindling very fast. Deforestation has scaled up the rate of siltation which is the direct cause of decline of Ahars.

14. Middlemen intervention: Apart from the above reasons, the local contractors and middlemen also discourage the local farmers from doing collective efforts towards repairing and maintenance of the system. They also mislead the farmers that such activities need to be done

through the Government so that they can make own monetary gains. There is lot of ignorance among the poor farmers and usually the upper class people take benefit of that.

19.7 NEED FOR GOING BACK TO INDIGENOUS SYSTEMS

Even if the area irrigated by traditional methods is on a decline and it has been mostly replaced by various modern techniques of surface and groundwater harvesting systems, it is essential that we revive the old systems. The major reasons for the same are furnished below:

1. Delays in major and minor irrigation projects: Pant (1982) reported that the technical evaluation cell of Planning Commission approved 529 projects (106 major and 423 minor) between 1971 and 1981, with an original outlay of Rs.6820 crore. The cost of all these schemes, according to latest estimates, compiled by ministry of irrigation has gone up by a staggering Rs. 3828 crore to Rs. 10648 crore. This indicated an overall increase of 56.13 % during the last 10 years. Statistics also reveal that 60 schemes, including four major ones, were completed in nine states, bringing the success ratio to only 11.34 %. One of the major reasons for the inadequate growth of the irrigation sector is the long time that it takes to commission major and minor irrigation projects. Delays in completion and increases in project costs of major and minor irrigation projects is a cause for concern. Instead of pumping in huge amount of money in schemes that are marred by technical, political and social glitches, it is time that we need to look into the already practiced systems at the grassroot level and focus on providing technical and monetary support to the already existing indigenous water management measures and that are much more suited for that particular place. For an example, there is enough literature to support that ahar-pyne system of irrigation was instrumental in saving all of Gaya district from the ravages of famine and drought. It is worth highlighting that through the 1866 famine of Orissa, the Bihar famine of 1873-74 and during the famine of 1886-87, Gaya district required practically no relief. Apart from irrigation, another useful purpose served by ahar-pyne system is to minimize the floods. Writing in the context of the then Gaya district, the collector (1947-49) observed that as long as these minor irrigation works were kept in a reasonable state of repair, floods in lower regions were well under control (Roy Choudhry, 1957). In 1949, a Flood Advisory Committee investigating continuous floods in Bihar's Gaya district came to the conclusion that "the fundamental reason for recurrence of floods was the destruction of the old irrigational system in the district."

2. Easy maintenance, cost and quality: The cost of ahar-pyne maintenance is quite low compared to canal maintenance which comes to about Rs. 5000 per ha. In case of ahar-pyne, it varies between Rs. 500 to Rs. 1000/ ha, depending on the extent to which *goam* is utilized. Further, the quality of construction is quite good because those who get engaged in the repairs are themselves the beneficiaries. Further, in some of the repairs the material used is the one which is locally and easily available. Pant (2004) noted that use of mozar which is obtained by mixing the wet mud with paddy straw quite effective in the repairs of embankment, including in raising its height. According to Pant (2004), as today's per ha cost of irrigation comes to about Rs. 80,000 and 46% of the total annual precipitation of 350 mhm in India is lost to the sea as river flow, the rejuvenation, development, and integration of ahar-pynes system with new diversion schemes present wide scope. The reason being, it mainly involves mobilisation of local material and man power resources with very little financial requirement (about Rs. 1000 per ha). This is especially

important at present times when financial crunch surrounds most states from all sides and participatory irrigation management is the rhetoric quick-fix.

3. Sustainability in the longer run: The sustainability of ahar-pyne system can be judged by the fact that these modes of irrigation are in existence for centuries. All the ahar-pyne systems that exist today are at least nearly hundred years old. The main reason of the sustainability of these indigenous systems is that the advantages emanating from them are twofold. First, these systems utilize water which otherwise would be wasted. Second, these systems, particularly in the past, saved the plains of South Bihar from the recurrent floods which otherwise would have devastated the countryside regularly. Lastly, if these indigenous systems are properly integrated with the recent canal irrigation schemes, the sustainability of both types of irrigation systems will enhance manifold (Pant, 2004). Proper utilisation of natural resources requires proficient consideration in many different aspects related with it. The real difference between the so-called modern and traditional methods is that the former, with an independent start, need gradual attainment of the proficiency, while the latter must have perfected those over centuries, otherwise those would not have survived. Detailed knowledge of traditional water resource management method therefore, may not only help in better formulation of new development projects but hasten the gradual rectification process which most of the existing problem projects are facing at present (Sengupta, 1985).

19.8 ATTEMPTS FOR REVIVAL

Some villages in Bihar have taken up the initiative to re-build and re-use the system. One such village is Dihra. It is a small village 28 km southwest of Patna city. In 1995, some village youths realised that they could impound the waters of the Pachuhuan (a seasonal stream passing through the village that falls into the nearby river Punpun) and use its bed as a reservoir to meet the village's irrigation needs. Essentially, this meant creating an ahar-pyne system. After many doubts, the village powers-that-be gave the go-ahead. Money was collected and work began in May 1995. After a month of shramdaan (voluntary labour) the villagers completed their work mid-June. Their efforts have borne fruit. By 2000 AD, the ahar was irrigating 80 ha of land. The people grow two cereal crops and one crop of vegetables every year. The returns from the sale of what they produce are good and the village is no longer a poor one. Even now community work of irrigation (*goam*) occurs every year in several villages of Bihar (www.rainwaterharvesting.org).

19.9 FUTURE STRATEGY

Revival of this traditional irrigation system could be one of the major activities for livelihood security. Ahar-pyne system is based on a minute understanding of the topography so that even at such mild slopes, pynes carrying water over several hundreds of metres could be constructed. Pynes also diverted water from the streams over long distances, irrigating large areas. According to a report by Ministry of Rural development (MoRD, 2006) revival of this system and ensuring their proper maintenance through community action should be a major plank of watershed projects in South Bihar.

For revival of traditional water harvesting systems, the most critical thing which needs to be done is the integration between new and old schemes. In the decade of 1950's, particularly during the first and the second five year plans, a number of diversion schemes were undertaken in South

Bihar. In most of the cases, the area brought under the command of these schemes had very elaborate system of indigenous irrigation network through ahars and pynes, particularly in the upper reaches. The planners realizing the valuable contribution of this indigenous system in subsidiary storage and water distribution; dovetailed it in their plan and thereby increased the capability of the run-of-the-river scheme on a rainfed river proposed to serve an area subject to fitful monsoon. They relied on the contribution of the existing ahars so much that they planned about two-third of the command was to be irrigated during the critical hathia period through the ahars which were to be filled up from canal networks by drawing maximum possible water during favourable period of river flow. However, the envisaged integration of ahar-pynes with the new schemes could not be done in a large number of cases and this indigenous system was made to languish over time. A recent study shows that the number of ahars in the command of Upper Mohar Irrigation Project covering the districts of Gaya and Aurangabad had dwindled to 44 in post project period from 109 in pre-project period (Metaplanner, 1994), consequently affecting the irrigation in an adverse manner. Had due attention been given to proper maintenance of these Indigenous systems and integrated management of new canal networks and old ahar-pynes was devised, all these new diversion schemes would have been grand success stories (Pant, 2004).Presently, the possible avenues of repair and revival are: hard manual labour during drought period, NREGA (National Rural Employment Guarantee Act), some relief schemes, food for work programme and also Minor Irrigation department which can spend some planned funds in the name of renovation of these systems.

Knowledge of water management handed over from generation to generation is extensive. One way to use it is to undertake extensive studies by the experts and then reflect it in their works. The other is to channelize it through peoples' participation in the projects themselves (Sengupta, 1985). Collective action is essential for the repair and maintenance of the system. For that, Olson (1982) argues that collective action is likely to be more feasible (i) The smaller the groups, (ii) the more homogenous the origin of the group (iii), the longer the members of the group have been associated with one another or the group has been in existence, (iv)the closer the social and physical proximity among group members, (v) the more differentiated (in a complementary way) the goals of different members of subgroups, (vi) the greater the sensitivity of the group to a threatened loss due to inaction and (vii) the more unequal the distribution of wealth and power among members. Keeping in mind the points discussed above, it is essential that concrete measures are taken at the earliest as the rate at which the traditional systems are declining; it would not be long that they would be left only for academic and historical importance.

References

Aggarwal, A., and Narain, S. (1997). *Dying Wisdom: Rise, Fall and the Potential of India's Traditional Water Harvesting System.*, State of India's Environment A Citizen's Report No. 4, Centre for Science and Environment, New Delhi.

Buchanan, F. (1939). An Account of District of Bhagalpur in 1810-11, Bihar and Orissa Research Society, Patna.

CSE (1997). *State of India's Environment, Fourth Citizens' Report [SOE-4], Dying Wisdom : Rise, Fall and Potential of India's Traditional Harvesting Systems*, pp. 87. New Delhi: Centre For Science and Environment.

Government of Bihar (1972). Season and Crop Report (table-III A), 1970-71, Patna.

Leach, E. R. (1961). *PulEliya: A village in Ceylon*. UK: Cambridge University Press.

Metaplanner (1990). Report of Post Facto Study of Upper Morhar Irrigation Project, Bihar: XII. Metaplanner and Management Consultant.

MoRD (2006). *From Hariyali to Neeranchal*. Report of the Technical Committee on Watershed Programmes in India. Department of Land Resources Ministry of Rural Development Government of India, January 2006.

Olson, M. (1982). *The Rise and Decline of Nations*. Yale University Press, New Haven.

O'Malley, L. S. S (1919). Bengal District Gazetters – Gaya Superintendent, Government Printing, Bihar and Orissa, Calcutta, pp.146-147.

Pandey, A. (2006). Ethnography of Participatory Irrigation Management in Vidisha District of Madhya Pradesh. Bhopal: Indian Institute of Forest Management.

Pandey, M. P. (1979). *Impact of Irrigation on Rural Development*. New Delhi: Concept Publication Co.

Pant, N. (1982). Major and medium irrigation projects: Analysis of cost escalation and delay in completion. *Economic and Political Weekly*. *17*(26), 34-43.

Pant, N. (2004). Indigenous Irrigation in South Bihar: A Case of Congruence of Boundaries, Accessed from http://www.indiana.edu/~iascp/Final/pant.pdf

Roy Choudhry, P.C. (1957). Bihar District Gazetters: Gaya, Government of Bihar, Patna, pp. 205

Sengupta, N. (1985). Irrigation: Traditional vs Modern. *Economic and Political Weekly*, *20* (45- 47), 1919-1938.

Sengupta, N. (1993). Land records and irrigation rights. In B. N. Yugandhar and K. GopalIyer (Eds.) *Land reforms in India, Volume 1, Bihar-institutional constraints*. New Delhi: Sage Publications.

Singh, R. (2010) .Water Management: Restoring India's Indigenous Knowledge Systems. *Farmers Forum*. Accessed from http://farmersforum.in/input/water-management-restoring-india%E2%80%99s-indigenous-knowledge-systems/.

Siy, R. Y. (1982). *Community resources management: Lessons from Zanjera*. Quizon city: University of Philippines Press.

Steward, J. H.(1949). Cultural Casuality and Law. *American Anthropologist, 51*(1), 1-27.

Wittfogel, K. (1957). *Oriental Despotism: A comparative Study of Total Power*. New Haven: Yale University Press.

Agroforestry Model to Improve Economic and Ecological Viability:
A case study of degraded tea lands of Sri Lanka

Prasad Dharmasena and M. S. Bhat

20.1 INTRODUCTION

Tea estates marginalization is not an abruptly raised problem in Sri Lanka. This has been widened with the beginning of commercial plantations in Sri Lanka in 1867 with 8 hectares at Loolkandura estate of Hewaheta in mid country. When 1869, tea lands were approximately 400 ha under cultivations. The lands extent had been expended to 90,000 ha in 1890. Presently it is recorded total extent in hectare as 2,21,969 on the elevation range of High grown, mid grown and low grown estates are recorded as 41,137, 71,018 and 1,09,814 accordingly (Sri Lanka Tea Board, 2009).

When British planters started commercial tea plantations in the country, well planned soil conservations methods or erosion mitigation technologies were not followed due to priorities of hasty profits making. Example, line plantings of seedling tea plants instead to contour plantings is one reason for acceleration of soil erosion. Secondly wider spacing between up and down planting rows (1.2m ×0.9m).The history of plantations industry in Sri Lanka reveals that tea industry have been started by clearing of virgin forest/untouched lands and Patna lands. Before, tea and coffee plantations in Sri Lanka, these virgin forest and Patna area would be rich with biodiversity, virgin soils and ecosystems. It is assumed that ecosystem and biodiversity were disturbed with the starting of plantations industry as well as deterioration of virgin soil was another ecological problem identified. The results of degradation of virgin soil have been physical, chemical, biological and social values with the exposure of forest soil for plantations. It was one reason to increase extent of marginal lands in plantations sector later on.

In additions to these reasons, planters from British period to present have been failed to undertake proper infilling programs on annual requirements. Some manual weeding controlling systems such as usage of scrapers also was responsible for accelerated soil erosion and time of replanting of old seedling fields top soil loss is substantially high. (Tea Research Institute-

Disaster Management and Risk Reduction: Role of Environmental Knowledge; Editors Anil K. Gupta, Sreeja S. Nair, Florian Bemmerlein-Lux and Sandhya Chatterji; Copyright © 2013, Narosa Publishing House, New Delhi

Thalawakelle, 2002). With the time these factors have resulted for marginalization of tea lands in Sri Lanka as removal of top soils and its' fertility.

Some policy failures also can be identified as a major factor lead to the marginalisation tea lands. For example, when the lands clearing was under taken prior to nationalization of plantations sector in Sri Lanka, there was not a proper national environment and forestry acts, policies or actions plans pertaining to the land clearing or environmental conservation. Hence, tea lands marginalization problem is not unexpectedly risen one. It has been developed gradually due to mismanagement of the plantations sector due to number of reasons before and after nationalization and also unplanned establishment of tea without sufficient environmental attention by British planters and local planters.

Land degradations have been a major issue in tea estates of Sri Lanka compared to rubber and coconut plantations. Nearly about 80 per cent of the land is old seedling tea which is often poorly managed. Large tracts of these old seedling tea plantations have been either neglected or left fallow. It is estimated that about 30 per cent of the entire tea land is marginal or uneconomic. Long steeps and poor management practices are responsible for severe soil erosion on tea lands *(UMWMP-1998)*.Early plantations industry was under the management of British planters and there were no other parties in the industry with entitlement for the plantations. But tea industry of Sri Lanka today depends on three parties on management systems (i) Private estates (ii) Government estates and (iii) Small holdings.

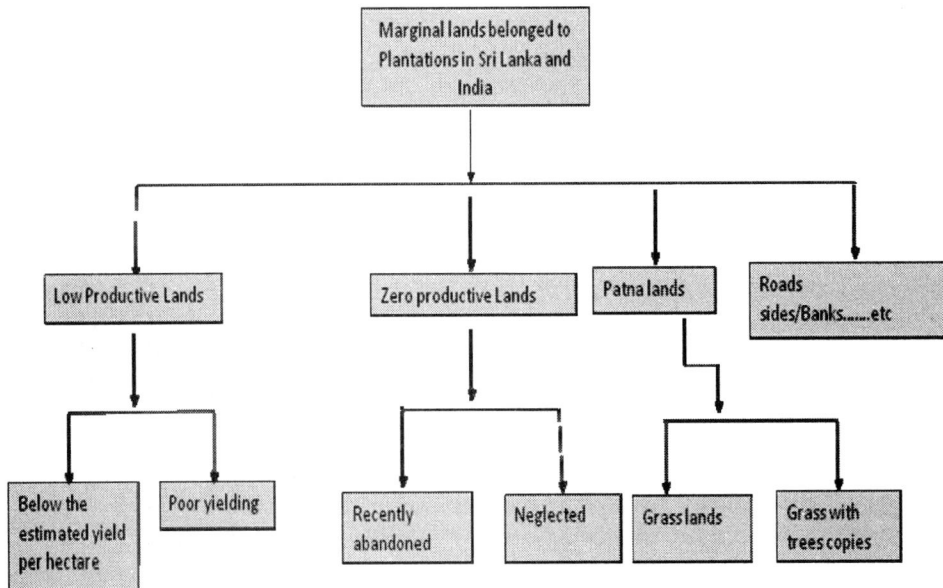

Figure 20.1: Types of marginal lands belonged to Plantation in Sri Lanka and India (Dharmasena,2011)

Present study is focused mainly on large scale plantations where it can be seen there is more marginal lands and consider as immediate rehabilitation and management is required. Marginal tea lands of the large scale tea plantations (Special reference to the Regional Plantations Company in Sri Lanka- RPCs) have four types of marginal lands on the definition of Dharmasena (2011). "Marginal tea land which ere has been giving low/negligible yields or zero yields per hectare due to environment or/and management limitation". Marginal tea lands are identified as high soil loss area. So, proper soil conservations systems and soil rehabilitations programs are important factor to be viable for the productivity of lands with replanting. But, Krishnarajah (1984) has recommended diversification of marginal tea lands to mix cropping gardens for improving export agricultural crops. Special feature of these lands are:

(i)   No or negligible No. of tea bushes
(ii)   All fields are very old seedling blocks
(iii)   Tea bush frame is not healthy
(iv)   Top soil is almost removed
(v)   No or very less shade trees
(vi)   No or negligible soil conservation adoption
(vii)   Soil erosion is high

20.2 STUDY AREA

Watawala/Ginigathena tea growing region one of controversial region in environment management of the country, is rested in western slope of the central hills of the country. It is proved that more abandoned lands in the NuwaraEliya district are observed in Watawala/Ginigathena region compared with other tea growing areas of the district where boarded to Kandy district (See Figure 20.2). Elevation of the region is 900m-1500m. This section is recognized as highest rainfall recorded sites of Sri Lanka. Even though the region receives average annual rainfall 4000m-4500mm (Punniyawardena, 2008) selected sites for studies are belonged to intermediate zones or mid grown plantations in tea classifications. Humidity of the Watawala/Ginigathena region is 80-85 per cent and widely spreading soils are Red yellow podzolic (RYP).

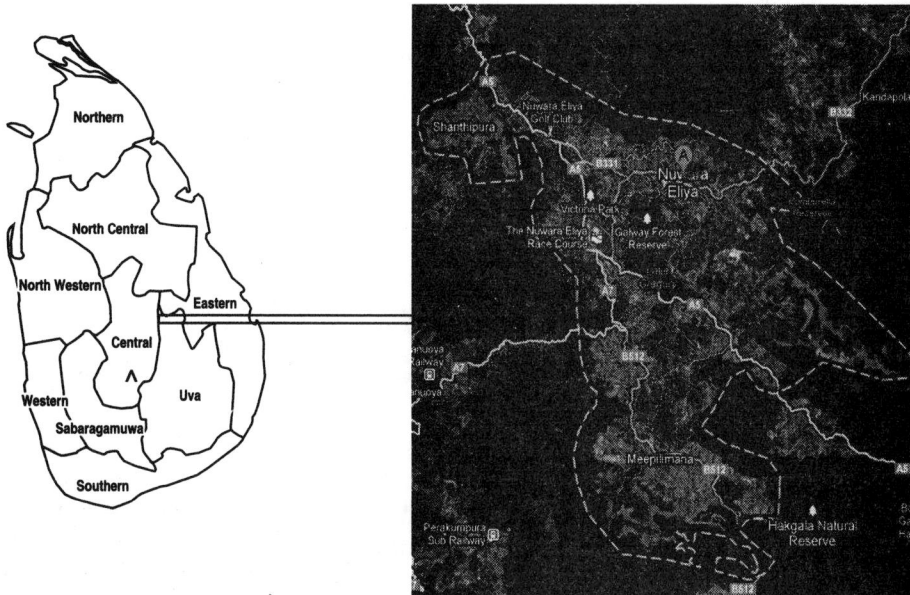

Figure 20.2: Study Area Map

20.3 METHODOLOGY

Two types of methods have been used to analyse existing condition of marginal lands and benefits of the agro forestry model

    (1) Economic approach
    (2) Ecological approach

20.3.1. Economic Analysis

Discounting measures were used as analytical tools

    (i)   CBR -Cost Benefits Ratios
    (ii)   NPV- Net Present Value
    (iii) IRR - Internal Rate of Return

20.3.2. Ecological Approach

    (i)   Installed 3 soil erosion measuring units to cover one hectare of degraded tea lands closed to the agroforestry model and 3 soil erosion measuring units to cover one hectare of the model under similar geological condition.

    (ii)   Analyzed soil samples to check soil building capacity under different land uses (pH values, SOM, Soil moisture, soil texture etc.)

    (iii) Analyzed soil samples of eroded soil to check for analyzing annual loss of N, P, K and other important soil nutrients.

20.4 RESULTS AND DISCUSSION

Watawalatea estates are under the mid and high grown tea mark of Sri Lanka and based in western slopes of the central hills of the country. There are about 35,000 hectares in central hills of the country which are totally unproductive lands belonged to 23 Regional Plantations Companies (RPCs)-Dharmasena P. (2008). It is required to minimise the soil erosion and concerning of the reduction of fertilizer cost is must. The Agroforestry model is contributing to enhance soil fertility by mulching. Mulching is very important event for building up soil fertility as top soil is eroded due to land degradation. Then selection of the suitable species is an important event for Agroforestry systems and it should be based on the quantity of the biomass availability and environmental and climatic conditions. This particular model five important crops were selected according to the land and climatic condition. Total extent of the model is recorded as 8 hectares. Elevation of the model site is between 850m to 1100m MSL and slope of the land is 45°. This particular model has been practiced in a zero productive which has been neglected before 08 years ago (2003).

    The nutrients provided by agroforestry through mulching of biomass can be monetarily valued as shown in following table. This study reveals that nutrient content of Gliricidia are immense and it may be vary due to elevation of the place. Gliricidia give maximum yield in Watawala region on favourable environmental conditions which is elevated 700m to 1400m elevation range. Following Table 20.1 give clear image about environmental benefits of applying nitrogen fixing species as Agroforestry plants.

Table 20.1: Species practicing in the model

Varieties	Planting distances	No of plants per ha
E-Eucalypt grandis A-Aricunut L-Lemmon grass G.-Gliricidia/.Vanilla	6m × 6m 2.6m × 2.66m - 2.6m × 2.6m	550 1600 15000 1600/1600

Table 20.2: Analysis of plant biomass of different species and soil improvement through mulching (By hedge rows) (Kg/ha/yr)

Species	N	P	K
Gliricidiasepuim	540.1	35.1	379.35
Cassiaspectabilis	377.0	20.0	108.0
Calliandracalothysus	387.0	10.0	83.0
Flemingiacongesta	280.25	28.9	204.85
Desmodiumresonii	281.25	15.0	62.25

Source: Mantrithilaka and Makadawara-1994

Conversion of above table to economic value may reflect environmental valuation of applying different species as agro forestry crops (Prices are based in 1996).

Table 20.3: Economic values of plant nutrients (Rs/ha/yr)

Species	N	P	K	Total
GliricidiaSepuim	14374.8	418.04	7105.23	21898.1
Cassia Spectabilis	9123.4	238.2	2022.84	11384.4
CalliandraCalothysus	9365.4	119.2	1545.6	11039.2
FlemingiaCongesta	6782.05	344.2	3836.8	10963.1
DesmodiumResonii	6806.25	178.65	1165.9	8150.84

Source: Mantrithilaka and Makadawara-1994

Experiments conducted by TRI of Sri Lanka shows that 100 plants of Gliricidia, 2-3 years old could provide 500kg green lopping with 2 pruning per year and this amount increases year by year. The nutrient supplied by this quantity of lopping is equivalent to about 13 kg Urea, 4 kg, sap phosphate, 6 kg muriate of potash, and 5 kg dolomite.

Experiments conducted in Mahailuppallama agricultural research institute of Sri Lanka shows that Gliricidia produces 150kg N /ha /yr. Gliricidia is known as to have insect repellent action especially against white ants. So, environmental and economic benefits from Gliricidia or Calliandra may be given comparable results where ever it practiced.

This particular Agroforestry model supply average 4800 kg/ha/yr green manure to the field by pruning 2 pruning cycles per year. According to the calculations of TRI and Mahailluppallama research station; this particular model contributes to enhance the quality of the land as follows.

Table 20.4: Type of manure and Quantity suggested based on the calculations of TRI

Type of manure	Quantity (Kg)
Urea	124
Sap phosphate	38
Muriate of potash	57
Dolomite	48

4.1 Prevention of Soil Erosion by Agroforestry model without Projects (Adjoining zero productive field to the model)

It was examined during 2003-2010, soil loss was analysed by installing physical soil erosion measuring units. This system urges us to compare and analyse the project viability. Without a project system on a total of 1162.4 tonnes of soil loss over six years was recorded. Soil losses on the agro forestry model were only 22 tonnes over the same period (Figure 20.3). This is almost 58 times less erosion. The annual average soil losses of 3.4 tonnes are negligible within the acceptable range for soil loss of 10 to 12 tonnes per year/ha. As well as we can easily see, soil loss of agro forestry project reach peak with starting point and after establishment of the project is to decline the soil loss. But, soil loss of non-project is constantly increasing with increasing rate. An agroforestry system wherever it practices, will contribute to control soil erosion but, magnitude may be fluctuated according to agro climatic conditions. Following graphs (Figures 20.3-20.5) gives authentic picture about with and without environmental benefits of the agroforestry models. It explains 1-8 years soil loss tons/ha with an agroforestry project and without a project.

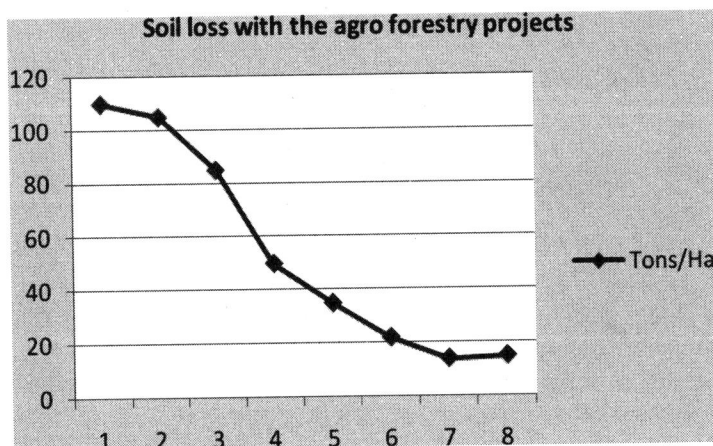

Figure 20.3: Soil loss with agroforestry projects

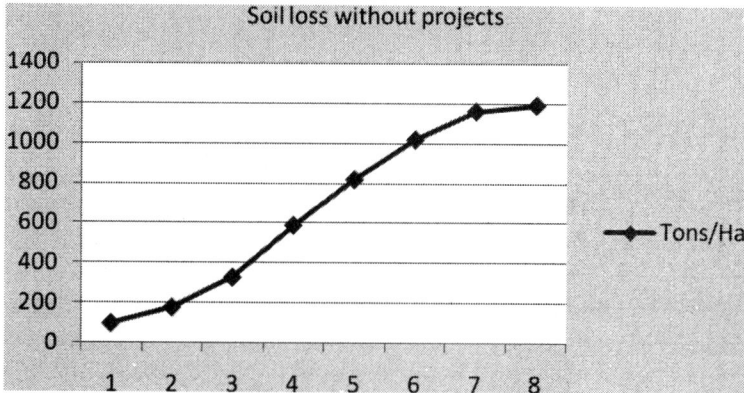

Figure 20.4: Graph depicting Soil losses without project (Cumulative value old seedling tea)

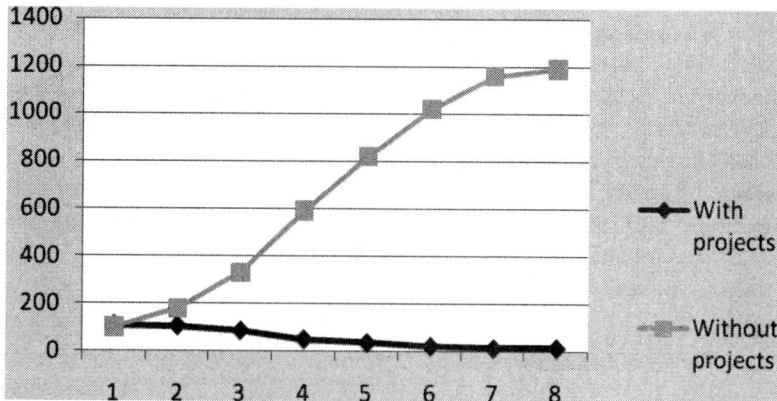

Figure 20.5: Graph showing comparative Soil loss comparison with projects and without projects

20.4.2 Discounting Measures of the project worth and project results

The techniques of discounting permit us to convert future streams of benefits and costs to their present values to determine whether to accept for implementing the projects that have variously time streams-that is pattern of when costs and benefits fall during the life of the project that differ from one another, and that are of different durations, the most common means of doing this is to subtract year by year the cost from the benefits to arrive at the incremental net benefits stream. So, called cash flow and then to discount that. This approach will give one of three discounted cash flow measures of projects value. The net present value, internal rate of return or the net benefits investment ratio and another discounted measure of projects worth is to find the present worth of cost and benefits. Streams are separately to divide the present worth of the cost stream to obtain the benefits. Cost ratio is a discounted measure of project worth. But because the net benefits and cost streams are discounted separately rather than subtracted from one another years by year. To be able to use discounted measures of project worth, we must decide upon the discount rate to be used for calculating the net present value, the cost benefits ratio, the net benefits investment ratio or rate below which it will be unacceptable for internal rate of return to fall.

For financial analysis, the discounted rate is usually the marginal cost of money to the farm or plantations firm for which the analysis is being done. This often is the rate at which the enterprise or company is able to borrow money. If the incremental capital to be obtained is a mixture of equity and borrowed capital, the discount rate will have to be weighted to take account of return necessary to attract equity capital on the one band and the borrowing rate on the other.

In most developing countries, especially agricultural countries, it is assumed to be somewhere between 8 and 15 percent in real terms, but here it used to be 10% discount rate for agro forestry project. It is purposely selected because; leading bankers of Sri Lanka are arranging their loan facilities for companies and farmer at 10 per cent of interest rate and it is assumed that project validity period is 18 years. Information used was 2001 to 2009 actual and data used for 2010 to 2018 years were projected.

20.4.2.1 Cost Benefits ratio-CBR

Discounted measure of project value is benefit cost ratio. This is the ratio obtained when the present worth of the benefit stream is divided by the present worth of cost stream. The cost benefit ratio is not commonly used in developing countries; this is because the value of the ratio will change depending on where the netting out in the cost benefits stream occurs. By the time discounted measure of project worth began to be applied in developing countries. Discounted cash flow measures of net present value and internal rate of return had become well known and were being widely used for privet investment.

$$\text{Cost Benefits Ratio} = \frac{\textbf{Present value of benefits}}{\textbf{Present value of cost}}$$

$$C_{BR} \quad \frac{\sum\limits_{t=1}^{n} \dfrac{B_n}{(1+i)^t}}{\sum\limits_{t=1}^{n} \dfrac{C_n}{(1+i)^t}}$$

B_n – benefits in each year
C_n – cost in each year
    t - Number of years
    i - Interest rate

Cost benefits ratio @ 10 %
$1715734.2/132757.87 = 12.92$

CBR was obtained for the Agroforestry project using reliable data. The projects life was considered as 18 years. Present value of costs and benefits of the projects were Rs. 128632.87 and Rs.1715734.2 respectively. Thus the CBR of the project was 12.92. According to CBR ratio, the project is extremely successful.

20.4.2.2 Net Present Value –NPV

This is the most straightforward discounted cash flow measure of project worth. It is simple. All projects with a positive NPV are theoretically acceptable.

Although NPV may be computed by subtracting the total discounted of the cost from the benefits it is easier and normal practice is to compute it by discounting the cash flow. An advantage of the NPV measure as compared with benefits cost ratio is that, it makes no difference at all, at what points in the computation process the netting out takes place.

The formal selection criterion for NPV measure of projects with a positive NPV, when discounted that the opportunity cost of capital, Net present value in the difference between discounted total benefits and discounted total costs of the project this could be calculated as,

$$NPV = \sum_{t=1}^{n} \frac{B_t}{(1+i)^t} - \sum_{t=1}^{n} \frac{C_t}{(1+i)^t}$$

where
B_t benefits in each year
C_t cost in each year
t - Number of years
i - interest rate

Though any project with positive NPV could be considered for implementation, large NPV better for the project in, The NPV calculated for the project was Rs. 1582976.33 (Table 20.4) and thus the project is extremely accepted on Net Present value tool.

Net present value = present value of benefits - present value of cost.
NPV @ 10 %=1715734.20 –132757.87 = 1582976.33
NPV @ 10 % of agro forestry project is Rs: 1582976.33.

20.4.2.3. Internal Rate of Return (IRR)

Another way of using the discounted cash flow procedure for measuring the worth of a project is to find the discount rate, which makes the NPV of the cash flow equal to zero. This discount rate is termed as the internal rate of return and in a sense represents the average earnings capacity of the capital invested in the project over the project life. The formal selection criterion in using the IRR is to accept all projects having an internal rate of return above the opportunity cost of capital.

The following formula will be useful in estimating the internal rate of return.

$$IRR = L_1 + D_{r1} * P_{w1} / A_{d1}$$

Where,

IRR-Internal rate of return

L_1- Lower discount rate

D_{rl}- Difference between the discount rates

P_{wl}- Present worth of cash flow at lower discount rate.

A_{dl}-Absolute difference between the present worth of the cash flow at the two discount rates.

Two discounted rates used for this analysis were 40 per cent and 50 per cent, (table20. 4)

IRR = 40 +10*(1028.45/49606.9)

$\quad$ = 40 + 10(0.020)

$\quad$ = 40 + 0.20

$\quad$ = 40.20%

IRR is more significant discounting measure. The project has 40.20 of IRR value, which is higher than the opportunity cost of capital. So, the project can be accepted according to IRR economic tool.

20.5 RECOMMENDATION

Analysis depicts that CBR, NPV and IRR according to these discounting techniques are acceptable rate. Further it was found that positive attitudes were there toward the agro forestry system by all type of responders and they prefer agroforestry system as a soil conservation method rather than traditional methods. Even a company invest for an agro forestry project by 40 per cent of an interest rate. It will be making profits after 18 years, because the project is having strong IRR value.

Local tree species can be recommended for agroforestry plantations for Watawala region which gives higher ecological values and Gliricidia hedgerows and one-Lemmon grass hedgerows are more suitable species for the project which can contribute to enhance the ecological value of the project. Because Gliricidia is nitrogen fixing tree so it is more suitable to hill country eroded lands and bio mass capacity is also very high. As the region is suffering less fertility, and this type of project maintains the quality of soil conditions. As well as aricunut and Vanilla plantations of the projects will contribute to enhance the economic return of the land. Eucalyptus trees will contribute to enhance the economic return in directly and works as a wind belts too.

For a sustainable and environment friendly culture, collaborative practices are required from past and present scientific researches and formulate appropriate agro forestry practices suitable for this region. For a sustainable culture, both organic farming and well maintained agro forestry models give environmental, economic and social benefits. This type of System is emphasized along with the varied environmental factors for improving marginal/abandoned lands of the region.

References

APAN Field Document No.4 and RAP Publication (1996/20). Bogor, Indonesia, Asia-Pacific Agroforestry Network and Bangkok, Thailand, Food and Agriculture Organization of the United Nations.

Ariyadasa, K.P. (1996). Sri Lanka profile. In *Asia-Pacific Agroforestry Profiles.* 2nd edition.

Bandaratillake, H.M. (2001). Impacts and effectiveness of logging bans in natural forests: Sri Lanka. In Patrick B. Durst, Thomas R. Waggener, Thomas EntersandTan Lay Cheng (eds.), Asia Pacific

Forestry Commission Report, *Forest out of Bounds: Impacts and effectiveness of logging bans in Asia-Pacific.* Bangkok, Thailand: Food and Agriculture Organization of the United Nations. Accessed from http://www.fao.org/docrep/003/X6967E/x6967e08.htm#bm08.

Central Bank of Sri Lanka (2000). Annual Report of Central Bank of Sri Lanka.

Evans, D.O., and Szott, L.T. (eds) (1995).Nitrogen Fixing Trees for Acid Soils. *Proceedings of a Forestry Sector Master Plan of Sri Lanka.* Ministry of Lands, Agriculture and Forestry, Sri Lanka.

Jacob, V. J. and Alles, W. S. (1987). Kandyan gardens of Sri Lanka. *Agroforestry Systems 5*(2), 123-137.

Jewell, N. (1995). The use of Landsat TM data for estimating the area of home gardens. *The Sri Lanka Forester (The Ceylon Forester,20 (3-4)*, 79-86.

Kass, D.L. (1995). Are nitrogen fixing trees a solution for acid soils? In D. O. Evans and L.T. Szott (eds). *Nitrogen Fixing Trees for Acid Soils.* Proceedings of a workshop organized by NTFA and CATIE held in Costa Rica July 3–8 1994 in Turrialba. NFT Research Report, Arkansas, USA.

Legg, C., and Jewell, N. (1995). A 1:50,000-scale forest map of Sri Lanka: The basis for a national NFT Research Report, Arkansas, USA.

Punniyawardena, B.V. (2008). *Rainfall and Agroecological Zones in Sri Lanka.* 2nd Ed., Peradeniya-Sri Lanka. Department of Agriculture.p12-13.

Szott, L.T. (1995). Growth and biomass production of nitrogen fixing trees on acid soils. In

Tonye, J., Ibewiro, B., and Duguma, B. (1997). Residue management of the planted fallow on an acidsoil in Cameroon: Crop yields and soil organic matter fractions. *Agroforestry Systems 37*, 199–207.

Young, A. (1997). *Agroforestry for Soil Management.*2nd edition. Nairobi, Kenya: ICRAF.

21

Vulnerability of Water and Electricity Supply to Natural Hazards

Claudia Bach and Jörn Birkmann

21.1 INTRODUCTION

In the light of climate change and climate variability, temperature and precipitation patterns are changing in intensity and duration. Additionally, the number of extreme weather events including heat waves/droughts and heavy precipitation is likely to increase due to climate change (see IPCC 2012). At the same time, a rising number of persons and values are being affected by such events due to socio-economic changes such as urbanization, population and/or economic growth. The accumulation of values and population particularly in urban areas also involves infrastructures that have to be built to allow the functioning of densely populated areas. In case of a hazardous event, these infrastructures can also be adversely affected.

Besides the accumulation of infrastructures in urban areas, the mentioned socio-economic changes (also including globalization and technological changes) require more complex systems of infrastructures that depend on each other as for example in the case of electricity and information and communication technology (ICT). In addition, most of these infrastructures are operated by privately owned companies, a situation that challenges the governance of such complex systems.

Against this background, the vulnerability of critical infrastructure services, such as electricity or water supply, towards natural hazards and the consequences of such shortfalls in critical infrastructure services on the population are key subjects for research and policy making. Many sectors such as transport, health services and economic activities depend on electricity. A major black-out for example could result in chaotic situations as lighting of public places, such streets, banks, stores, supermarkets, stations or airports to name just a few would stop functioning. The same accounts for information and communication systems, water supply etc.. Although peoples' lives and economic operations are highly dependent on the functioning of these services, there have been solely few attempts to assess the vulnerability of critical infrastructures.

Developing a comprehensive assessment for infrastructures based on the understanding that vulnerability can be described as a function of exposure, susceptibility and coping capacity in the context of sustainable development (as proposed by the BBC framework) (see Birkmann, 2006),

Disaster Management and Risk Reduction: Role of Environmental Knowledge; Editors Anil K. Gupta, Sreeja S. Nair, Florian Bemmerlein-Lux and Sandhya Chatterji; Copyright © 2013, Narosa Publishing House, New Delhi

requires a two steps approach. First, the vulnerability of selected infrastructure systems has to be evaluated. Second, the dependency of other systems and on these infrastructure services has to be assessed. Referring to these steps, the paper will introduce opportunities and challenges of vulnerability assessments of critical infrastructures. In this regard the paper introduces an approach to capture second- and third-order effects of a black-out of critical infrastructures, which can be used for improvements in the planning process when developing infrastructure systems but also for improving disaster risk management by taking effects of infrastructure failure systematically into account. The findings shown below were developed within the scope of a German project on critical infrastructures and civil protection funded by the Federal Office for Civil Protection (BBK).

21.2 EXTREME WEATHER EVENTS

In the last decade it became evident, that climate is and will be changing in shorter periods of time around the globe (see e.g. IPCC 2001 and 2007). This means, first of all, a regional increase in temperatures (Rosenzweig et al., 2007). Regarding hot days for example, several scenarios such as changes in the mean temperature and temperature variability can be assumed. An increase by the overall land temperature of 0.4 to 0.8 °C since the late 19th century can already be reported (see Folland and Karl, 2001) and especially an increase in regional temperatures can be expected with very high confidence (Rosenzweig et al., 2007). An increase of the mean temperature then means, that what is defined as an extreme event today (e.g. as a certain percentile of temperatures being measured) will occur more often.

Additionally, extreme events as well as specific patterns of development have been identified as major drivers for disaster risk (IPCC 2012). At the global level, costs of losses due to extreme weather events are rapidly increasing since the 1970s (see Münchener Rückversicherungs-Gesellschaft, 2011). Although this development is being influenced by changes such as economic and population growth leading to an increase in values and people exposed, there remains a rising trend when losses are normalized (Rosenzweig et al., 2007).

In the context of disaster risk management, infrastructures play a crucial role. Not only in terms of hazard impact but also regarding the resumption of services in the aftermath of a natural hazard or disaster (UN/ISDR 2007, p. 101 f). The magnitude of damages and losses of infrastructures can for example by demonstrated by the Elbe flooding in 2002 where damages on infrastructures amounted up to 50% of the overall damage caused and accounted for almost 5 billion Euros (Mechler and Weichselgartner 2003, p. 18).

21.3 VULNERABILITY OF CRITICAL INFRASTRUCTURES

Since the damage caused by extreme events on infrastructures is already quite high and is expected to increase in the future, it is essential to improve the knowledge base regarding infrastructure systems and their underlying vulnerability towards different hazards in order to reduce risks. However, comprehensive approaches for assessing the vulnerability of critical infrastructures and the effects of their shortfall are basically lacking. Although there have been some attempts to conceptualize critical infrastructures and their vulnerabilities (see e.g. Rinaldi 2001, IRGC 2006, Hellström, 2007; Lauwe and Riegel, 2008; Kröger, 2008; Lenz, 2009), most of them remain at a very general level or focus on one specific detail, without providing a systematic overview of the overall vulnerability of the system, its components and processes.. The paper specifically refers to approaches which can be applied by local authorities in order to conduct a self-assessment of their municipality and the respective infrastructure services. Within the scope of this paper, we are referring only to meteorological hazards which will be influenced by climate change. Accordingly, new approaches to address this gap will be presented within the scope of the paper.

Before assessing vulnerability of critical infrastructures, it is however essential to also define the term critical infrastructure. In general, critical infrastructures mainly encompass infrastructures that provide crucial services and encompass key institutions for the functioning of a society (compare e.g. United States Department of Homeland 2009, p.109 or Australian Government 2010, p.8). According to a definition by the German Federal Ministry of the Interior (Federal MOI),

"Critical infrastructures (CI) are organizational and physical structures and facilities of such vital importance to a nation's society and economy that their failure or degradation would result in sustained supply shortages, significant disruption of public safety and security, or other dramatic consequences." (Federal MOI, 2009)

The definition differentiates between:

- Technical basic infrastructure, including: power and drinking water supply, information and communications technology, transport and sewage disposal and
- Socio-economic services infrastructure such as public health, food, emergency and rescue services including disaster control and management, finance and insurance business, parliament, government, public administration, law enforcement agencies as well as the media and cultural objects (Federal MOI, 2009)

This definition however, is too broad to be operationalized by an in-depth vulnerability assessment. Therefore, within the scope of this work, it will only be referred to drinking water and electricity supply in terms of critical infrastructures.

Regarding vulnerability, definitions and frameworks are numerous and differ strongly. Correspondingly, the term has to be specified and will be used according to the United Nations International Strategy for Disaster Reduction (UN/ISDR). Vulnerability will be understood as *"The characteristics and circumstances of a community, system or asset that make it susceptible to the damaging effects of a hazard."* (UN/ISDR, 2009)

To conceptualize this definition, a further component will be added. *Coping capacity* can thereby be understood as the flipside of susceptibility and is defined as *"The ability of people, organizations and systems, using available skills and resources, to face and manage adverse conditions, emergencies or disasters"* (UN/ISDR, 2009).

Thus, vulnerability can be described as a function of a certain hazard exposure and the susceptibility and coping capacity of the exposed elements (see among others Birkmann, 2006):

f (Vulnerability) = Exposure × Susceptibility × Coping Capacity

To assess the vulnerability of infrastructure services, there are two steps that have to be taken into account. In a first step, the physical vulnerability of infrastructure components is assessed (see chapter 4). This step is especially relevant for hazards that have clear and precise exposure patterns, thus hazards which allow for a spatial boundary. In addition it is expected that the hazard physically (e.g. by the intrusion of water) damages components of the infrastructure. This applies for example to sudden onset hazards such as flooding, flash floods and storms.

In a second step (IIa and b), an analysis of the root causes and effects of an infrastructures failure have to be considered. This step is especially relevant for creeping hazards such as heat waves and dry spells which do hardly cause any physically damage to infrastructure components but rather influence many factors that might pose additional stress to such infrastructures, such as increases in the demand of drinking water and electricity due to heat stress, while at the same time the production has to be reduced (limited cooling water) (Birkmann et al., 2012). Additionally, the vulnerability of the population towards critical infrastructure failures is assessed, thus their susceptibility and coping capacity (self-capacity and capacity by the state) with regard to the failure of critical infrastructure services (see Figure 21.1).

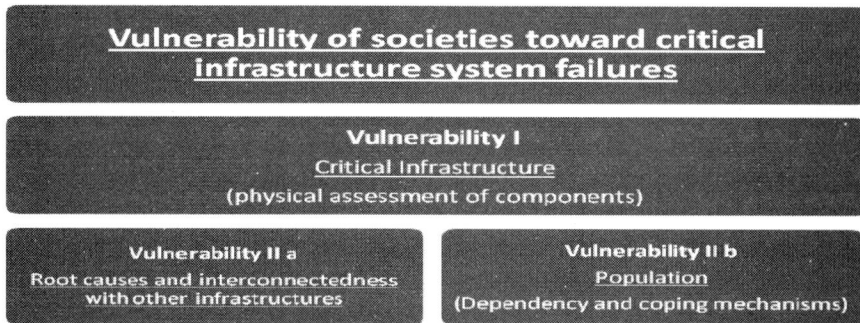

Figure 21.1: Framework for the vulnerability assessment of societies toward critical infrastructure system failures

21.4. VULNERABILITY ASSESSMENT OF CRITICAL INFRASTRUCTURE COMPONENTS

As described above, the vulnerability of critical infrastructures can be carried out in two steps, starting with a physical assessment of an infrastructure's components. Referring to drinking water supply, components possibly affected by extreme weather events are listed in Table 21.1. These components could be affected by an extreme weather event such as flooding or flash floods. To assess the vulnerability of the components, a concept was developed by Krings (2010) which is shown in Figure 21.2.

Table 21.1: Components of water supply

Process	Component
Water extraction	Wells, surface water
Conditioning	Water works, blending station
Feeding of water into the system	Pumping station
Transfer of water	Feeder
Adjustment of pressure	Pumping station
Buffer store	(High-level) tank
Regulation	Control station
Transportation of water	Pipes

Source: Krings, 2012, p. 40.

According to the context of Krings, the assessment within its first phase needs to examine a certain (flooding) scenario which allows assessing those components and processes exposed to the hazard phenomena. After having identified all components (see Table 21.1), that might be spatially exposed to a hazard, the components' susceptibility towards flooding has to be analysed. This assessment phase also deals with the technical and organizational replaceability (coping capacity). The comparison and assessment of the selected components and processes and their vulnerability will finally also provide information about the overall infrastructure vulnerability (see Figure 21.2) (Krings, 2010).

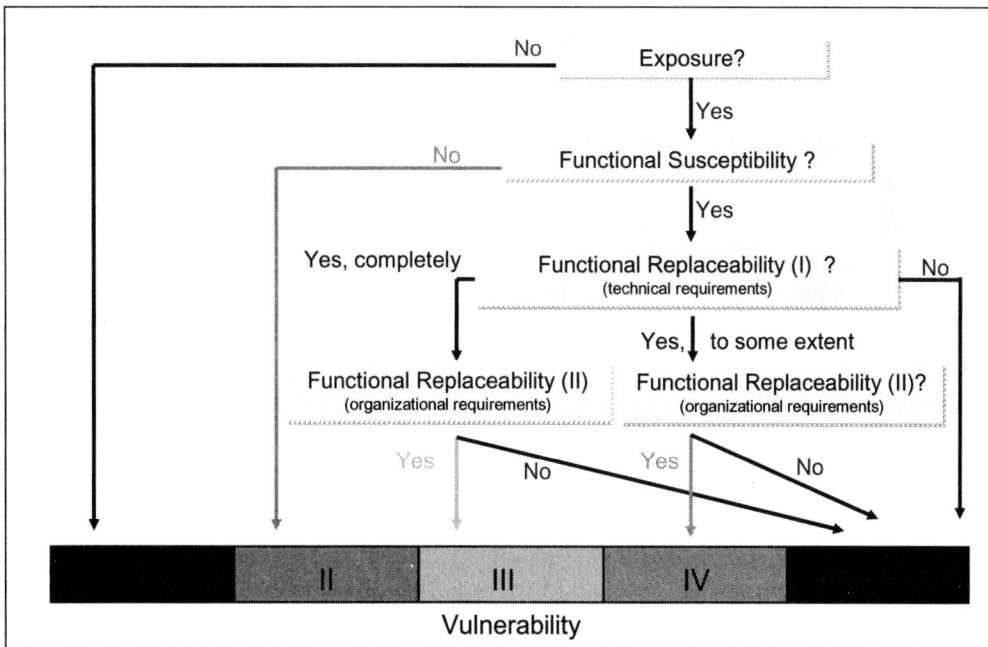

Figure 21.2: Concept for the vulnerability assessment of infrastructures towards flooding
Source: applied from Krings, 2012, p. 25.

This procedure also allows for the cooperation with privately owned companies operating infrastructure systems which might be reluctant to hand out information regarding details of components and systems. The provision of rather aggregated information on the final susceptibility and replaceability of different components and processes allows these companies to keep detailed information still confidential. This is even more important as an increasing number of former publicly owned companies are being privatized (see e.g. IRGC, 2006; Gheorghe et al., 2006 or Kröger, 2008).

Additionally, the concept can also be applied towards flash floods. By the use of Geo-Information-Systems (GIS) and a high resolution of data, local depressions can be identified as a kind of exposure zone for flash floods. In such depressions water will gather during flash floods also distant to rivers. The maps generated provide an overview of different exposure zones in a city (see Figure 21.3).

For hazards such as heat waves and dry periods, the described assessment methodology however is difficult to apply. The second assessment step (see Figure 21.1) instead plays a more important role for these types of hazards. For these hazards, the interaction of several factors (root causes) and the interaction of sectors play a more important role than the physical damage (see e.g. Birkmann et al., 2012).

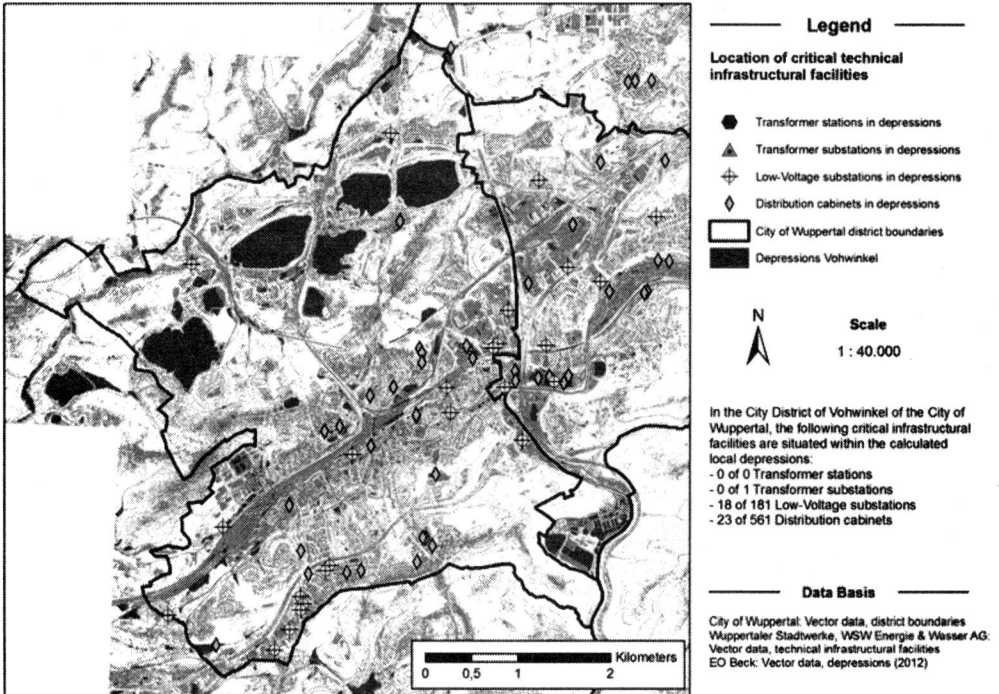

Figure 21.3: Exposure map for critical infrastructures by the example of the German city of Wuppertal

21.5   ASSESSMENT OF ROOT CAUSES, SECOND-ORDER EFFECTS AND CASUAL INTERRELATIONS

21.5.1 Fundamental Changes Influencing Vulnerability

For slow onset hazards such as heat waves and dry periods towards critical infrastructure, physical damage hardly plays a role. Instead many factors that might pose additional stress to such infrastructures have to be analysed in more detail. In Germany for example, many regions have to face an aging population and an overall reduction of the amount of people that have to be supplied with services (see German Federal Statistical Office 2010), in opposite, e.g. in Asia extremely rapid population growth and increasing concentration of people and values challenges respective critical infrastructure provision, such as drinking water supply (see e.g. Mustafa, 2010). For electricity supply for example, climatic and demographic factors, as well as economic factors and lifestyle effects play an important role in determining consumption patterns (Olonscheck, Hülsten and Kropp, 2011, p. 4795). These trends and shifts also affect the supply side or lead to a mismatch between demand and supply. Additionally, the natural hazard itself can shape the relation between demand and supply. During hot and dry periods e.g. an increased demand for water as well as for electricity caused by an increased use of air conditioning and fans has been observed (BMU 2008, Savonis et al., 2008).

In order to consider such influencing factors as well as interrelations of sectors, a second assessment step was developed (see also Figure 21.1) which can be described by the help of indicators and criteria respectively.

21.5.1.1 Interrelation between sectors

Besides the underlying factors which influence the overall operation of infrastructure services, the services itself might depend on each other (such as water and electricity supply) and likewise does the population depend on them. Thus, the relation between different infrastructure services and the population can be seen as a system. According to Vester (2002), systems can be understood as a formation of different elements being connected in structures, organizations or interdependencies. Against the background of this system understanding (e.g. Krieger, 1998 or Vester, 2004), it is important to examine the interconnectedness of infrastructures and the population. Thus, different infrastructure sectors cannot be assessed separately but have to be part of an overall understanding of a system of population and other infrastructures. Population in (urban) societies heavily depends on infrastructure services such as drinking water supply, transport and economic activities. These services are highly interrelated. Electricity supply for example is needed for all sectors such as transport, drinking water supply or information and communication technologies. At the same time, electricity supply is based on functioning information and communication technology and transport (e.g. for the delivery of coal) and a malfunction of one of these services can lead to cascading effects on other sectors. Therefore, an analysis of interrelations, especially between water and electricity supply and the population has to be part of a comprehensive infrastructure vulnerability assessment.

21.5.1.2 Water and electricity

In case of a major black-out, neither banks nor street light would work and stations as well as airports would perhaps face chaotic situations (Reichenbach et al., 2008). Likewise are water supply infrastructures affected in case of a blackout as the system components (also see Table 21.1) need electricity for their functioning. At the same time, electricity generation depends to major parts on the availability of water. Hydroelectric power plants require water but even other sources of electricity generation such as nuclear or coal-fired power plants depend on this natural resource. They are in need of water for the cooling of their rods and consume about 50l of fresh water for the production of one kilowatt hour of electricity (Lönker, 2003). To preserve the functioning of aquatic ecosystems in rivers, a further temperature increase of river water caused by discharging heated water from power plants has to be avoided and respective guidelines and laws exist for this issue. Besides ecosystem stress, problems in shipping might also lead to a lack of coal delivery for these power plants. This can have effects on electricity generation as coal for coal-fired power plants is mainly shipped via rivers. In periods of high temperatures and low precipitation, low tides may then lead to problems in water supply but also to disruptions in electricity supply.

In Germany, the major source of energy are coal-fired (lignite and stone coal) power plants which contribute to about 40% of the overall energy supply. They are followed by nuclear power plants (23%) and renewable energies (16%) of which only 3% are derived by hydroelectric power plants. The rest (ca. 20%) is provided by natural gas and other sources. (BDEW 2010) During the heat wave that hit Europe in 2003 major nuclear and coal-fired power stations in Germany had a reduced output by between − 15 % to − 100 % (BfG 2006; Lönker, 2003). Thus, power plants using ambient air for cooling could operate superiorly (BfG 2006). Furthermore, shipping traffic on the rivers Danube and Rhine was partly discontinued (Müller-Westmeier and Willing, 2007).

To assess the overall dependency of electricity supply on water, it is important to consider different sources of energy, their location (e.g. close to rivers or the sea) and the amount to which they contribute to the overall electricity supply. It is further important to take into account the quantities of energy that are or can be imported from other countries or which amounts could (technically) be transferred within a country in case of a regional shortfall in electricity production (BMWi, 2005). The respective figures and shares can be utilized as indicators determining the vulnerability of electricity supply towards heat waves, e.g. by dependency levels on water supply. Long-term monitoring of system shifts as well as the comparison with other countries/regions can thereby create a basis for developing vulnerability reduction plans. The latter ones might be combined with infrastructure development as well as mitigation plans.

21.5.1.3 Vulnerability of the population

Assuming a shortfall of drinking water supply/electricity, the vulnerability of the population towards a failure can be assessed by the mentioned formula of exposure, susceptibility and coping capacity. For some hazards such as floods, e.g. not only socio-economic features such as age and income play an important role regarding peoples vulnerability, but also the experience people have to effectively deal with such phenomena (see Birkmann et al., 2010). The relevant factors were thus translated into criteria which might be applied by municipalities in order to analyse their vulnerability towards blackouts qualitatively.

Applying the concept to failures in the electricity supply system, exposure can firstly be described as the severity (regarding the geographical as well as the time scale) of the failure or the cascading effect. Susceptibility instead describes the societies' weaknesses and the coping capacities its resources to deal with such failures. Society thereby does not only mean individual or household level (in-) capacities but includes civil protection agencies that should provide help in such adverse events particularly for most vulnerable groups. Susceptibility can for both (public and private) actors be translated into dependency on electricity supply. On a household level, this encompassed especially persons that are highly dependent such as those in need of certain electric health devices. On a municipal level, this criterion refers to the potential of maintaining the work during blackouts.

The coping capacity component can be translated into preparedness towards failures. It describes on a household level measures such as the storage of drinking water or the existence of an oven and or candles. At a municipal level, it analyses the preparedness of important facilities such as hospitals or retirement homes as well as of emergency services especially in terms of emergency power supply. Additionally, alternative communications methodologies such as satellite telephones should be considered.

21.5.2 Transferability

The vulnerability assessment shown above that includes a two-step approach has been developed within the scope of a German project on critical infrastructures and civil protection , KIBEX, funded by the German Federal Office of Civil Protection and Disaster Assistance (BBK) and refers to German/ European infrastructures systems. Hence, the transferability of the assessment method to other countries and regions around the world has to be critically discussed. Whereas the first step, mainly accounting for flooding and flash floods, can be applied by choosing a scenario and obtaining information of the functioning of infrastructure components and systems, the transferability of the second step (interdependency of systems and dependency of the population) is more complex and might be context specific. Consequently, specific information is needed to conduct the second step in countries in transition or in developing countries. Nevertheless, also in developing countries and countries in transition, critical infrastructure services play an increasing role.

Focusing on water supply, the underlying factors that influence the overall availability of water are different. Additionally, water cannot – compared to electricity – easily be transported (UBA, 2000) via long distances. Thus, the assessment has to consider water availability as well as other underlying factors shaping the supply such as population growth, urbanization and demand at a very local level which requires the availability of the respective data. For electricity supply set-up instead also the national level might serve as a level of reference where data might be easier available.

Besides this common problem in data availability, also the overall applicability of the assessment method around the world needs to be discussed. The method was developed by assuming a high level of supply security and interconnectedness of different critical infrastructures as well as very low probability of failure and a high dependency of the population on infrastructure services. We assumed:

i.   An overall availability of resources.
ii. A high level of the population has access to infrastructure services.
iii. The infrastructure services meet the demand and have a very low probability of failure.
iv. Major parts of the supplied population are not aware of the probability of a major failure of infrastructure systems and cascading effects and are hardly prepared. They are thus vulnerable to failures in supply.

However, these assumptions might not meet the conditions in other countries. Whereas in Germany more than 80% of the overall water resources of 188 bn. m³ remain unused (UBA 2009a), water is hardly available in other countries. Due to the excessive use of their renewable water resources many countries, particularly in North Africa, Middle-East and central Asia are or will be experiencing water scarcity (UNEP, 2008a).

Referring to the access to water, further differences can be found. Whereas in developed countries a majority of the population has access to water (in Germany an average of about 99% (UBA, 2009b)), only between 70% (in rural) and 90% (in urban areas) of the population in upper and lower middle income as well as least developed countries have access to improved drinking water (Gleick, 2009). By 2004 in many regions in Africa even less than 65% of the overall population had access to improved sources of drinking water (UNEP, 2008b).

The third assumption that has to be considered is the functioning of infrastructure services which could be disrupted. In some countries, societies are used to a high level of supply security regarding water and electricity supply. For Germany e.g., the electricity supply is constantly improving so that in 2008 only about 17 minutes (Bundesnetzagentur, 2010) (from about 23 minutes in 2004) (Bundesnetzagentur, 2006) of its failure per year and consumer could be reported. This does not account for a view major blackouts as e.g. occurred in 2005 in Münsterland, Germany but describes a general trend. Thus, major failures in electricity supply are not expected by the society. Additionally, hardly any major blackouts of several days are reported and blackouts are not perceived as threats (Palm, 2009) by the population. Accordingly, hardly any preparation measures are taken (Nye, 2010; Graham and Thrift, 2007).

At the same time, the malfunction of certain devices can be hypothetically assessed, however, the in-depth understanding of cascading effects to other sectors is still lacking, such as the impact of a black out on different water supply systems.

In this context it remains to be solved on how to assess the vulnerability of infrastructure services that are not reliable. One could assume that the population in regions where the malfunction of critical infrastructures belongs to everyday life has better coping strategies regarding their blackout. However, the absence of infrastructure services such as water and energy supply increases the overall susceptibility of the population towards natural hazards. This aspect has not been included so far into the presented methodology. It has to be discussed, whether the application of a critical infrastructure vulnerability assessment for countries with a lack of such functioning services is feasible or if it would rather be part of a vulnerability assessment of the population defining their susceptibility via a lack of access to infrastructures.

21.6 CONCLUSION

Natural hazards and so called extreme events can generate major stresses for critical infrastructures. However, a real risk only exists if critical infrastructures are vulnerable and if many population groups are strongly depending on these services of critical infrastructures.

To integrate not only the physical vulnerability of infrastructure components which might be exposed to a natural hazard or so called extreme event, an assessment encompassing two steps was outlined. In a first step, a methodology for the vulnerability assessment of infrastructure components which has been developed so far for flooding and flash floods is introduced. A second step then accounts for factors increasing the pressure on infrastructure systems such as urbanization and population growth which have direct effects on the demand side and thus influence the vulnerability of processes such as electricity generation. The second step however, should also encompass an assessment of cascading effects and the vulnerability of the population towards infrastructure failures. This step is important as not only the level of dependency and the accustomedness to failures are different but also the interdependency of infrastructures can vary. Referring to electricity supply and water resources for example, sources of power generation and sources used for cooling (e.g. air, rivers, sea) have an influence on the infrastructure vulnerability.

Besides quantitative approaches, the overall concept has included a variety of criteria that allow municipalities or private companies to assess the vulnerability of critical infrastructures and the dependency of the population on these infrastructure services. The assessment also considered respective management capacities and coping capacities of individual households.

Although the assessment provides a first systematic approach to assess the vulnerability of critical infrastructures and complexity of systems as well as the various dependencies of people on these services, there remains a major research task for the future. Particularly, the Fukushima crises in Japan underscored the necessity to better understand cascading effects of such critical infrastructure failures – especially in the context of existing governance structures. In addition, the framework presented is mainly based on countries with a high access to these infrastructure services. Hence, until now the framework accounts for developed countries with resources to which almost 100% of the population have access to. It also assumes that infrastructure services are running (mainly) without interruptions. Finally, a major source of insecurity and a false sense of security also might be determined by the fact that many population groups are unprepared to deal with such failures, particularly due to a lack of awareness about e.g. their strong dependency on these services and the potential of a breakdown.

References

Australian Government (2010). Critical Infrastructure Resilience Strategy. Accessed from: http://www.ag.gov.au/Documents/Australian%20Government%20s%20Critical%20Infrastructu re%20Resilience%20Strategy.PDF, last accessed: 01.06.2012.

Birkmann, J. (2006). Measuring vulnerability to promote disaster-resilient societies: Conceptual framework and definitions. In J. Birkmann (Ed.) *Measuring Vulnerability to Natural Hazards – Towards Disaster Resilient Societies* (pp. 9-54). Tokyo: United Nations University Press.

Birkmann, J., Bach, C., and Vollmer, M. (2012). Tools for Resilience Building and Adaptive Spatial Governance – Challenges for Spatial and Urban Planning in Dealing with Vulnerability. In *Raumforschung und Raumordnung, 70*(4), 293-308.

Birkmann, J., Bach, C., Setidadi, N., Olonscheck, M., Walther, C., Taubenböck, H., Wurm,M., and Roth, A. (unpublished). 3rd Progress Report to the BBK of the KIBEX Project, Bonn.

Bundesamt für Gewässerkunde (BfG) (2006). Niedrigwasserperiode 2003 in Deutschland. Ursachen – Wirkungen – Folgen. Koblenz.

Bundesnetzagentur (2006). Monitoring bericht 2006 der Bundesagentur für Elektrizität, Gas, Telekommunikation, Post und Eisenbahn. Accessed from http://www.bundesnetzagentur.de-/cae/servlet/contentblob/31292/publicationFile/1122/Monitoringbericht2006Id7263pdf.pdf Last accessed on 19.04.2011.

Bundesnetzagentur (2010). Versorgungsqualität-SAIDI-Wert2008. Accessed from http://www.bundesnetzagentur.de/DE/Sachgebiete/ElektrizitaetGas/Sonderthemen/SAIDIWertStro m2008/SAIDIWertStrom2008_node.html. Last accessed on15.04.2011.

Bundesverband der Energie- und Wasserwirtschaft (BDEW) (2010). Brutto-Stromerzeugung nach Energieträgern. Accessedfrom http://www.bdew.de/bdew.nsf/id/ DE_Bruttostromerzeugung_ in_Deutschland/$file/Bruttostromerzeugung%20in%20Deutschland%202009pdf. Last accessed on 14.04.2011.

Folland, C. K., and Karl, T. R. (2001). Observed Climate Variability and Change. In J. T. Houghton, Y. Ding, D. J. Griggs, M. Noguer, P. J. van der Linden, X. Dai, K. Maskell and C. A. Johnson (Eds.) *Climate Change 2001: The Scientific Basis.* Contribution of Working Group I the Third Assessment Report of the Intergovernmental Panel on Climate Change, Cambridge.

German Federal Ministry for the Environment, Nature Conservation and Nuclear Safety (BMU) (2008). Deutsche Anpassungsstrategie an den Klimawandel, Hintergrundpapier, accessible http://www.bmu.de/files/pdfs/allgemein/application/pdf/das_hintergrund.pdf. Last accessed on 16.04.2011.

German Federal Ministry of Economics and Technology (BMWi) (2005). Energiepolitik. Accessed from http://www.bmwi.de/BMWi/Navigation/Energie/energiepolitik.html.Last accessed on 14.04.2011.

German Federal Ministry of the Interior (Federal MOI) (2009). National Strategy for Critical Infrastructure Protection (CIP Strategy), Berlin. Accessed from http://www.bmi.bund.de/SharedDocs/Downloads/EN/Broschueren/cip_stategy.html?nn=441658. Last accessed on 15.04.2011.

German Federal Statistical Office (2010). Statistisches Jahrbuch 2010, Wiesbaden. Accessed from http://www.destatis.de/jetspeed/portal/cms/Sites/destatis/SharedContent/Oeffentlich/-B3/Publikation/Jahrbuch/StatistischesJahrbuch,property=file.pdf. Last accessed on 21.04.2011.

Gheroghe, A. V., Masera, M., Weijnen, M., and De Vries, L. (2006). *Critical Infrastructure at Risk : Securing the European Electric Power System*. Dodrecht:Springer.

Gleick, P. H. (2009). The world's water 2008-2009. Pacific Institute for Studies in Development, Environment, and Security, Island Press, Washington. Accessed from http://www.worldwater.org/data20082009/Table3.pdf. Last accessed on 20.04.2011.

Graham, S., and Thrift, N. (2007). Out of order: Understanding repair and maintenance.*Theory, Culture and Society 24* (3), 1-25.

Innenministerium Baden-Württemberg (IM Ba-Wü) und Bundesamt für Bevölkerungsschutz und Katastrophenhilfe (BBK)(Eds.) (2010). Krisenmanagement Stromausfall – Kurzfassung, Heidelberg. Accessed from http://www.bbk.bund.de/cln_028/nn_398890/-SharedDocs/Publikationen/Publikationen_20Kritis/Krisenhandbuch__Stromausfall_Kurzfassung_pdf,templateId=raw,property=publicationFile.pdf/Krisenhandbuch_Stromausfall_Kurzfassung_pdf .pdf. Last accessed on 20.04.2011.

Intergovernmental Panel on Climate Change (IPCC) (2001). *Climate Change 2001: Synthesis Report.* A Contribution of Working Groups I, II, and III to the Third Assessment Report of the Integovernmental Panel on Climate Change [Watson, R.T. and the Core Writing Team (eds.)]. Cambridge University Press, Cambridge,United Kingdom, and New York, NY, USA, 398 pp.

Intergovernmental Panel on Climate Change (IPCC) (2007). *Climate Change 2007: Synthesis Report.* Contribution of Working Groups I, II and III to the Fourth Assessment Report of the Intergovernmental Panel on Climate Change [Core Writing Team, Pachauri, R.K and Reisinger, A.(eds.)]. IPCC, Geneva, Switzerland, 104 pp.

Intergovernmental Panel on Climate Change (IPCC) (2012). *Managing the Risks of Extreme Event and Disasters to advance Climate Change Adaptation.* A Special Report of Working Groups I and II of the Intergovernmental Panel on Climate Change. In: Field, C.B., V. Barros, T. F. Stocker, D. Qin, D. J. Dokken, K. L. Ebi, M. D. Mastrandrea, K. J. Mach, G.-K. Plattner, S. K. Allen, M. Tignor, and P. M. Midgley (Eds.).Cambridge University Press, Cambridge, UK, and New York, NY, USA, 582 pp.

International Risk Governance Council (IRGC) (2006). White Paper on Managing and Reducing Social Vulnerabilities from Coupled Critical Infrastructures, Châtelaine.

Krieger, D.J. (1998). Einführung in die allgemeine Systemtheorie. 2. Aufl. UTB, Paderborn.

Krings, S. (2010). Verwundbarkeitsassessment der Strom- und Trinkwasserversorgung gegenüber Hochwasserereignissen. In J. Birkmann, S. Dech, M. Gähler, S. Krings, W. Kühling, K. Meisel K., Roth, A. Schieritz, A.,Taubenböck, H, Vollmer, M., Welle, T.,Wolfertz, J., Wurm, M. and H. Zwenzner (Eds.) Abschätzung der Verwundbarkeit gegenüber Hochwasserereignissen auf kommunaler Ebene. p. 21-47, BBK, Bonn.

Kröger, W. (2008). Critical infrastructures at risk: A need for a new conceptual approach and extended analytical tools. *Reliability Engineering and System Safety, 93*, 1781-1787.

Lauwe, P. and Riegel, C. (2008).Schutz Kritischer Infrastrukturen - Konzepte zur Versorgungssicherheit. Informationen zur Raumentwicklung (1/2), 113-125.

Lönker, O. (2003). Hitzefrei für Atomstrom. *Neue Energie 9*, 22– 23.

Mechler. R, and Weichselgartner, J. (2003). Disaster Loss Financing in Germany – The Case of the Elbe River Floods 2002. Interim Report IR-03-021 of the International Institute for Applied Systems Analysis. Luxemburg. Accessed from: www.iiasa.ac.at/Admin/PUB/-Documents/IR-03-021.pdf.

Müller-Westermeier, G., and Willing, P. (2007). Klimastatusbericht 2007. Deutscher Wetterdienst (DWD), Offenbach.

Münchener Rückversicherungs-Gesellschaft (2011). GroßeNaturkatastrophenweltweit 1950-2010, NatCatService, .Accessed from http://www.munichre.com/app_pages/www/@res/pdf/-NatCatService/great_natural_catastrophes/NatCatSERVICE_Great_1950_2010_number_touch_de .pdf. Last accessed on18.04.2011.

Nye, D.E. (2010). When the lights went out: A history of blackouts in America. Cambridge:MIT Press

Olonscheck, M., Hülsten, A., and Kropo, J. (2011). Heating and cooling energy demand and related emissions of the German residential building stock under climate change. *Energy Policy, 39*, 4795-4806.

Palm, J. (2009). Emergency Management in the Swedish electricity grid from a household perspective. *Journal of Contengencies and Crisis Management, 17* (1), 55-63.

Reichenbach, G., Wolff, H., Göbel, R., & Stokar von Neuforn, S. (2008). Risiken und Herausforderungen für die öffentliche Sicherheit in Deutschland. Berlin: Grünbuch des Zukunftsforum Öffentliche Sicherheit.

Rinaldi, S. M., Peerenboom, J. P., and Kelly, T. K. (2001). Identifying, understanding and analysing critical infrastructure interdependencies. *IEEE Control Systems Magazine, 21* (6), 12-25.

Rosenzweig, C., Casassa, G., Karolyn, D. J., Imeson, A., Liu, C., Menzel, A., Rawlins, S., Root, T. L., Seguin, B., and Tryjanowsk, P. (2007). Assessment of observed changes and responses in natural and managed systems.In M. L. Parry, O. F. Canziani, J. P. Palutikof, P. J. van der Linden and C. E. Hanson (Eds.). *Climate Change 2007: Impacts, Adaptation and Vulnerability*. Contribution of Working Group II to the Fourth Assessment Report of the Intergovernmental Panel on Climate Change, Cambridge.

Savonis, M. J., Burkett, V. R. and Potter, R. (2008). Impacts of Climate Change and Variability on Transportation Systems and Infrastructure: Gulf Coast Study, Phase 1. Synthesis and Assessment Product 4.7.

Umweltbundesamt (UBA) (2000). Liberalisierung der deutschenWasserversorgung .Accessed from http://www.umweltdaten.de/publikationen/fpdf-l/1888.pdf. Last accessed on 20.04.2011.

Umweltbundesamt (UBA) (2009a). Öffentliche Wasserversorgung, .Accessedfrom http://www.umweltbundesamt-daten-zur- nodeIdent=2302. Last accessed on 20.04.2011.

Umweltbundesamt (UBA) (2009b). Wasserdargebot und Wassernutzung 2007. Accessed from http://www.umweltbundesamt-daten-zur- umwelt.de/umweltdaten/public/document/download / Image.do?ident=17535. Last accessed on 20.04.2011.

United Nations Environment Programme (UNEP) (2008a). Excessive withdrawal of renewable water resources. Accessed from http://www.unep.org/dewa/vitalwater/article81.html. Last accessed on 20.04.2011.

United Nations Environment Programme (UNEP) (2008b). Toward a world of thirst?. Accessed from http://www.unep.org/dewa/vitalwater/index.html. Last accessed on 19.04.2011.

United Nations International Strategy for Disaster Reduction (UN ISDR) (2007). *Words Into Action: A Guide for Implementing the Hyogo Framework.* Geneva, available via: http://www.unisdr.org/files/594_10382.pdf. Last accessed: 11.01.2012.

United Nations International Strategy for Disaster Reduction (UN/ISDR) (2009). UNISDR Terminology on Disaster Risk Reduction. Accessed from http://www.unisdr.org/eng/-terminology/terminology-2009-eng.html last accessed on 14.04.2011.

United States Department of Homeland (2009). National Infrastructure Protection Plan – Partnering to enhance protection and resiliency. Accessed from: http://www.dhs.gov/-xlibrary/assets/NIPP_Plan.pdf. Last accessed: 01.06.2012.

Vester, F. (2002). Unsere Welt - ein vernetztes System. 2. Aufl. Deutscher Taschenbuch Verlag, München.

Vester, F. (2004). Die Kunst vernetzt zu denken. Ideen und Werkzeuge für einen neuen Umgang mit Komplexität. Ein Bericht an den Club of Rome. 4. Aufl. Deutscher Taschenbuch Verlag, München.

Index

Index   **289**

About the Editors

Dr. Anil K Gupta is working as Associate Professor of Policy Planning at NIDM since 2006. He was founder & Head/Director of Institute of Environment and Development Studies at Bundelkhand University since 2003. He obtained Ph.D. in the area of EIA/Environmental-health in 1994 and did post-doc work at NEERI (CSIR). He received Young Scientist Award of Govt. of Madhya Pradesh in 1996. He has the credit of over 100 publications as journal papers, books, chapters and training modules. He guided 4 Ph.D. & 25 M.Phil./M.Tech and M.Sc. research thesis. His contributions pioneer to National Human Resource Plan for DRM, National Strategy on Climate Change, eco-DRR, National Action Plan for Chemical Disaster Management, University education in DM. He is project director of ekDRM (GIZ-NIDM), PI of Bundelkhand Drought Vulnerability & Mitigation Analysis (ICSSR), Integrating Climate-change Adaptation with Flood DRR Gorakhpur Study (CDKN) and Coastal Andhra and Tamil Nadu (EU). He was invited by UNESCO, UNU, UNEP, ICIMOD, World Water Forum, Italian Space Agency for academic deliberations in many countries.

Sreeja S. Nair is working as Assistant Professor at NIDM since 2007. She is a disaster management professional having more than 12 years of experience in the field. Before joining NIDM she was working with Disaster Risk Management Programme of UNDP during 2004-07. She published 10 papers in peer reviewed national and international journals, authored 3 training modules and 2 books (Editor). She is the coordinator of Indo German Cooperation on Environmental Knowledge for Disaster Risk Management and Co-Principal Investigator of ICSSR Research project on Drought Vulnerability and Mitigation Analysis and also involved in the GIZ-European Union pilot project on integrating climate-change adaptation with disaster management planning process coastal districts of Andhra Pradesh and Tamil Nadu. Her areas of expertise include geoinformatics applications in disaster management, environmental law, ecosystem approach to disaster risk management, chemical disaster management, and disaster data management. She has been associated as resource person with UNSPIDER, UNEP, UNDP, ICIMOD etc in India and other countries.

Florian Bemmerlein-Lux is Advisor, GIZ Germany for Programme on Environmental Planning and Disaster Risk Management. He is Director of the IFANOS consortium Germany and Ifnaos Concept & Planning. He is an ecologist and biologist (Germany and USA) and served as senior advisor for environment and planning for various international organisations. As director of ifanos concept and planning in Germany, his current projects are in Europe, Asia and southern Africa. He is adjunct professor of the Institute of Environmental Engineering of the University of Beijing teaching Institutional Development in Environment (IDE). With over 27 years of practical experience he is specialised in landscape ecology and instruments of environmental planning, management and protection of protected areas and integration of environmental issues into spatial planning.

Dr. Sandhya Chatterji is Advisor, GIZ India for Programme on Environmental Planning and Disaster Risk Management. She is Director of the IFANOS consortium India including Ifnaos Concept and Planning. She has obtained her Ph.D degree from Jawaharlal Nehru University for the pioneering work done on planning development action in Zanskar tehsil, Ladakh in 1986. She has over 25 years of experience with planning, implementing and evaluation of projects on natural resource management including agriculture, irrigation, forestry, watershed management, as well as livelihood and social development projects focused on gender, self help group promotion, and environment and disaster risk management.